Challenges in Poultry Production Systems and Nutritional Interventions

Challenges in Poultry Production Systems and Nutritional Interventions

Guest Editor

Janghan Choi

MDPI

Basel • Beijing • Wuhan • Barcelona • Belgrade • Novi Sad • Cluj • Manchester

Guest Editor
Janghan Choi
Department of Animal and
Food Sciences
Texas Tech University
Lubbock
United States

Editorial Office
MDPI AG
Grosspeteranlage 5
4052 Basel, Switzerland

This is a reprint of the Special Issue, published open access by the journal *Animals* (ISSN 2076-2615), freely accessible at: www.mdpi.com/journal/animals/special_issues/2PX449WH83.

For citation purposes, cite each article independently as indicated on the article page online and using the guide below:

Lastname, A.A.; Lastname, B.B. Article Title. *Journal Name* **Year**, *Volume Number*, Page Range.

ISBN 978-3-7258-3668-0 (Hbk)
ISBN 978-3-7258-3667-3 (PDF)
https://doi.org/10.3390/books978-3-7258-3667-3

Contents

animals

Editorial

Challenges in Poultry Production Systems and Nutritional Interventions

Janghan Choi

Department of Animal and Food Sciences, Texas Tech University, Lubbock, TX 79409, USA; janghan.choi@ttu.edu

Chicken is the most produced and consumed meat in the world [1]. Poultry production yields affordable and highly nutritious (e.g., protein, vitamins, irons, omega 3, etc.) meat and eggs all over the world. There was a great achievement in poultry production that increased the meat and egg yields, production efficiency, which was resistant against specific challenges over recent decades [2,3]. However, there are still many challenges that can negatively impact productivity, efficiency, food safety, and animal welfare in poultry production [4]. Challenges in poultry production, such as feed costs, parasitic and bacterial infections, heat stress, muscle myopathies, mycotoxin contamination, and issues related to rearing conditions and stocking density, have prompted extensive research into diverse nutritional and systemic interventions to address these problems.

This Special Issue collects valuable research articles and reviews describing the current challenges in poultry production systems, and nutritional and systemic interventions to cope with these issues. Parasitic and bacterial infections are some of the major issues that negatively influence production, welfare, and food safety issues in poultry production, and the withdrawal of anticoccidial and antibacterial drugs has aggravated the negative impacts of the challenges in poultry production [5,6]. *Eimeria* infection alone, or with *Clostridium perfringens* (e.g., necrotic enteritis) cause tremendous economic losses in poultry production [7,8]. In this Special Issue, Choi et al. [9] demonstrated that different *Eimeria* infection doses negatively impact productivity and efficiency in broiler chickens raised in floor pens by negatively impacting their feed intake, nutrient digestibility, gut microbiota, and foot pad dermatitis. Furthermore, this study provided information on *Eimeria* infection models (e.g., infection doses and their responses) for broilers in the floor pen condition. Most of the previous *Eimeria* research in broiler chickens was conducted in cage studies, while most of the broilers are normally raised in the floor pen condition in the US [10]. Sharma et al. [10] showed that inflammatory response in *Eimeria* infection and necrotic enteritis could disrupt the growth and development of bone via regulating RANKL-RANK pathway in chickens. Zhang et al. [11] demonstrated that the supplementation of 1000 mg/kg tannic acid (plant extract), which is known to have antimicrobial, antioxidant, and anti-inflammatory effects [4,12,13], improved gut integrity, attenuated the inflammatory response, and enhanced antioxidant capacity in broiler chickens under the necrotic enteritis challenge. Moreover, Yang et al. [14] showed that microencapsulated sodium butyrate (e.g., an organic acid) enhanced immune status, intestinal morphology, and volatile fatty acid production, mainly by beneficially affecting gut microbiota in broiler chickens infected with *Clostridium perfringens*. Zhi et al. [15] showed that the supplementation of sea-buckthorn flavonoids (e.g., plant extracts) improved growth performance, systemic immune responses, and gut barrier integrity in broiler chickens challenged with bacterial lipopolysaccharides. Our Special Issue highlighted the detrimental effects of

Received: 5 February 2025
Revised: 11 February 2025
Accepted: 12 February 2025
Published: 13 February 2025

Citation: Choi, J. Challenges in Poultry Production Systems and Nutritional Interventions. *Animals* **2025**, *15*, 530. https://doi.org/10.3390/ani15040530

parasitic and bacterial infections on broiler chickens, while presenting variety of nutritional strategies to mitigate these challenges.

Improving productivity and efficiency by nutritional interventions is also important in general conditions in poultry production. Phosphorus, the third-most expensive nutrient in the poultry diet, plays important roles in regulating energy metabolism, cell homeostasis, bone growth, gut health, and overall performance in chickens [16,17]. Decreasing phosphorous levels would be beneficial to decrease feed costs and to reduce phosphorus excretion to the environment, while it can decrease the productivity and efficiency of the poultry production [18]. The supplementation of phytase could increase the utilization of phosphorus in phosphorus-deficient diets. In this Special Issue, Yu et al. [19] demonstrated that the supplementation of an *Escherichia coli*-derived 6-phytase improved growth performance and bone health in broilers fed phosphorus-deficient diets by improving the utilization of phosphorus. These results suggest that the reduction of phosphorous with the addition of the phytase would not negatively impact the growth of broilers via enhancing phosphorus utilization. Corn–soybean meal diets have high contents of non-starch polysaccharide (NSP) including arabinoxylan, β-glucan, cellulose, xylose, mannose, and pectin, which can lead to poor productivity and efficiency by decreasing feed intake and compromising nutrient digestion and absorption, mainly through increasing the digesta viscosity in the gastrointestinal tract of chickens [20,21]. Shuaib et al. [22] showed that the addition of a β-mannanase with the inclusion of 10% soyhull, which is rich in NSP, increased the growth performance of laying hens during the late peak production without affecting the egg production and quality. Their study demonstrated that supplementation of phytase and β-mannanase could improve the nutrient utilization of poultry.

Rearing temperature and environments are crucial factors, and as important as the nutritional and genetic factors in poultry production. In the global warming era, the temperature in the poultry house and in the feed mill is increasing. Poultry is sensitive to heat stress, which can negatively impact animal welfare, growth performance and egg production, and quality of chickens [23]. In this Special Issue, Sumanu et al. [24] demonstrated that the supplementation of a probiotic (*Saccharomyces cerevisiae*) and ascorbic acid improved growth performance, intestinal morphology, and antioxidant capacity heat stress in broilers subjected to heat stress. The incidence and severity of mycotoxins contamination in the ingredients and feeds are increasing with the global warming [25]. Lee et al. [26] showed that the inclusion of the two toxin binders, namely (1) clay minerals (85% bentonite and 12% clinoptilolite) with 3% charcoal and (2) clay minerals (66.1% aluminosilicates), with natural components (0.8% artichoke and rosemary plant extracts, 7% yeast extract, 0.5% beta-glucans, and 25.6% carriers), attenuated the negative impacts of feeding diets contaminated by ochratoxin A (a highly toxic mycotoxin produced by several fungi, including the Aspergillus and Penicillium genera) in broiler breeders. Low rearing temperatures (e.g., cold stress) could influence the physiology of chickens [27]. Kim et al. [28] demonstrated that low rearing temperatures impaired the antioxidant system and modulated lipid metabolism in laying hens without affecting egg production and quality in layers. Al-Baadani et al. [29] and Kang et al. [30] demonstrated that environmental factors could affect animal welfare, productivity, efficiency, and systemic health in chickens. Al-Baadani et al. [29] demonstrated that high stocking density could compromise productivity, efficiency, meat yield, and gut health, and the supplementation of gum Arabic and commercial prebiotics improved growth performance, production efficiency, and gut health of broiler chickens, potentially by beneficially modulating gut microbiota. Kang et al. [30] demonstrated that stimulated voluntary activities by placing enrichment huts in the light enriched broiler house improved animal welfare and systemic health in fast growing modern broilers. Our Special Issue provides variable strategies to cope with diverse environmental strategies.

Taken together, diverse challenging conditions in poultry production and effective nutritional and systemic coping strategies were researched and discussed in this Special Issue. There is still not a single bullet to resolve the current challenges in poultry production yet. Continuous studies with novel ideas are required to understand the challenges, and to elucidate the coping strategies in poultry production.

Funding: This research received no external funding.

Conflicts of Interest: The author declares no conflict of interest.

References

1. Uzundumlu, A.S.; Dilli, M. Estimating chicken meat productions of leader countries for 2019–2025 years. *Ciência Rural* **2022**, *53*, e20210477. [CrossRef]
2. Collins, K.; Kiepper, B.; Ritz, C.; McLendon, B.; Wilson, J. Growth, livability, feed consumption, and carcass composition of the Athens Canadian Random Bred 1955 meat-type chicken versus the 2012 high-yielding Cobb 500 broiler. *Poult. Sci.* **2014**, *93*, 2953–2962. [CrossRef] [PubMed]
3. Lordelo, M.; Cid, J.; Cordovil, C.M.; Alves, S.P.; Bessa, R.J.; Carolino, I. A comparison between the quality of eggs from indigenous chicken breeds and that from commercial layers. *Poult. Sci.* **2020**, *99*, 1768–1776. [CrossRef]
4. Choi, J.; Kim, W.K. Dietary application of tannins as a potential mitigation strategy for current challenges in poultry production: A review. *Animals* **2020**, *10*, 2389. [CrossRef] [PubMed]
5. Cervantes, H.M. Antibiotic-free poultry production: Is it sustainable? *J. Appl. Poult. Res.* **2015**, *24*, 91–97. [CrossRef]
6. Kadykalo, S.; Roberts, T.; Thompson, M.; Wilson, J.; Lang, M.; Espeisse, O. The value of anticoccidials for sustainable global poultry production. *Int. J. Antimicrob. Agents* **2018**, *51*, 304–310. [CrossRef]
7. Gilbert, W.; Bellet, C.; Blake, D.P.; Tomley, F.M.; Rushton, J. Revisiting the economic impacts of Eimeria and its control in European intensive broiler systems with a recursive modeling approach. *Front. Vet. Sci.* **2020**, *7*, 558182. [CrossRef] [PubMed]
8. Goo, D.; Choi, J.; Ko, H.; Choppa, V.S.R.; Liu, G.; Lillehoj, H.S.; Kim, W.K. Effects of Eimeria maxima infection doses on growth performance and gut health in dual-infection model of necrotic enteritis in broiler chickens. *Front. Physiol.* **2023**, *14*, 1269398. [CrossRef] [PubMed]
9. Choi, J.; Goo, D.; Sharma, M.K.; Ko, H.; Liu, G.; Paneru, D.; Choppa, V.S.R.; Lee, J.; Kim, W.K. Effects of different Eimeria inoculation doses on growth performance, daily feed intake, gut health, gut microbiota, foot pad dermatitis, and Eimeria gene expression in broilers raised in floor pens for 35 days. *Animals* **2023**, *13*, 2237. [CrossRef] [PubMed]
10. Sharma, M.K.; Regmi, P.; Applegate, T.; Chai, L.; Kim, W.K. Osteoimmunology: A link between gastrointestinal diseases and skeletal health in chickens. *Animals* **2023**, *13*, 1816. [CrossRef] [PubMed]
11. Zhang, Z.; Xu, P.; Liu, C.; Chen, J.; Ren, B.; Du, E.; Guo, S.; Li, P.; Li, L.; Ding, B. Effect of Tannic Acid on Antioxidant Function, Immunity, and Intestinal Barrier of Broilers Co-Infected with Coccidia and Clostridium perfringens. *Animals* **2024**, *14*, 955. [CrossRef] [PubMed]
12. Choi, J.; Yadav, S.; Vaddu, S.; Thippareddi, H.; Kim, W.K. In vitro and in vivo evaluation of tannic acid as an antibacterial agent in broilers infected with Salmonella Typhimurium. *Poult. Sci.* **2023**, *102*, 102987. [CrossRef] [PubMed]
13. Choi, J.; Tompkins, Y.H.; Teng, P.-Y.; Gogal Jr, R.M.; Kim, W.K. Effects of tannic acid supplementation on growth performance, oocyst shedding, and gut health of in broilers infected with eimeria maxima. *Animals* **2022**, *12*, 1378. [CrossRef]
14. Yang, T.; Sun, Y.; Dai, Z.; Liu, J.; Xiao, S.; Liu, Y.; Wang, X.; Yang, S.; Zhang, R.; Yang, C. Microencapsulated sodium butyrate alleviates immune injury and intestinal problems caused by clostridium perfringens through gut microbiota. *Animals* **2023**, *13*, 3784. [CrossRef]
15. Zhi, K.; Gong, F.; Chen, L.; Li, Z.; Li, X.; Mei, H.; Fu, C.; Zhao, Y.; Liu, Z.; He, J. Effects of Sea-Buckthorn Flavonoids on Growth Performance, Serum Inflammation, Intestinal Barrier and Microbiota in LPS-Challenged Broilers. *Animals* **2024**, *14*, 2073. [CrossRef]
16. Choi, J.; Lee, J.; Kim, W.K. Alterations in the gut microbiota of Eimeria infected broiler chickens fed diets supplemented with varying levels of dietary calcium and phosphorus, along with 25-hydroxycholecalciferol. *Poult. Sci.* **2024**, *103*, 104223. [CrossRef]
17. Shi, H.; Lopes, T.; Tompkins, Y.H.; Liu, G.; Choi, J.; Sharma, M.K.; Kim, W.K. Effects of phytase supplementation on broilers fed with calcium and phosphorus-reduced diets, challenged with Eimeria maxima and Eimeria acervulina: Influence on growth performance, body composition, bone health, and intestinal integrity. *Poult. Sci.* **2024**, *103*, 103511. [CrossRef] [PubMed]
18. Knowlton, K.; Radcliffe, J.; Novak, C.; Emmerson, D. Animal management to reduce phosphorus losses to the environment. *J. Anim. Sci.* **2004**, *82*, E173–E195. [CrossRef] [PubMed]

19. Yu, M.; Oketch, E.O.; Hong, J.-S.; Chathuranga, N.C.; Seo, E.; Park, H.; Vasanthakumari, B.L.; Lee, H.; Heo, J.-M. Efficacy and Equivalency of Phytase for Available Phosphorus in Broilers Fed an Available Phosphorus-Deficient Diet. *Animals* **2023**, *14*, 41. [CrossRef]
20. Smeets, N.; Nuyens, F.; Van Campenhout, L.; Delezie, E.; Niewold, T. Interactions between the concentration of non-starch polysaccharides in wheat and the addition of an enzyme mixture in a broiler digestibility and performance trial. *Poult. Sci.* **2018**, *97*, 2064–2070. [CrossRef] [PubMed]
21. Chen, X.; Zhang, G.-m.; Wang, W.-w.; Liu, G.-h.; Cai, H.-y.; Purba, A.; Zheng, A.-j. Compound non-starch polysaccharide enzymes improve growth performance, slaughter performance, immune function, and apparent utilization rate of nutrients in broiler chickens fed a low-metabolizable energy diet. *Front. Vet. Sci.* **2023**, *10*, 1162811. [CrossRef] [PubMed]
22. Shuaib, M.; Hafeez, A.; Tahir, M.; Sufyan, A.; Ullah, O.; Shams, M.A.; Siddiqui, S.A.; Swelum, A.A. Effects of β-Mannanase Supplementation and Soyhull Inclusion on Production Performance, Economics, Egg Quality, Blood Biochemicals, Nutrient Digestibility, and Intestinal Morphology in Golden Brown Hens (RIR× Fayoumi) during Late Peak Production. *Animals* **2024**, *14*, 2047. [CrossRef] [PubMed]
23. Lara, L.J.; Rostagno, M.H. Impact of heat stress on poultry production. *Animals* **2013**, *3*, 356–369. [CrossRef] [PubMed]
24. Sumanu, V.O.; Naidoo, V.; Oosthuizen, M.; Chamunorwa, J.P. A Technical Report on the Potential Effects of Heat Stress on Antioxidant Enzymes Activities, Performance and Small Intestinal Morphology in Broiler Chickens Administered Probiotic and Ascorbic Acid during the Hot Summer Season. *Animals* **2023**, *13*, 3407. [CrossRef]
25. Medina, Á.; González-Jartín, J.M.; Sainz, M.J. Impact of global warming on mycotoxins. *Curr. Opin. Food Sci.* **2017**, *18*, 76–81. [CrossRef]
26. Lee, J.; Cho, H.; Song, D.; Chang, S.; An, J.; Nam, J.; Lee, B.; Kim, S.; Kim, W.K.; Cho, J. Effects of Combinations of Toxin Binders with or without Natural Components on Broiler Breeders Exposed to Ochratoxin A. *Animals* **2023**, *13*, 2266. [CrossRef]
27. Hangalapura, B.; Nieuwland, M.; Buyse, J.; Kemp, B.; Parmentier, H. Effect of duration of cold stress on plasma adrenal and thyroid hormone levels and immune responses in chicken lines divergently selected for antibody responses. *Poult. Sci.* **2004**, *83*, 1644–1649. [CrossRef] [PubMed]
28. Kim, D.-H.; Song, J.-Y.; Park, J.; Kwon, B.-Y.; Lee, K.-W. The Effect of Low Temperature on Laying Performance and Physiological Stress Responses in Laying Hens. *Animals* **2023**, *13*, 3824. [CrossRef] [PubMed]
29. Al-Baadani, H.H.; Alhotan, R.A.; Azzam, M.M. Evaluation of the interaction between gum Arabic addition and stocking density on growth performance, carcass characteristics, and general health parameters of broiler chickens. *Animals* **2023**, *13*, 3024. [CrossRef]
30. Kang, S.W.; Christensen, K.D.; Jr, M.T.K.; Orlowski, S.K. Effects of Environmental Enrichments on Welfare and Hepatic Metabolic Regulation of Broiler Chickens. *Animals* **2024**, *14*, 557. [CrossRef] [PubMed]

animals

MDPI

Article

Effects of Different *Eimeria* Inoculation Doses on Growth Performance, Daily Feed Intake, Gut Health, Gut Microbiota, Foot Pad Dermatitis, and *Eimeria* Gene Expression in Broilers Raised in Floor Pens for 35 Days

Janghan Choi [1,2], Doyun Goo [1], Milan Kumar Sharma [1], Hanseo Ko [1], Guanchen Liu [1], Deependra Paneru [1], Venkata Sesha Reddy Choppa [1], Jihwan Lee [1] and Woo Kyun Kim [1,*]

1 Department of Poultry Science, University of Georgia, Athens, GA 30602, USA; choij@uga.edu (J.C.); dgoo@uga.edu (D.G.); milan.sharma@uga.edu (M.K.S.); hsko@uga.edu (H.K.); guanchen.liu@uga.edu (G.L.); dpaneru@uga.edu (D.P.); venkatasesharedy.choppa@uga.edu (V.S.R.C.); jihwan.lee@uga.edu (J.L.)

2 US National Poultry Research Center, United States Department of Agriculture Agricultural Research Service, Athens, GA 30605, USA

* Correspondence: wkkim@uga.edu; Tel.: +1-706-542-1346

Simple Summary: Coccidiosis, which is induced by *Eimeria* spp., is one of the most predominant diseases and causes tremendous economic losses in the world. The effects of *Eimeria* infection on broilers at slaughter ages in floor pen conditions should be elucidated to conduct further studies investigating the effects of feed additives and bioactive compounds as alternatives for anti-coccidial drugs in broilers. The study was aimed to investigate the effects of different inoculation doses of *E. acervulina*, *E. maxima*, and *E. tenella* with different doses on growth performance, gut ecosystem, and oocyst shedding in broilers raised in floor pens for 35 days. *Eimeria* infection decreased body weight (BW) in the acute phase (D 21), and this effect was prolonged to the final day (D 35). *Eimeria* oocysts were observed in the litter until D 35, which may indicate that *Eimeria* spp. reinfected broilers. *Eimeria* infection dramatically reduced crude fat (CF) digestibility in the acute phase, which may be associated with reduced fat content in the broilers on D 35. Gut microbiota was negatively affected by *Eimeria* infection in both acute phase and on D 35. In conclusion, *Eimeria* infection negatively affected the growth performance and gut ecosystem in broilers, and the negative effects were prolonged to D 35 in floor pen conditions.

Abstract: The study was conducted to investigate the effects of different *Eimeria* inoculation doses on the growth performance, gut ecosystem, and body composition of broilers in floor pens for 35 days. A total of 750 15-day-old broilers were allocated to five experimental groups with six replicate pens. The five experimental groups included unchallenged control (CON); *Eimeria* dose 1 (ED1): *E. acervulina*: 31,250/*E. maxima*: 6250/*E. tenella*: 6250; *Eimeria* dose 2 (ED2): *E. acervulina*: 62,500/*E. maxima*: 12,500/*E. tenella*: 12,500; *Eimeria* dose 3 (ED3): *E. acervulina*: 125,000/*E. maxima*: 25,000/*E. tenella*: 25,000; and *Eimeria* dose 4 (ED4): *E. acervulina*: 250,000/*E. maxima*: 50,000/*E. tenella*: 50,000. On D 21, BW were linearly reduced by increased *Eimeria* inoculation doses ($p < 0.01$). On D 35, the *Eimeria* challenge groups had significantly lower BW compared to the CON group. Increased *Eimeria* inoculation doses linearly decreased crude fat (CF) ($p < 0.01$) on D 21. Increased *Eimeria* inoculation doses tended to increase the relative abundance of the phylum Proteobacteria ($p = 0.098$) on D 21. On D 35, lean:fat was linearly reduced by increased *Eimeria* inoculation doses ($p < 0.05$). *Eimeria* infection negatively influenced growth performance and gut health in broilers in the acute phase, and the negative effects were prolonged to D 35 in floor pen conditions.

Keywords: *Eimeria*; apparent ileal digestibility; gut microbiota; body composition; floor pen; broilers

check for updates

Citation: Choi, J.; Goo, D.; Sharma, M.K.; Ko, H.; Liu, G.; Paneru, D.; Choppa, V.S.R.; Lee, J.; Kim, W.K. Effects of Different *Eimeria* Inoculation Doses on Growth Performance, Daily Feed Intake, Gut Health, Gut Microbiota, Foot Pad Dermatitis, and *Eimeria* Gene Expression in Broilers Raised in Floor Pens for 35 Days. *Animals* **2023**, *13*, 2237. https://doi.org/10.3390/ani13132237

Academic Editor: Christian Matthias Bauer

Received: 1 June 2023
Revised: 22 June 2023
Accepted: 5 July 2023
Published: 7 July 2023

1. Introduction

Coccidiosis, which is induced by *Eimeria* spp., is one of the most predominant diseases and causes tremendous economic losses estimated to be approximately USD 3 billion to 10.4 billion pounds annually in worldwide broiler production [1]. There are seven known *Eimeria* spp. in chickens, and *Eimeria acervulina*, *E. tenella*, and *E. maxima* are known to be most predominant *Eimeria* spp. in broiler production [2]. The *Eimeria* life cycle contains the exogenous (e.g., sporulation in the excreta) and the endogenous (asexual and sexual reproduction in the host intestine) life cycles [3]. During the endogenous phase, the activities of *Eimeria* spp. disrupt the functionality and integrity of the gastrointestinal tract of chickens, which induces watery feces and negatively influences nutrient utilization, growth performance, and animal welfare in the acute phase [4,5]. Watery feces can increase the moisture content of the litter, which can increase the incidence and severity of foot pad dermatitis (FPD) by increasing the litter ammonia level [6]. Decreased nutrient digestion and absorption due to *Eimeria* infection in the acute phase have a potential to alter body composition of broiler chickens at slaughter ages [7].

Traditionally, anti-coccidial drugs such as ionophores and synthetic drugs have been utilized for the control of coccidiosis in broilers [8]. However, due to public concerns regarding the spread of antibiotic-resistant microbes, the utilization of anti-coccidial drugs has been restricted in many countries [9]. Therefore, it has become essential to explore alternatives to anti-coccidial drugs in poultry production. Diverse bioactive compounds including amino acids [10], essential oils [11], and plant extracts [12] were studied to use as alternatives for anticoccidial drugs in broiler production. However, most of the *Eimeria* studies were conducted in cage conditions. *Eimeria* trials conducted in cage conditions have following limitations: (1) reinfection of *Eimeria* via the oral–fecal route is unlikely to occur because feces are regularly removed before they can build up in cages; (2) some cages are not suitable for accommodating broilers during the finisher phase (D 28 to 42); and (3) cage conditions cannot completely mimic the field conditions because broilers are normally raised in the litter condition in the poultry industry [13]. Proper research settings for *Eimeria* studies in floor pens are required, and the effects of *Eimeria* infection with different inoculation doses on growth performance, gut health, gut microbiota, and body composition in broilers at slaughter ages in the floor pen conditions should be elucidated to conduct further studies investigating the effects of feed additives and bioactive compounds as alternatives for anti-coccidial drugs in broilers. Therefore, this study was designed to investigate the effects of different *Eimeria* inoculation doses on the growth performance, litter moisture content, oocyst shedding, *Eimeria* gene expression, nutrient digestion and absorption, gut microbiota, incidence and severity of FPD, and body composition in broilers raised in floor pens for 35 days.

2. Materials and Methods

2.1. Experimental Design, Animals, Experimental Infection, Diets, and Growth Performance

The Institutional Animal Care and Use Committee at the University of Georgia approved animal care and use protocols of the current study. A total of 750 fifteen-day-old Cobb 500 male broilers were randomly distributed to 5 experimental groups with 6 replicate pens of 25 birds per pen. The 5 experimental groups were different inoculation doses of *E. acervulina*, *E. maxima*, and *E. tenella* as follows: unchallenged control (CON); *Eimeria* dose 1 (ED1): *E. acervulina*: 31,250/*E. maxima*: 6250/*E. tenella*: 6250; *Eimeria* dose 2 (ED2): *E. acervulina*: 62,500/*E. maxima*: 12,500/*E. tenella*: 12,500; *Eimeria* dose 3 (ED3): *E. acervulina*: 125,000/*E. maxima*: 25,000/*E. tenella*: 25,000; and *Eimeria* dose 4 (ED4): *E. acervulina*: 250,000/*E. maxima*: 50,000/*E. tenella*: 50,000. One milliliter of PBS containing freshly prepared *Eimeria* spp. (stored in the 5% potassium dichromate solution at 4 °C for less than 3 months) was administered to individual birds on D 15 [12]. The doses for *Eimeria* spp. were determined according to our previous study [14]. Birds were reared in floor pens (width: 1.52 m, length: 1.22 m, height: 0.61 m) equipped with one feeder and several nipple drinkers, and birds had free access to water and feed during the entire experimental period. The front side was blocked by wire, while the sides are blocked by hardboard, indicating the front side was more vulnerable for cross contamination. The CON group was distributed in a way that it could be located across from the front side to minimize the cross contamination, and *Eimeria* infected groups were randomly allocated. Temperature and light were managed according to Cobb 500 broiler management guide (2018). Diets were formulated to meet or exceed the recommended level according to Cobb 500 nutrient requirement guide (2018) and contained 0.3% titanium dioxide (Acros Organics, Morris Plains, NJ, USA) as an inert marker as shown in Table 1. The diets for the CON group in all phases contained 0.05% anticoccidial drug (Coban 90, Elanco Animal Health, Greenfield, IN, USA) to prevent cross contamination to the CON group. The feeding phases were divided into the grower (D 15 to 28) and finisher phases (D 28 to 35). Growth performance parameters including body weight (BW), average daily gain (ADG), average daily feed intake (ADFI), and feed conversion ratio (FCR) were measured on D 21 (6 days post inoculation (dpi)), D 28, and D 35, and daily feed intake (DFI) was measured through the entire experiment period.

Table 1. Ingredients and nutrient compositions of diets (as-fed basis).

Items	D 15 to 28	D 28 to 35
Ingredients (kg/ton)		
Corn	675.26	693.66
Soybean meal (480 g crude protein/kg)	262.75	238.77
Soybean oil	17.70	23.05
Dicalcium phosphate	12.62	12.81
Limestone	10.58	10.66
Sand [1]	7	7
Titanium dioxide [2]	3	3
Common Salt	3.45	3.47
DL-Methionine 99%	2.65	2.48
L-Lysine HCl 78%	2.01	2.15
Vitamin Premix [3]	1	1
Mineral Premix [4]	0.8	0.80
Choline 60%	0.70	0.60
L-threonine	0.48	0.54
Total	1000	1000
Calculated energy and nutrient value, %		
Metabolizable energy, kcal/kg	3100	3150
Crude fat	4.39	4.972
Crude protein	18	17
SID [5] methionine	0.544	0.515
SID total sulfur amino acids	0.800	0.760
SID lysine	1.02	0.970
SID threonine	0.660	0.630
Total calcium	0.760	0.760
Available phosphate	0.380	0.380

[1] The diet for the CON group contained 0.05% anticoccidial drug (Coban 90, Elanco Animal Health, Greenfield, IN, USA) in the sand part. [2] Titanium dioxide 3 g/kg (Acros Organics, Morris Plains, NJ, USA). [3] Vitamin mix included the following per kg: vitamin A (IU): 3,527,160; vitamin D3 (ICU): 1,399,921; Vitamin E (IU): 19,400; Vitamin B12 (mg): 8.8; Menadione (mg): 1102; Riboflavin (mg): 3527; d-Pantothenic acid (mg): 5467; Thiamin (mg): 970, Niacin (mg): 20,282; Vitamin B6 (mg): 1455; Folic acid (mg): 573; and Biotin (mg): 79. [4] Mineral mix included the following per kg: Ca (g): 0.72; Mn (g): 3.04; Zn (g): 2.43; Mg (g): 0.61; Fe (g): 0.59; Cu (g): 22.68; I (g): 22.68; Se (g): 9.07. [5] SID: standard ileal digestible amino acid.

2.2. Lesion Score, Oocyst Shedding, Gut Permeability, and Digesta and Litter Moisture Content

On D 21 (6 dpi), D 28, and D 35, duodenum, jejunum-ileum section, and ceca were collected, and lesion scores for each section were evaluated according to the 4-score scale [15]. On 22 (7 dpi), D 28, and D 35, cloaca content from one sacrificed bird per pen and approximately 30 g of litter samples were collected for oocyst shedding. Litter samples were collected in the different areas of the pen except areas near water nipples, and the litter samples were thoroughly mixed. Tap water (40 mL) were added to the samples (5 g), and the samples were placed at room temperature overnight to dissolve hard fecal particles. Afterwards, the samples were vortexed and diluted 10 times with saturated salt solution. *E. acervulina*, *E. maxima*, and *E. tenella* in the solution were counted using a hemocytometer (Hausser Scientific Company, Horsham, PA, USA). Ileal content and litter samples were oven-dried (75 °C) until a constant weight was achieved to determine their moisture content [16]. Gut permeability by using fluorescein isothiocyanate–dextran (molecular weight: 4 kDa; FITC-D4; Sigma-Aldrich Co., St. Louis, MO, USA) was determined on D 20 (5 dpi) and D 27 according to Choi et al. [12]. The concentration of FITD-F4 in the serum was determined using a prepared standard curve.

2.3. Foot Pad Dermatitis (FPD) and Body Composition Analysis

On D 35, severity and incidence of FPD were recorded according to Sorin et al. [17], as shown in Table 2.

Table 2. Footpad dermatitis (FPD) scoring system according to Sorin et al. [17].

FPD Score	Description
0	No lesion
1	FPD covers less than 50% of the food pad
2	FPD covers more than 50% of the food pad

One bird per pen was euthanized via cervical dislocation and scanned using a dual-energy X-ray absorptiometry (DEXA, GE Healthcare, Madison, WI, USA) as shown in Figure 1.

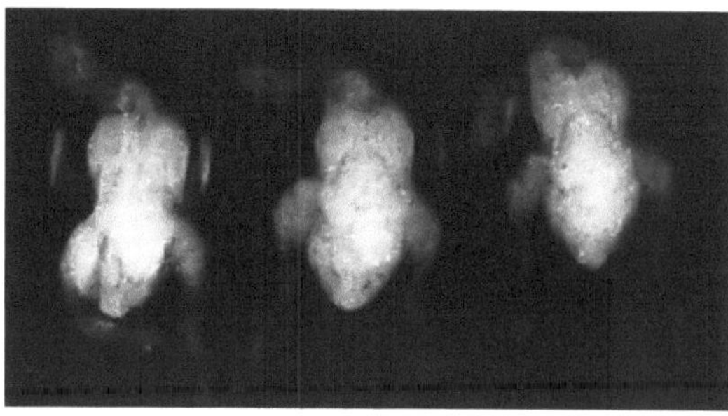

Parameters
Tissue weight (g)
Lean weight (g)
Fat weight (g)
Fat percentage (%)
Lean percentage (%)
Lean : fat (g/g)

Figure 1. Scanned images of dual-energy X-ray absorptiometry (DEXA) of broiler chickens and corresponding parameters.

2.4. Apparent Ileal Digestibility (AID) of Nutrients

On D 21 (6 dpi) and D 35, ileal content was collected between 10 cm below of the Meckel's diverticulum and 10 cm upper of the ileo-ceca-colonic junction. Feed (0.5 g) and ileal samples (0.3 g) were ashed at 600 °C overnight, and concentrations of titanium dioxide were analyzed according to Short et al. [18]. The concentrations of CP and crude fat (CF) were determined using nitrogen combustion analyses according to AOAC international (2000) analytical method 990.03 and analytical method 942.05, respectively. Apparent ileal digestibility (AID) of dry matter (DM), organic matter (OM), ash, CP, and CF were calculated according to Lin and Olukosi [19].

2.5. Intestinal Morphology

On D 21 (6 dpi) and D 35, the duodenum (mid-part of the duodenal loop), jejunum (10 cm upper of the Meckel's diverticulum), and mid-ceca were collected, and remaining digesta was rinsed with PBS and placed into 10% neutral buffered formalin solution. After 72 h fixation, samples were embedded in paraffin and sliced into 4 μm sections, and hematoxylin and eosin (H&E) staining was performed. The stained slides were pictured using a microscope (BZ-X810; Keyence, Osaka, Japan). The villus height (VH) and crypt depth (CD) were measured for the duodenum and jejunum samples, and CD was measured for the ceca samples using ImageJ (National Institutes of Health, Bethesda, MD, USA).

2.6. Activities of Jejunal Brush Border Digestive Enzymes and Alkaline Phosphatase in the Serum

On D 21 (6 dpi) and D 35, the jejunum (located 10 cm above the Meckel's diverticulum) was collected and immediately snap-frozen in liquid nitrogen. It was then stored at −80 °C for further analysis. Around 100 mg of the jejunum samples (10 cm upper of the Meckel's diverticulum) were homogenized in 1.8 mL PBS by using a beads beater (Biospec Products, Bartlesville, OK, USA). The homogenized samples were centrifuged at 12,000× *g* and 4 °C for 15 min, and the protein concentration of the supernatant was determined using Pierce BCA Protein Assay Kits (Thermo Fisher Scientific, Waltham, MA, USA). Activities of maltase and sucrase were evaluated according to Fan et al. [20]. Activities of alkaline phosphatase in the intestine and serum were determined according to Lackeyram et al. [21]. Activities of aminopeptidase N (APN) were measured according to the method by Maroux et al. [22]. Lipase activities were analyzed according to the method of Elgharbawy et al. [23]. The activities of digestive enzymes were expressed as values per mg protein.

2.7. RNA Extraction and Quantitative Real-Time Reverse Transcription PCR (qRT-PCR)

On D 21 (6 dpi), whole-tissue samples of the duodenum (mid-part of the duodenal loop), jejunum (10 cm upper of the Meckel's diverticulum), and mid-ceca were collected and immediately snap-frozen in liquid nitrogen. They were then stored at −80 °C for further analysis. Around 100 mg of the whole-tissue of the duodenum (mid-part of the duodenal loop), jejunum (10 cm upper of the Meckel's diverticulum), and mid-ceca was homogenized in QIAzol lysis reagents (Qiagen, Valencia, CA, USA) using a beads beater (Biospec Products, Bartlesville, OK, USA). RNA was extracted according to the manufacturer's procedure, and RNA quantity was measured using a NanoDrop 2000 spectrophotometer (Thermo Fisher Scientific, Waltham, MA, USA). One microgram of RNA was utilized to synthesize the first-strand cDNA using high-capacity cDNA synthesis kits (Applied Biosystems, Foster City, CA, USA). Primers used in the study are presented in Table 3.

Quantitative real-time reverse transcription PCR (qRT-PCR) was conducted using SYBR Green Master Mix with a Step One thermocycler (Applied Biosystems, Foster City, CA, USA). The final PCR volume (10 μL) contained 5 μL of SYBR Green Master Mix, 1.5 μL of cDNA, 0.5 μL of forward and reverse primers (10 μM), and 2.5 μL of water. Thermal cycle conditions for all reactions were as follows: 95 °C denature for 10 min, 40 cycles at 95 °C for 15 s and 60 °C for 1 min, 95 °C for 15 s, 60 °C for 1 min and 95 °C for 15 s. The melting curve of each gene was checked to confirm the specificity of each PCR product. Several PCR products from each gene were stained with 6 × DNA loading dye (Thermo Fisher Scientific, Waltham, MA, USA), electrophoresed on a 3% agarose gel in a Tris-acetate-EDTA buffer, and visualized by adding ethidium bromide to confirm the specificity of each PCR product. Relative abundance of *Eimeria 18s* genes was normalized by using host reference genes (geometric mean of beta actin and glyceraldehyde 3-phosphate dehydrogenase) to quantify *Eimeria* spp. in the host tissue, and relative abundance of *Eimeria* genes was normalized with a housekeeping gene (18s) for each *Eimeria* spp. [24]. Relative mRNA abundance was determined by using the $2^{-\Delta\Delta Ct}$ method [25]. The negative control, containing no cDNA, was included in each run, and each sample was run in duplicate.

Table 3. Primers used in the study.

Genes	Sequence, 5′ to 3′	Amplicon	Accession Number
Beta actin	F: CAACACAGTGCTGTCTGGTGGTA R: ATCGTACTCCTGCTTGCTGATCC	205	NM_205518.2
GAPDH [1]	F: GCTAAGGCTGTGGGGAAAGT R: TCAGCAGCAGCCTTCACTAC	161	NM_204305.2
	Eimeria acervulina		
18s	F: CTGCGAATGGCTCATTAAAA R: AATAAACACAGCCCCTCCAG	123	KT184333.1
APN [2]	F: CGTCTCCTCATTGATGCTGT R: GCCTTCTGTCTCCTTTCAGG	109	XM_013396512.1
Flagella-related protein	F: AGTGTTTAGCCCATCGAACC R: TTGCCACATTAACGACAGGT	123	XM_013394342.1
EF1 [3]	F: CCCAACAGCTCTGAGACAAA R: TGTATAGGCCGAAAGCAATG	118	XM_013397273.1
EF2	F: GATTCCGTTGATGGTGTTTG R: AGAAGTGCGCGATCTACCTT	107	XM_013392455.1
GAM56 [4]	F: ATCAGCAGCAGTCCTACACG R: GGTGTGCATAGCCATAGGTG	103	MK519446.1
GAM82	F: GGCAGCACTTGAAGCAATAA R: CGACCGACAATGTTTCTGAC	117	MK519447.1
	Eimeria maxima		
18s	F: TCGCGTCTCTAATGATCGTC R: TCTGCAATTCACAATGCGTA	110	AF027724.1
APN	F: TTTCGCCGTTGATTCTGTAG R: CTCCCCATTCAAGACCAAGT	124	XM_013483142.1
Flagella-related protein	F: GAGTTCCAGTCGTGGGATTT R: ACTGCCTGAAGCAGAAAGGT	116	XM_013478390.1
EF1A	F: TTGATGATAGCAGCGTCTCC R: GATGCCCTACAGCACACATT	124	GO305089.1
EF2	F: GATGGAAAGGGAGAACAGGA R: ACAGAATCCACGACGACAAG	125	GO305837
GAM56	F: CTTCCCTGAAACCCCTATGA R: TGAGGCTACGAAATGTGAGC	121	AY129951.2
GAM82	F: AGGTACCCCAGCTATGATGC R: CACGCGAGTATATGCTGGAT	109	AY179510.2
	Eimeria tenella		
18s	F: GTGCAAGGTTACGGAAGGAT R: CTACTGCTCTGTGGGTTGCT	113	XM 013373467.1
APN	F: TTCAAGACAGTTTGCCGAAG R: GCACAACCTCTGCACCTTTA	119	ETH_00013105
Flagella-related protein	F: GAGACAGGGCATTTGCTTC R: TGGTAGAAGCCGTAGGCAAT	120	XM_013378497.1
EF1A	F: TGATCGTGGGGATAAACAAA R: GGGTTGTAGCCCACTGTCTT	115	CD665486.1
GAM22	F: TAGCCACCCTAGTCGGTTTC R: ATCGCTTCTGGATCGATTTC	100	MH445412.1
GAM56	F: AGATGGGCACTTACCAGGAG R: AAGTTCTCCAGCCACTGGTC	112	XM_013376832.1

[1] *GAPDH*: glyceraldehyde 3-phosphate dehydrogenase. [2] *APN*: aminopeptidase N. [3] *EF*: elongation factor. [4] *GAM*: gametocyte protein.

2.8. DNA Extraction and Microbiome Analysis

On D 21 (6 dpi) and D 35, the cecal content was collected and immediately snap-frozen in liquid nitrogen. It was then stored at −80 °C for further analysis. DNA was extracted from the cecal content by employing QIAamp® DNA stool mini kit (Qiagen GmbH, Hilden, Germany) according to manufacturer's protocol. Quality and quantity of extracted DNA were checked using a NanoDrop 2000 spectrophotometer (Thermo Fisher Scientific), and samples were shipped to LC sciences (Houston, TX, USA) for 16s rRNA gene sequencing.

Qimme2 (version 2022.02) was used to process and analyze 16s rRNA gene sequences [26]. Using Qiime2's built-in functions, alpha diversity, beta diversity, and the phylum and family level composition were analyzed and presented.

2.9. Statistical Analyses

Statistical analyses and graph construction were performed using SAS (version 9.4; SAS Inst. Inc., Cary, NC, USA) and GraphPad Prism (Version 9.1.0; GraphPad Software, San Diego, CA, USA), respectively. All experimental groups were compared using PROC MIXED in a completely randomized design followed by the Tukey's HSD (honestly significant difference) test. For quantitative beta diversity measurement, each experimental group was placed as the control group, and experimental groups were compared using PROC MIXED with Dunnett's post hoc test. *Eimeria* lesion score data were compared using Kruskal–Wallis test followed by the Dwass–Steel–Critchlow–Fligner test. Orthogonal polynomial contrasts were performed to analyze the significance of linear or quadratic effects of different *Eimeria* doses, and the inoculation doses of *E. acervulina*, *E. maxima*, and *E. tenella* were normalized by using the base 2 logarithm of the number of sporulated Eimeria oocyst number for orthogonal polynomial contrasts. Statistical significance was set at $p < 0.05$, and trends ($0.05 \leq p \leq 0.1$) were also shown.

3. Results

3.1. Growth Performance and Daily Feed Intake (DFI)

The results of the growth performance are shown in Table 4. On D 21, BW was linearly ($p < 0.01$) and quadratically ($p < 0.05$) decreased due to increased *Eimeria* inoculation doses, the ED1 and ED2 groups had significantly lower BW compared to the CON group, and the ED4 group had significantly lower BW compared to the ED1 and ED2 groups. Increased *Eimeria* inoculation doses linearly and quadratically reduced ADFI ($p < 0.01$), and the CON group and the ED4 group had the highest and lowest ADFI among the experimental groups, respectively ($p < 0.05$). The FCR was linearly increased by increased *Eimeria* inoculation doses ($p < 0.01$), and the ED4 group and the CON group had the highest and lowest FCR, respectively, among the experimental groups ($p < 0.01$).

On D 28, increased *Eimeria* inoculation doses linearly ($p < 0.01$) and quadratically ($p < 0.05$) decreased BW, ADG, and FCR, and the *Eimeria* challenge groups had lower BW, ADG, and ADFI compared to the CON group. On D 35, *Eimeria* challenged groups had significantly lower BW compared to the CON group, and increased *Eimeria* doses linearly ($p < 0.05$) and quadratically ($p < 0.05$) reduced BW. In the whole phase, increased *Eimeria* inoculation doses linearly ($p < 0.01$) and quadratically ($p < 0.01$) reduced ADG, and the *Eimeria* infected groups had lower ADG compared to the CON group ($p < 0.01$). The ADFI were linearly ($p < 0.01$) and quadratically (tendency; $p = 0.076$) decreased by *Eimeria* infection, and the *Eimeria* challenged groups had lower ADFI compared to the CON group in the whole phase ($p < 0.01$). The CON group and the ED4 group had the lowest and highest FCR, respectively, among the experimental groups ($p < 0.01$), and increased *Eimeria* inoculation doses resulted in a linear increase in FCR in the whole phase ($p < 0.01$).

The DFI during the entire experimental period is presented in Figure 2. On D 19, DFI was linearly reduced by increased *Eimeria* inoculation doses ($p < 0.01$). From D 20 to 23, DFI was linearly ($p < 0.01$) and quadratically ($p < 0.05$) decreased by increased *Eimeria* inoculation doses. Increased *Eimeria* inoculation doses linearly decreased DFI on D 26 ($p < 0.01$) and D 27 ($p < 0.05$). On D 28, increased *Eimeria* inoculation doses tended to linearly ($p = 0.055$) and quadratically ($p = 0.075$) reduce DFI. On D 29, the *Eimeria* challenge tended to linearly reduce DFI ($p = 0.057$). No differences were observed in the DFI from D 30 to 35 ($p > 0.1$).

Table 4. Effects of different *Eimeria* infection doses on growth performance parameters including body weight (BW; g), average daily gain (ADG; g/d), average daily feed intake (ADFI; g/d), and feed conversion ratio (FCR, g/g) in broilers [1].

	CON	ED1	ED2	ED3	ED4	SEM [2]	*p* Value	Linear	Quadratic
Initial BW	440.8	442.4	443.0	444.3	442.8	18.9	0.998	0.809	0.845
D 15 to 21									
BW	859.76 ^{a}	769.97 ^{b}	747.12 ^{b}	719.24 ^{bc}	689.27 ^{c}	30.13	<0.001	<0.001	0.020
ADG	69.83 ^{a}	54.60 ^{b}	50.68 ^{bc}	45.82 ^{cd}	41.08 ^{d}	2.88	<0.001	<0.001	<0.001
ADFI	99.99 ^{a}	88.44 ^{b}	87.30 ^{bc}	82.44 ^{cd}	80.63 ^{d}	3.51	<0.001	<0.001	0.007
FCR	1.43 ^{d}	1.62 ^{c}	1.72 ^{bc}	1.80 ^{b}	1.97 ^{a}	0.080	<0.001	<0.001	0.598
D 21 to 28									
BW	1478.8 ^{a}	1312.0 ^{b}	1301.3 b	1241.7 ^{b}	1230.6 ^{b}	51.37	<0.001	<0.001	0.003
ADG	88.43 ^{a}	77.43 ^{b}	79.17 ^{b}	74.63 ^{b}	77.33 ^{b}	4.98	<0.001	<0.001	0.01
ADFI	145.83 ^{a}	134.45 ^{b}	133.06 ^{b}	126.48 ^{b}	129.52 ^{b}	6.23	<0.001	<0.001	0.02
FCR	1.65	1.74	1.68	1.70	1.68	0.060	0.211	0.847	0.136
D 28 to 35									
BW	2237.9 ^{a}	2049.5 ^{b}	2057.6 ^{b}	1975.8 ^{b}	1961.7 ^{b}	64.59	<0.001	<0.001	0.015
ADG	108.44	105.37	108.04	104.87	104.45	5.02	0.519	0.202	0.946
ADFI	180.15	181.55	182.48	180.80	180.50	4.48	0.902	0.992	0.389
FCR	1.66	1.73	1.69	1.73	1.73	0.06	0.314	0.110	0.599
Whole phase									
ADG	89.86 ^{a}	80.36 ^{b}	80.73 ^{b}	76.57 ^{b}	75.95 ^{b}	2.92	<0.001	<0.001	0.006
ADFI	144.09 ^{a}	137.13 ^{b}	136.63 ^{b}	132.28 ^{b}	132.70 ^{b}	3.85	<0.001	<0.001	0.076
FCR	1.59 ^{c}	1.70 ^{b}	1.70 ^{b}	1.74 ^{ab}	1.78 ^{a}	0.04	<0.001	<0.001	0.146

[1] CON (unchallenged control); ED1 (*Eimeria* dose 1): *E. acervulina*: 31,250/*E. maxima*: 6250/*E. tenella*: 6250; ED2 (*Eimeria* dose 2): *E. acervulina*: 62,500/*E. maxima*: 12,500/*E. tenella*: 12,500; ED3 (*Eimeria* dose 3): *E. acervulina*: 125,000/*E. maxima*: 25,000/*E. tenella*: 25,000; ED4 (*Eimeria* dose 4): *E. acervulina*: 250,000/*E. maxima*: 50,000/*E. tenella*: 50,000. Oral gavage of *Eimeria* spp. was performed on D 15. Experimental groups were compared using PROC MIXED followed by the Tukey's HSD (honestly significant difference) test. Different letters in the same row means significant differences ($p < 0.05$) among the experimental groups. Orthogonal polynomial contrasts were conducted to see linear and quadratic patterns among the experimental groups. [2] Standard errors of the means.

3.2. Lesion Score and Gut Permeability

On D 21, the CON group had significantly lower lesion scores for *E. acervulina*, *E. maxima*, and *E. tenella* compared to the Eimeria infected groups as shown in Figure 3. The ED4 group had significantly higher *E. tenella* lesion scores compared to the ED1 group. On D 28, although no statistical differences in *E. acervulina*, *E. maxima*, and *E. tenella* lesion scores were observed among the experimental groups ($p > 0.1$), lesion scores of *E. acervulina* and *E. maxima* were observed in all experimental groups, and *E. tenella* lesion score was observed in the ED2, ED3, and ED4 groups. On D 35, *E. acervulina* lesion scores were observed only in the ED3 group, and *E. maxima* lesion scores were observed in all experimental groups. *E. tenella* lesion scores were observed in all experimental groups except the ED2 group.

As shown in Table 5, on D 20, the ED4 group had significantly higher gut permeability compared to the CON and ED1 groups, and the ED3 group had significantly higher gut permeability compared to the ED2 group. However, no significant differences were observed in gut permeability on D 27 ($p > 0.1$).

3.3. Digesta and Litter Moisture Content and Foot Pad Dermatitis (FPD)

As shown in Table 6, on D 21 and D 35, increased *Eimeria* inoculation doses resulted in a linear increase in ileal moisture content ($p < 0.01$), and the ED4 group had significantly higher ileal moisture content compared to the CON group. On D 21, litter moisture content tended to be increased in a linear trend by increased *Eimeria* inoculation doses ($p = 0.098$). Increased *Eimeria* inoculation doses caused a quadratic increase in the litter moisture content on D 35 ($p < 0.05$).

Table 5. Effects of different *Eimeria* infection doses on gut permeability concentration (mg/mL) of fluorescein isothiocyanate–dextran (molecular weight: 4 kDa; FITC-D4) in broilers on D 20 and D 27 [1].

	CON	ED1	ED2	ED3	ED4	SEM [2]	*p* Value	Linear	Quadratic
D 20	0.002 [c]	0.057 [c]	0.125 [bc]	0.355 [a]	0.326 [ab]	0.125	<0.001	<0.001	0.976
D 27	0.195	0.551	0.530	0.387	0.462	0.3	0.286	0.361	0.157

[1] CON (unchallenged control); ED1 (*Eimeria* dose 1): *E. acervulina*: 31,250/*E. maxima*: 6250/*E. tenella*: 6250; ED2 (*Eimeria* dose 2): *E. acervulina*: 62,500/*E. maxima*: 12,500/*E. tenella*: 12,500; ED3 (*Eimeria* dose 3): *E. acervulina*: 125,000/*E. maxima*: 25,000/*E. tenella*: 25,000; ED4 (*Eimeria* dose 4): *E. acervulina*: 250,000/*E. maxima*: 50,000/*E. tenella*: 50,000. Oral gavage of *Eimeria* spp. was performed on D 15. Experimental groups were compared using PROC MIXED followed by the Tukey's HSD (honestly significant difference) test. Different letters in the same row means significant differences ($p < 0.05$) among the experimental groups. Orthogonal polynomial contrasts were conducted to see linear and quadratic patterns among the experimental groups. [2] Standard errors of the means.

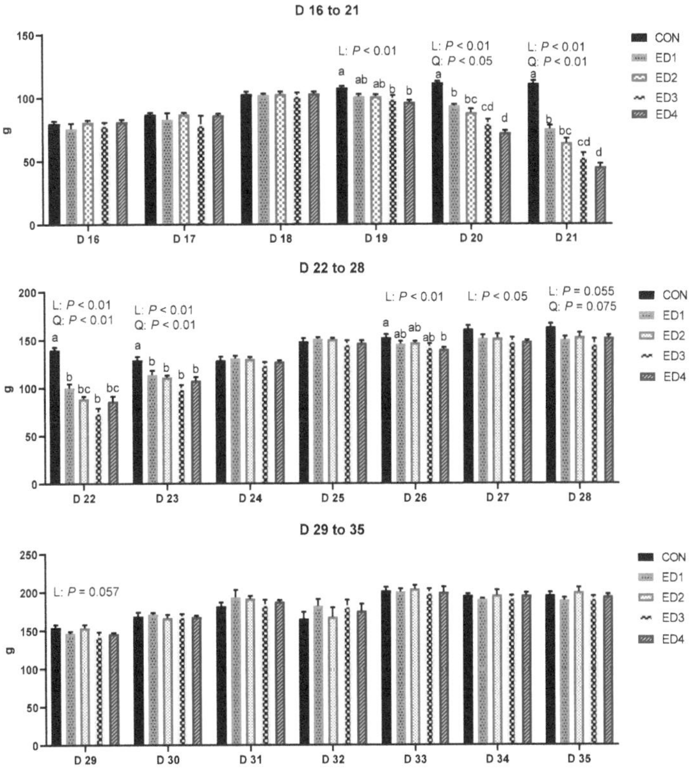

Figure 2. Daily feed intake (g) during the entire experimental period in the unchallenged control (CON); *Eimeria* dose 1 (ED1): *E. acervulina*: 31,250/*E. maxima*: 6250/*E. tenella*: 6250; *Eimeria* dose 2 (ED2): *E. acervulina*: 62,500/*E. maxima*: 12,500/*E. tenella*: 12,500; *Eimeria* dose 3 (ED3): *E. acervulina*: 125,000/*E. maxima*: 25,000/*E. tenella*: 25,000; and *Eimeria* dose 4 (ED4): *E. acervulina*: 250,000/*E. maxima*: 50,000/*E. tenella*: 50,000 groups. Oral gavage of *Eimeria* spp. was performed on D 15. Experimental groups were compared using PROC MIXED followed by the Tukey's HSD (honestly significant difference) test, and different letters mean significant differences ($p < 0.05$) among the experimental groups on the same day. Orthogonal polynomial contrasts were conducted to see linear and quadratic patterns among the experimental groups on the same day.

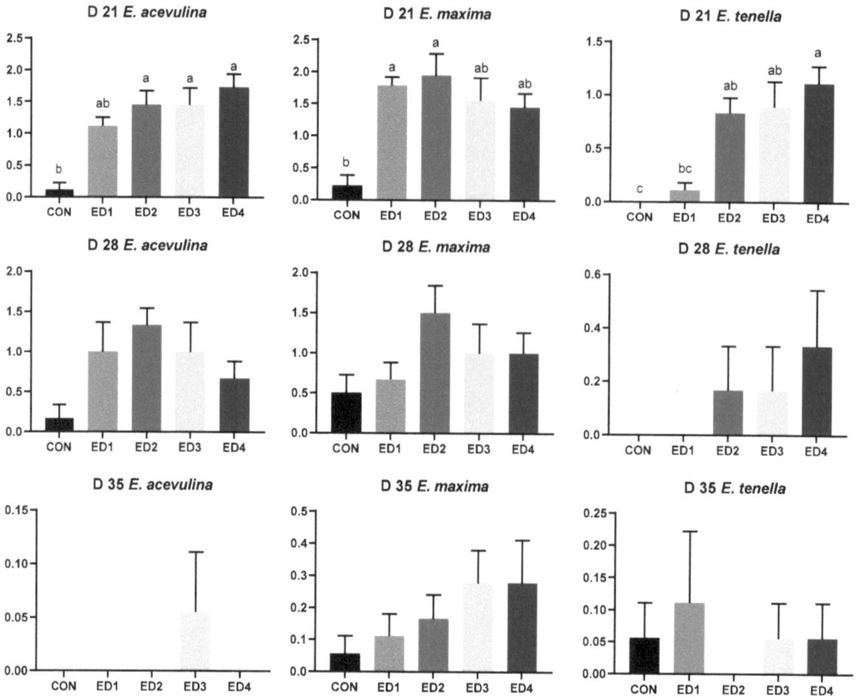

Figure 3. Duodenal, jejunal, and cecal lesion score in the unchallenged control (CON); *Eimeria* dose 1 (ED1): *E. acervulina*: 31,250/*E. maxima*: 6250/*E. tenella*: 6250; *Eimeria* dose 2 (ED2): *E. acervulina*: 62,500/*E. maxima*: 12,500/*E. tenella*: 12,500; *Eimeria* dose 3 (ED3): *E. acervulina*: 125,000/*E. maxima*: 25,000/*E. tenella*: 25,000; and *Eimeria* dose 4 (ED4): *E. acervulina*: 250,000/*E. maxima*: 50,000/*E. tenella*: 50,000 groups on D 21, D 28, and D 35. Oral gavage of *Eimeria* spp. was performed on D 15. Lesion scores for each section were evaluated according to the 4-score scale [15]. Experimental groups were compared using Kruskal–Wallis test followed by the Dwass–Steel–Critchlow–Fligner test. Different letters mean significant differences ($p < 0.05$) among the experimental groups.

Table 6. Effects of different *Eimeria* infection doses on ileal digesta and litter moisture content (%) in broilers [1].

	CON	ED1	ED2	ED3	ED4	SEM [2]	*p* Value	Linear	Quadratic
Ileum digesta									
D 21	81.3 [b]	85.1 [ab]	86.3 [ab]	86.6 [ab]	89.9 [a]	3.04	0.002	<0.001	0.697
D 35	86.3 [b]	89.5 [ab]	87.9 [ab]	90.2 [ab]	93.0 [a]	3.62	0.040	0.005	0.545
Litter									
D 22	26.8	32.8	30.0	33.0	37.0	9.16	0.415	0.098	0.891
D 28	20.1	25.6	17.9	22.5	22.7	6.82	0.383	0.816	0.873
D 35	24.1	28.4	25.8	25.0	23.0	3.35	0.100	0.200	0.047

[1] CON (unchallenged control); ED1 (*Eimeria* dose 1): *E. acervulina*: 31,250/*E. maxima*: 6250/*E. tenella*: 6250; ED2 (*Eimeria* dose 2): *E. acervulina*: 62,500/*E. maxima*: 12,500/*E. tenella*: 12,500; ED3 (*Eimeria* dose 3): *E. acervulina*: 125,000/*E. maxima*: 25,000/*E. tenella*: 25,000; ED4 (*Eimeria* dose 4): *E. acervulina*: 250,000/*E. maxima*: 50,000/*E. tenella*: 50,000. Oral gavage of *Eimeria* spp. inoculum (1 mL of PBS) was performed on D 15. Experimental groups were compared using PROC MIXED followed by the Tukey's HSD (honestly significant difference) test. Different letters in the same row means significant differences ($p < 0.05$) among the experimental groups. Orthogonal polynomial contrasts were conducted to see linear and quadratic patterns among the experimental groups. [2] Standard errors of the means.

The severity and incidence of FPD on D 35 were quadratically increased by increased *Eimeria* inoculation doses ($p < 0.05$; Figure 4).

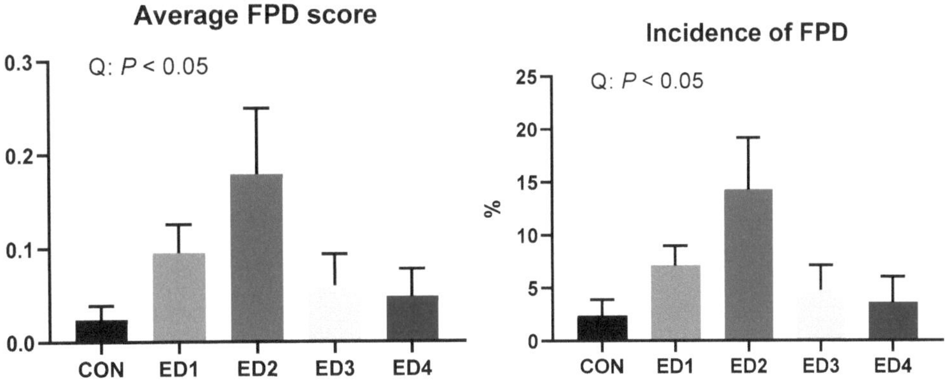

Figure 4. Foot pad dermatitis (FPD) in the unchallenged control (CON); *Eimeria* dose 1 (ED1): *E. acervulina*: 31,250/*E. maxima*: 6250/*E. tenella*: 6250; *Eimeria* dose 2 (ED2): *E. acervulina*: 62,500/*E. maxima*: 12,500/*E. tenella*: 12,500; *Eimeria* dose 3 (ED3): *E. acervulina*: 125,000/*E. maxima*: 25,000/*E. tenella*: 25,000; and *Eimeria* dose 4 (ED4): *E. acervulina*: 250,000/*E. maxima*: 50,000/*E. tenella*: 50,000 groups on D 35. Oral gavage of *Eimeria* spp. inoculum (1 mL of PBS) was performed on D 15. Severity (0 to 2), and incidence of FPD was recorded according to Sorin et al. [17]. Experimental groups were compared using Kruskal–Wallis test followed by the Dwass–Steel–Critchlow–Fligner test. Orthogonal polynomial contrasts were conducted to see linear and quadratic patterns among the experimental groups.

3.4. Oocyst Shedding

Table 7 shows that increased inoculation doses of *E. acervulina*, *E. maxima*, and *E. tenella* linearly increased the number of oocysts of *E. acervulina* and *E. tenella* ($p < 0.05$) and tended to linearly increase *E. maxima* oocyst in the cloaca content ($p = 0.081$) on D 22. The oocyst number of *E. maxima* in the cloaca were quadratically modulated by increased doses of *E. maxima* ($p < 0.05$). Linearly increased number of oocysts of *E. acervulina* and *E. tenella* in the litter were observed by increased infection doses of *Eimeria* spp. ($p < 0.01$), and the ED3 and ED4 groups had significantly higher *E. acervulina* oocysts in the litter compared to the ED1 group, and the ED2 and ED3 groups had significantly higher *E. tenella* oocysts compared to the ED1 group on D 22. However, no statistical differences were observed in the number of *Eimeria* oocysts among the *Eimeria* challenged groups ($p > 0.1$), *Eimeria* oocysts were observed either in the cloaca content and litter in the *Eimeria* infected groups on D 28 and 35.

3.5. Apparent Ileal Digestibility (AID) of Nutrients

As shown in Table 8, increased *Eimeria* inoculation doses linearly decreased the AID of CP and EE ($p < 0.01$) on D 21, and the AID of EE tended to be quadratically reduced by increased *Eimeria* inoculation doses ($p = 0.056$). The ED4 group had a significantly lower AID of CP compared to the CON group. The ED3 and ED4 groups had a significantly lower AID of EE compared to the CON group, and the ED2, ED3, and ED4 groups had minus values for the AID of CF. On D 35, the AID of ash was linearly decreased by increased *Eimeria* inoculation doses ($p < 0.05$).

Table 7. Effects of different doses of *Eimeria* spp. on oocyst shedding of *Eimeria acervulina*, *Eimeria maxima*, and *Eimeria tenella* in the cloaca content and litter of broilers on D 22, D 28, and D 35 [1].

	CON	ED1	ED2	ED3	ED4	SEM [2]	*p* Value	Linear	Quadratic
D 22/Cloaca									
E. acervulina	0	161,677	619,889	716,744	1,497,462	1,003,493	0.12	0.011	0.583
E. maxima	0 [b]	125,476 [ab]	227,080 [a]	100,460 [ab]	129,101 [ab]	99,216	0.018	0.081	0.012
E. tenella	0	80,162	407,110	203,800	466,214	329,393	0.101	0.02	0.751
D 22/Litter									
E. acervulina	0 [b]	14,626 [b]	38,062 [ab]	86,810 [a]	87,311 [a]	34,489	<0.001	<0.001	0.956
E. maxima	0	27,237	25,777	23,110	12,657	16,497	0.0429	0.329	0.005
E. tenella	0 [b]	1201 [b]	25,437 [a]	24,141 [a]	18,550 [ab]	13,424	0.004	0.002	0.068
D 28/Cloaca									
E. acervulina	1074 [b]	185,269 [a]	35,218 [ab]	66,520 [ab]	35,088 [ab]	104,594	0.048	0.71	0.13
E. maxima	0	982.17	982.17	982.17	982.17	3449	0.041	0.12	0.018
E. tenella	0 [b]	6111.7 [ab]	11,395 [a]	2804.88 [ab]	2008.89 [ab]	5956	0.025	0.927	0.005
D 28/Litter									
E. acervulina	2444	939	2681	2679	5044	3341	0.352	0.12	0.251
E. maxima	602	496	146	0	1001	932	0.39	0.804	0.102
E. tenella	1593	1753	1167	2922	1437	2253	0.706	0.771	0.785
D 35/Cloaca									
E. acervulina	31,641	11,185	14,544	27,598	1013	31,505	0.457	0.281	0.958
E. maxima	22,004	1909.96	0	552	7407	21,874	0.394	0.29	0.104
E. tenella	3745	37,915	992	1485	305	31,115	0.195	0.291	0.49
D 35/Litter									
E. acervulina	26,572	22,367	15,392	10,743	19,730	30,159	0.908	0.522	0.539
E. maxima	6500	0	1409	3480	2347	5556	0.347	0.507	0.191
E. tenella	2524.98	14,995	2117.5	8523.45	7920.53	14,321	0.531	0.817	0.756

[1] CON (unchallenged control); ED1 (*Eimeria* dose 1): *E. acervulina*: 31,250/*E. maxima*: 6250/*E. tenella*: 6250; ED2 (*Eimeria* dose 2): *E. acervulina*: 62,500/*E. maxima*: 12,500/*E. tenella*: 12,500; ED3 (*Eimeria* dose 3): *E. acervulina*: 125,000/*E. maxima*: 25,000/*E. tenella*: 25,000; ED4 (*Eimeria* dose 4): *E. acervulina*: 250,000/*E. maxima*: 50,000/*E. tenella*: 50,000. Oral gavage of *Eimeria* spp. inoculum (1 mL of PBS) was performed on D 15. Experimental groups were compared using PROC MIXED followed by the Tukey's HSD (honestly significant difference) test. Different letters in the same row means significant differences ($p < 0.05$) among the experimental groups. Orthogonal polynomial contrasts were conducted to see linear and quadratic patterns among the experimental groups. [2] Standard errors of the means.

Table 8. Effects of different *Eimeria* infection doses on apparent ileal digestibility (%) of dry matter (DM), organic matter (OM), ash, crude protein (CP), and crude fat (CF) in broilers on D 21 and D 35 [1].

	CON	ED1	ED2	ED3	ED4	SEM [2]	*p* Value	Linear	Quadratic
D 21									
DM	52.49	56.45	48.09	47.06	47.14	13.99	0.726	0.295	0.985
OM	55.16	58.38	50.38	49.79	49.35	13.29	0.716	0.268	0.996
Ash	7.87	24.24	9.76	1.42	10.17	28.19	0.72	0.64	0.84
CP	67.92 [a]	62.04 [ab]	53.82 [ab]	53.15 [ab]	46.50 [b]	11.3	0.037	0.002	0.738
CF	53.47 [a]	11.22 [ab]	−8.54 [b]	−9.23 [b]	−11.21 [b]	31.34	0.013	0.002	0.056
D 35									
DM	68.25	71.46	67.24	69.68	68.25	2.65	0.358	0.761	0.687
OM	69.20	72.74	68.73	71.38	70.53	7.75	0.398	0.83	0.751
Ash	46.56	45.21	36.75	36.40	25.61	6.57	0.045	0.005	0.504
CP	76.20	78.40	75.72	77.16	76.98	2.71	0.764	0.957	0.923
CF	45.90	58.08	54.81	56.71	48.69	15.12	0.875	0.899	0.347

[1] CON (unchallenged control); ED1 (*Eimeria* dose 1): *E. acervulina*: 31,250/*E. maxima*: 6250/*E. tenella*: 6250; ED2 (*Eimeria* dose 2): *E. acervulina*: 62,500/*E. maxima*: 12,500/*E. tenella*: 12,500; ED3 (*Eimeria* dose 3): *E. acervulina*: 125,000/*E. maxima*: 25,000/*E. tenella*: 25,000; ED4 (*Eimeria* dose 4): *E. acervulina*: 250,000/*E. maxima*: 50,000/*E. tenella*: 50,000. Oral gavage of *Eimeria* spp. inoculum (1 mL of PBS) was performed on D 15. Experimental groups were compared using PROC MIXED followed by the Tukey's HSD (honestly significant difference) test. Different letters in the same row means significant differences ($p < 0.05$) among the experimental groups. Orthogonal polynomial contrasts were conducted to see linear and quadratic patterns among the experimental groups. [2] Standard errors of the means.

3.6. Intestinal Morphology

As shown in Table 9, duodenal VH were linearly reduced by increased *Eimeria* inoculation doses ($p < 0.01$), and the ED3 and ED4 groups had significantly lower duodenal VH compared to the CON and ED1 groups. Duodenal CD was linearly deepened by increased *Eimeria* inoculation doses ($p < 0.01$), and the CON group had lower duodenal CD compared to the ED2, ED3, and ED4 groups ($p < 0.01$). Duodenal VH:CD were linearly reduced by increased *Eimeria* inoculation doses ($p < 0.01$), and the *Eimeria* challenged groups has significantly lower duodenal VH:CD compared to the CON group ($p < 0.01$). Jejunal VH was linearly decreased by increased *Eimeria* inoculation doses ($p < 0.01$), and the ED3 and ED4 groups had significantly lower jejunal VH compared to the CON and ED1 groups. Jejunal CD was linearly deepened by increased *Eimeria* inoculation doses ($p < 0.01$), and the CON group had the lowest jejunal CD among the experimental groups ($p < 0.01$). Jejunal VH:CD was linearly reduced by increased *Eimeria* inoculation doses ($p < 0.01$), and the ED3 and ED4 groups had the lowest jejunal VH:CD among the experimental groups ($p < 0.01$), and the CON group had the highest jejunal VH:CD among the experimental groups ($p < 0.01$). Cecal CD was linearly deepened by increased *Eimeria* inoculation doses ($p < 0.01$), and the ED3 and ED4 groups had significantly deeper cecal CD compared to the CON group ($p < 0.05$). On D 35, duodenal CD was quadratically deepened by increased *Eimeria* inoculation doses ($p < 0.01$), and the ED2 group had significantly higher duodenal CD compared to the CON group. Increased *Eimeria* inoculation doses resulted in a quadratic increase in duodenal VH:CD ($p < 0.05$).

Table 9. Effects of different *Eimeria* infection doses on intestinal morphology parameters including villus height (VH, μm), crypt depth (CD, μm), and VH:CD in the duodenum, jejunum, and ceca of broilers on D 21 and D 35 [1].

	CON	ED1	ED2	ED3	ED4	SEM [2]	*p* Value	Linear	Quadratic
D 21									
Duodenal VH	2480.12 [a]	2381.18 [a]	2067.76 [ab]	1790.54 [b]	1814.41 [b]	248.14	<0.001	<0.001	0.464
Duodenal CD	228.78 [d]	295.56 [cd]	380.47 [bc]	474 [ab]	509.76 [a]	66.48	<0.001	<0.001	0.603
Duodenal VH:CD	11.85 [a]	8.32 [b]	5.76 [bc]	3.88 [c]	3.88 [c]	1.601	<0.001	<0.001	0.004
Jejunal VH	1014.28 [a]	831.82 [ab]	733.49 [bc]	599.69 [c]	553.8 [c]	120.69	<0.001	<0.001	0.209
Jejunal CD	191.18 [b]	359.39 [a]	385.2 [a]	454.07 [a]	444.97 [a]	65.84	<0.001	<0.001	0.005
Jejunal VH:CD	5.45 [a]	2.45 [b]	2.08 [bc]	1.40 [c]	1.29 [c]	0.575	<0.001	<0.001	<0.001
Cecal CD	292.8 [b]	381.5 [ab]	502.3 [ab]	581.3 [a]	583.0 [a]	155.41	0.011	<0.001	0.372
D 35									
Duodenal VH	2432.73	2319.83	2387.54	2596.8	2541.9	299.81	0.507	0.212	0.579
Duodenal CD	270.09 [b]	358.24 [ab]	393.32 [a]	313.91 [ab]	298.64 [ab]	71.07	0.045	0.89	0.007
Duodenal VH:CD	9.50	6.72	6.63	9.05	8.97	1.91	0.032	0.609	0.012
Jejunal VH	1352.43	1358.23	1232.07	1237.21	1313.67	368.98	0.955	0.68	0.633
Jejunal CD	211.67	286.91	251.41	251.31	249.53	76.7	0.585	0.689	0.321
Jejunal VH:CD	6.86	5.33	5.25	5.30	5.51	1.44	0.278	0.154	0.113
Cecal CD	348.9	432.7	385.7	423.0	386.6	66.88	0.257	0.468	0.15

[1] CON (unchallenged control); ED1 (*Eimeria* dose 1): *E. acervulina*: 31,250/*E. maxima*: 6250/*E. tenella*: 6250; ED2 (*Eimeria* dose 2): *E. acervulina*: 62,500/*E. maxima*: 12,500/*E. tenella*: 12,500; ED3 (*Eimeria* dose 3): *E. acervulina*: 125,000/*E. maxima*: 25,000/*E. tenella*: 25,000; ED4 (*Eimeria* dose 4): *E. acervulina*: 250,000/*E. maxima*: 50,000/*E. tenella*: 50,000. Oral gavage of *Eimeria* spp. inoculum (1 mL of PBS) was performed on D 15. Experimental groups were compared using PROC MIXED followed by the Tukey's HSD (honestly significant difference) test. Different letters in the same row means significant differences ($p < 0.05$) among the experimental groups. Orthogonal polynomial contrasts were conducted to see linear and quadratic patterns among the experimental groups. [2] Standard errors of the means.

3.7. Activities of Jejunal Brush Border Digestive Enzymes

As shown in Table 10, activities of APN were lower in the ED1 group compared to the CON group on D 21 ($p < 0.05$). Activities of serum alkaline phosphatase (SAP) were linearly reduced by increased *Eimeria* inoculation doses ($p < 0.01$), and the ED3 and ED4 groups had significantly lower activities of SAP compared to the CON group. The ED2 group had significantly higher sucrase activities compared to the ED3 group. Increased *Eimeria* inoculation doses linearly decreased maltase activities ($p < 0.01$), and the ED1 and ED4 groups had significantly lower maltase activities compared to the CON group. On D 35, activities of APN were linearly decreased by increased *Eimeria* infection doses ($p < 0.05$).

Table 10. Effects of different *Eimeria* infection doses on activities of jejunal brush border digestive enzymes including maltase (nmol glucose released/mg protein/min), sucrase (nmol glucose released/mg protein/min), aminopeptidase N (APN; nmol p-nitroaniline liberated/mg protein/min), lipase (mmol p-nitrophenyl phosphate liberated/mg protein/min), and intestinal alkaline phosphatase (IAP; µmol p-nitrophenol liberated/mg protein/min) and serum alkaline phosphatase (SAP; µmol p-nitrophenol liberated/mL serum/min)] in broilers on D 21 and D 35 [1].

	CON	ED1	ED2	ED3	ED4	SEM [2]	p Value	Linear	Quadratic
D 21									
APN	12.95 [a]	10.83 [b]	12.71 [ab]	11.49 [ab]	11.46 [ab]	1.20	0.023	0.144	0.559
Lipase	0.702	0.481	0.545	0.499	0.539	0.199	0.358	0.245	0.189
Maltase	3.508 [a]	2.254 [b]	3.149 [ab]	2.297 [ab]	2.007 [b]	0.728	0.006	0.004	0.872
Sucrase	0.521 [ab]	0.496 [ab]	0.657 [a]	0.399 [b]	0.483 [ab]	0.118	0.016	0.273	0.275
SAP	2.575 [a]	1.842 [ab]	2.08 [ab]	1.583 [b]	1.528 [b]	0.544	0.017	0.003	0.463
D 35									
APN	14.54	14.37	12.37	13.97	12.20	1.67	0.054	0.027	0.880
Lipase	0.734	0.549	0.555	0.638	0.61	0.152	0.229	0.420	0.101
Maltase	4.300	3.022	3.574	3.761	3.269	1.294	0.500	0.435	0.546
Sucrase	0.499	0.394	0.524	0.453	0.457	0.206	0.844	0.927	0.962
IAP	0.157	0.13	0.16	0.149	0.163	0.029	0.328	0.445	0.358
SAP	1.536	1.506	1.734	1.639	1.465	0.414	0.795	0.989	0.345

[1] CON (unchallenged control); ED1 (*Eimeria* dose 1): *E. acervulina*: 31,250/*E. maxima*: 6250/*E. tenella*: 6250; ED2 (*Eimeria* dose 2): *E. acervulina*: 62,500/*E. maxima*: 12,500/*E. tenella*: 12,500; ED3 (*Eimeria* dose 3): *E. acervulina*: 125,000/*E. maxima*: 25,000/*E. tenella*: 25,000; ED4 (*Eimeria* dose 4): *E. acervulina*: 250,000/*E. maxima*: 50,000/*E. tenella*: 50,000. Oral gavage of *Eimeria* spp. inoculum (1 mL of PBS) was performed on D 15. Experimental groups were compared using PROC MIXED followed by the Tukey's HSD (honestly significant difference) test. Different letters in the same row means significant differences ($p < 0.05$) among the experimental groups. Orthogonal polynomial contrasts were conducted to see linear and quadratic patterns among the experimental groups. [2] Standard errors of the means.

3.8. Alpha Diversity of the Cecal Microbiome Communities

No differences were observed in the alpha diversity parameters (biodiversity of the samples) on D 21 ($p > 0.1$; Table 11). On D 35, alpha diversity parameters including pielou evenness (evenness; $p < 0.05$), faith phylogenetic diversity (biodiversity based on phylogeny; $p < 0.05$), shannon entropy (richness and evenness; $p < 0.05$), and observed features (richness; $p < 0.01$) were linearly reduced by increased *Eimeria* inoculation doses.

3.9. Beta Diversity of the Cecal Microbiome Communities

As shown in Figure 5, no differences were observed in unweighted unifrac distance (dissimilarity among samples without considering abundance information) to each experimental group both on D 21 and D 35 ($p > 0.1$).

Table 11. Effects of different *Eimeria* infection doses on alpha diversity parameters of the cecal microbial communities in broilers on D 21 and D 35 [1].

	CON	ED1	ED2	ED3	ED4	SEM [2]	*p* Value	Linear	Quadratic
D 21									
Pielou evenness	0.610	0.646	0.575	0.586	0.634	0.108	0.756	0.923	0.526
Faith phylogenetic diversity	16.91	15.29	18.11	16.51	16.07	3.037	0.592	0.907	0.659
Shannon entropy	5.015	5.208	4.694	4.880	5.205	0.957	0.864	0.966	0.516
Observed features	298.8	267.3	303.0	316.7	295.5	56.93	0.663	0.567	0.988
D 35									
Pielou evenness	0.680	0.743	0.671	0.631	0.625	0.076	0.09	0.034	0.377
Faith phylogenetic diversity	20.5	19.26	20.15	17.31	18.12	2.338	0.129	0.036	0.914
Shannon entropy	5.94	6.506	5.656	5.328	5.259	0.752	0.051	0.015	0.520
Observed features	424.5	435.3	346.3	351.0	340.5	63.51	0.03	0.005	0.604

[1] CON (unchallenged control); ED1 (*Eimeria* dose 1): *E. acervulina*: 31,250/*E. maxima*: 6250/*E. tenella*: 6250; ED2 (*Eimeria* dose 2): *E. acervulina*: 62,500/*E. maxima*: 12,500/*E. tenella*: 12,500; ED3 (*Eimeria* dose 3): *E. acervulina*: 125,000/*E. maxima*: 25,000/*E. tenella*: 25,000; ED4 (*Eimeria* dose 4): *E. acervulina*: 250,000/*E. maxima*: 50,000/*E. tenella*: 50,000. Oral gavage of *Eimeria* spp. inoculum (1 mL of PBS) was performed on D 15. Experimental groups were compared using PROC MIXED followed by the Tukey's HSD (honestly significant difference) test. Orthogonal polynomial contrasts were conducted to see linear and quadratic patterns among the experimental groups. [2] Standard errors of the means.

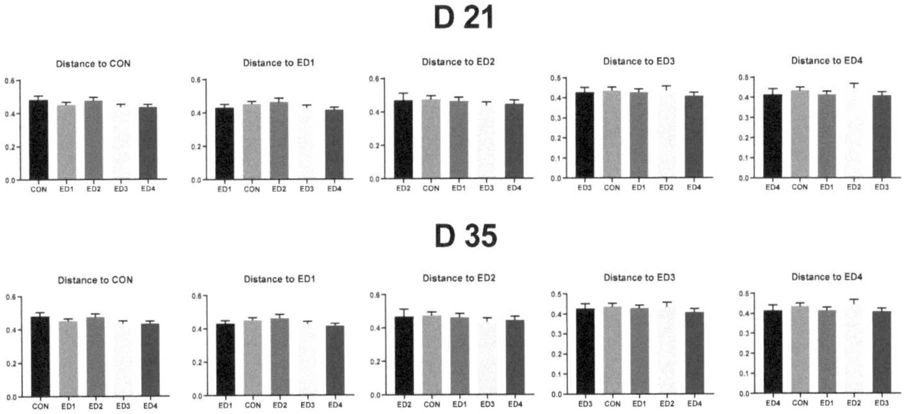

Figure 5. Unweighted unifrac measurement (quantitative beta diversity) in the unchallenged control (CON); *Eimeria* dose 1 (ED1): *E. acervulina*: 31,250/*E. maxima*: 6250/*E. tenella*: 6250; *Eimeria* dose 2 (ED2): *E. acervulina*: 62,500/*E. maxima*: 12,500/*E. tenella*: 12,500; *Eimeria* dose 3 (ED3): *E. acervulina*: 125,000/*E. maxima*: 25,000/*E. tenella*: 25,000; and *Eimeria* dose 4 (ED4): *E. acervulina*: 250,000/*E. maxima*: 50,000/*E. tenella*: 50,000 groups. groups. Oral gavage of *Eimeria* spp. inoculum (1 mL of PBS) was performed on D 15. Each experimental group was placed as the control group, and experimental groups were compared by using one-way PROC MIXED with Dunnett's post hoc test.

Figure 6 shows that on D 21, the ED3 group had significantly higher weighted unifrac distance (dissimilarity among samples with considering abundance information) to compared to the CON and ED1 groups ($p < 0.05$). On D 35, the ED2 ($p < 0.01$) and ED4 ($p < 0.05$) groups had significantly higher weighted unifrac distance compared to the CON and ED1 groups. The ED2 group had significantly higher weighted unifrac distance compared to the ED3 group ($p < 0.01$).

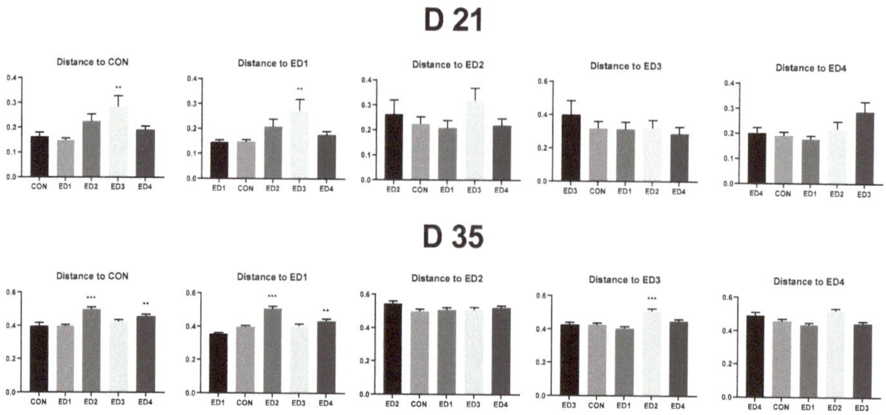

Figure 6. Weighted unifrac measurement (quantitative beta diversity) in the unchallenged control (CON); *Eimeria* dose 1 (ED1): *E. acervulina*: 31,250/*E. maxima*: 6250/*E. tenella*: 6250; *Eimeria* dose 2 (ED2): *E. acervulina*: 62,500/*E. maxima*: 12,500/*E. tenella*: 12,500; *Eimeria* dose 3 (ED3): *E. acervulina*: 125,000/*E. maxima*: 25,000/*E. tenella*: 25,000; and *Eimeria* dose 4 (ED4): *E. acervulina*: 250,000/*E. maxima*: 50,000/*E. tenella*: 50,000 groups. Oral gavage of *Eimeria* spp. inoculum (1 mL of PBS) was performed on D 15. Each experimental group was placed as the control group, and experimental groups were compared by using one-way PROC MIXED with Dunnett's post hoc test. ** represents when $p < 0.05$, and *** represents when $p < 0.01$.

As shown in Figure 7, no visual differences were observed in beta diversity parameters including unweighted unifrac and weighted unifrac on D 21 and D 35.

Figure 7. Visualized beta diversity indices including unweighted and weighted emperor in the unchallenged control (CON); *Eimeria* dose 1 (ED1): *E. acervulina*: 31,250/*E. maxima*: 6250/*E. tenella*: 6250; *Eimeria* dose 2 (ED2): *E. acervulina*: 62,500/*E. maxima*: 12,500/*E. tenella*: 12,500; *Eimeria* dose 3 (ED3): *E. acervulina*: 125,000/*E. maxima*: 25,000/*E. tenella*: 25,000; and *Eimeria* dose 4 (ED4): *E. acervulina*: 250,000/*E. maxima*: 50,000/*E. tenella*: 50,000 groups.

3.10. Taxa Abundance of the Cecal Microbiome Communities

As shown in Figure 8, increased *Eimeria* inoculation doses tended to increase the relative abundance of the phylum Proteobacteria ($p = 0.098$). Increased *Eimeria* inoculation doses tended to linearly enlarge the relative abundance of the family Enterobacteriaceae (Figure 9; $p = 0.091$). The relative abundance of the family Bacillaceae was enlarged in a linear trend by increased *Eimeria* inoculation doses ($p < 0.05$). Increased *Eimeria* inoculation doses linearly ($p < 0.05$) and quadratically (tendency; $p = 0.074$) reduced the relative abundance of the family Christensenellaceae, and the ED2 group had significantly lower the relative abundance of the family Christensenellaceae compared to the CON group. Increased *Eimeria* inoculation doses linearly reduced the relative abundance of the family Peptostreptococcaceae, and the ED4 group had a significantly lower relative abundance of the family Peptostreptococcaceae compared to the CON group.

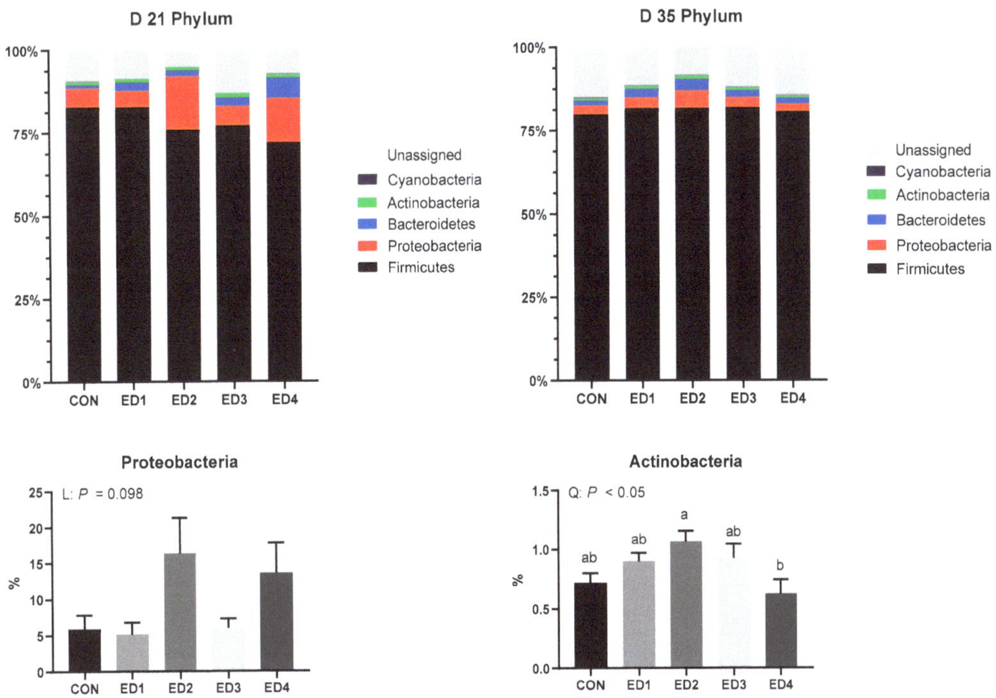

Figure 8. Phylum-level composition of the cecal microbial communities in the unchallenged control (CON); *Eimeria* dose 1 (ED1): *E. acervulina*: 31,250/*E. maxima*: 6250/*E. tenella*: 6250; *Eimeria* dose 2 (ED2): *E. acervulina*: 62,500/*E. maxima*: 12,500/*E. tenella*: 12,500; *Eimeria* dose 3 (ED3): *E. acervulina*: 125,000/*E. maxima*: 25,000/*E. tenella*: 25,000; and *Eimeria* dose 4 (ED4): *E. acervulina*: 250,000/*E. maxima*: 50,000/*E. tenella*: 50,000 groups. Oral gavage of *Eimeria* spp. Inoculum (1 mL of PBS) was performed on D 15. The phylum with statistical differences were presented. Experimental groups were compared using PROC MIXED followed by the Tukey's HSD (honestly significant difference) test, and different letters mean significant differences ($p < 0.05$) among the experimental groups. Orthogonal polynomial contrasts were conducted to see linear and quadratic patterns among the experimental groups.

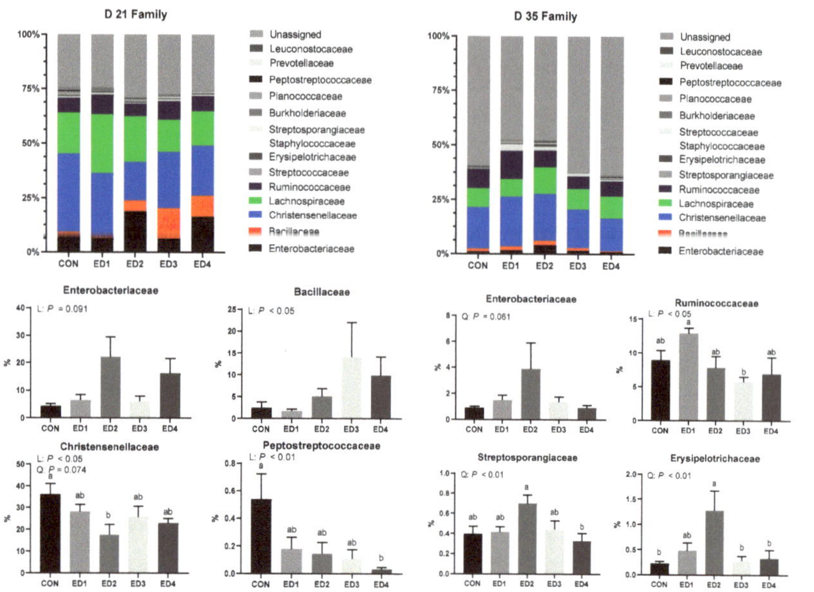

Figure 9. Family-level composition of the cecal microbial communities in the unchallenged control (CON); *Eimeria* dose 1 (ED1): *E. acervulina*: 31,250/*E. maxima*: 6250/*E. tenella*: 6250; *Eimeria* dose 2 (ED2): *E. acervulina*: 62,500/*E. maxima*: 12,500/*E. tenella*: 12,500; *Eimeria* dose 3 (ED3): *E. acervulina*: 125,000/*E. maxima*: 25,000/*E. tenella*: 25,000; and *Eimeria* dose 4 (ED4): *E. acervulina*: 250,000/*E. maxima*: 50,000/*E. tenella*: 50,000 groups. Oral gavage of *Eimeria* spp. inoculum (1 mL of PBS) was performed on D 15. The family with statistical differences were presented. Experimental groups were compared using PROC MIXED followed by the Tukey's HSD (honestly significant difference) test, and different letters mean significant differences ($p < 0.05$) among the experimental groups. Orthogonal polynomial contrasts were conducted to see linear and quadratic patterns among the experimental groups.

On D 35, the relative abundance of the phylum Actinobacteria were quadratically enlarged by increased *Eimeria* inoculation doses ($p < 0.05$), and the ED4 group had a significantly lower relative abundance of the phylum Actinobacteria compared to the ED2 group (Figure 8). Increased *Eimeria* inoculation doses tended to increase the relative abundance of the family Enterobacteriaceae with a linear trend ($p = 0.061$; Figure 9). The relative abundance of the family Ruminococcaceae was linearly reduced by *Eimeria* infection ($p < 0.05$), and the ED3 group had a significantly lower relative abundance of the family Ruminococcaceae compared to the ED1 group. Increased *Eimeria* inoculation doses quadratically increased the relative abundance of the family Streptosporangiceae ($p < 0.01$), and the ED2 group had significantly higher relative abundance of the family Streptosporangiceae compared to the ED4 group. Increased *Eimeria* inoculation doses quadratically increased the relative abundance of the family Erysipelotrichaceae ($p < 0.01$), and the ED2 group had significantly higher relative abundance of the family Streptosporangiceae compared to the CON, ED3, and ED4 groups.

3.11. *Eimeria* Gene Expression

As shown in Table 12, different *Eimeria* doses did not modulate mRNA expression of *E. acervulina* and *E. tenella* genes including *APN*, flagella-related protein, elongation factors (*EF*), and gametocyte proteins (*GAM*) in broilers on D 21 ($p > 0.1$). However, higher *Eimeria* inoculation doses linearly upregulated the gene expression of *E. maxima APN* ($p < 0.05$), and the ED3 group had significantly higher *E. maxima APN* compared to the ED1 and ED2 groups. Higher *E. maxima* doses linearly increased the gene expression of *E. maxima* EF2 ($p < 0.05$). Higher *E. maxima* doses linearly upregulated the gene expression of *E. maxima GAM56* ($p < 0.01$), and the ED3 group had significantly higher gene expression of *E. maxima GAM56* compared to the ED1 and ED2 groups. Gene expression of *E. maxima GAM82* was linearly increased by higher doses of *Eimeria* ($p < 0.01$), and the ED3 group had a significantly higher gene expression of *E. maxima GAM82* compared to the ED1 group.

Table 12. Effects of different doses of *Eimeria* spp. on the relative mRNA expression of genes relating to viability and sexual reproduction of *E. acervulina*, *E. maxima*, and *E. tenella* in broilers on D 21 [1,2].

	ED1	ED2	ED3	ED4	SEM [3]	*p* Value	Linear	Quadratic
E. acervulina								
18s	1.023	4.913	1.245	3.918	2.457	0.048	0.277	0.551
APN	1.038	1.058	1.238	1.038	0.824	0.969	0.908	0.753
Flagella-related protein	1.056	1.295	1.841	1.066	1.215	0.664	0.806	0.331
EF1	1.04	0.947	1.83	1.068	0.715	0.277	0.556	0.356
EF2	1.008	0.957	1.47	0.857	0.721	0.483	0.965	0.363
GAM56	1.022	0.889	2.36	1.1	1.559	0.358	0.573	0.398
GAM82	1.010	0.950	1.238	1.067	0.734	0.916	0.748	0.858
E. maxima								
18s	1.444	1.370	0.652	0.592	1.062	0.373	0.107	0.989
APN	1.059 [b]	1.096 [b]	2.151 [a]	1.775 [ab]	0.62	0.015	0.01	0.424
Flagella-related protein	1.044	0.972	1.041	0.92	0.339	0.904	0.628	0.861
EF1A	1.027	1.355	0.973	1.199	0.477	0.51	0.879	0.796
EF2	1.024	1.225	1.481	1.335	0.308	0.104	0.048	0.184
GAM56	1.035 [c]	1.241 [bc]	2.022 [ab]	2.144 [a]	0.534	0.003	<0.001	0.85
GAM82	1.021 [b]	1.102 [ab]	1.819 [a]	1.595 [ab]	0.457	0.017	0.008	0.422
E. tenella								
18s	4.11	68.12	58.98	27.34	77.12	0.401	0.646	0.115
APN	2.361	0.743	0.598	1.548	1.954	0.44	0.496	0.133
Flagella-related protein	1.310	1.904	1.048	1.852	1.259	0.610	0.752	0.848
EF1A	1.326	1.534	1.267	0.979	1.055	0.838	0.524	0.581
GAM22	1.218	1.087	1.007	1.852	1.092	0.542	0.393	0.299
GAM56	3.682	2.623	1.868	6.219	4.912	0.461	0.473	0.204

[1] CON (unchallenged control); ED1 (*Eimeria* dose 1): *E. acervulina*: 31,250/*E. maxima*: 6250/*E. tenella*: 6250; ED2 (*Eimeria* dose 2): *E. acervulina*: 62,500/*E. maxima*: 12,500/*E. tenella*: 12,500; ED3 (*Eimeria* dose 3): *E. acervulina*: 125,000/*E. maxima*: 25,000/*E. tenella*: 25,000; ED4 (*Eimeria* dose 4): *E. acervulina*: 250,000/*E. maxima*: 50,000/*E. tenella*: 50,000. Oral gavage of *Eimeria* spp. inoculum (1 mL of PBS) was performed on D 15. Experimental groups were compared using PROC MIXED followed by the Tukey's HSD (honestly significant difference) test. Different letters in the same row means significant differences ($p < 0.05$) among the experimental groups. Orthogonal polynomial contrasts were conducted to see linear and quadratic patterns among the experimental groups. [2] Relative abundance of *Eimeria* 18s genes was normalized by using host reference genes (beta actin and glyceraldehyde 3-phosphate dehydrogenase), and relative abundance of *Eimeria* genes including aminopeptidase N *(APN)*, elongation factor *(EF)*, and gametocyte protein *(GAM)* were normalized by using *Eimeria* 18s genes. [3] Standard errors of the means.

3.12. Body Composition

The body composition of broilers infected with *Eimeria* spp. on D 21, 28, and 35 is shown in Table 13. On D 21, total weight was linearly reduced by increased *Eimeria* inoculation doses ($p < 0.05$), and the ED4 group had a significantly lower total weight compared to the CON group. Increased *Eimeria* infection doses linearly reduced fat weight ($p < 0.05$) and lean weight ($p < 0.01$). No differences were observed on D 28 ($p > 0.1$). On D 35, fat weight was linearly reduced by increased *Eimeria* inoculation doses ($p < 0.05$), and the ED4 group tended to have lower fat weight compared to the CON group ($p = 0.055$). The lean weight tended to be linearly decreased by increased *Eimeria* inoculation doses ($p = 0.068$). Fat percentage tended to be reduced with a linear trend by increased *Eimeria* inoculation doses ($p = 0.060$), and lean percentage tended to be increased with a linear trend by increased *Eimeria* inoculation doses ($p = 0.059$). Lean:fat ratio was linearly reduced by increased *Eimeria* inoculation doses ($p < 0.05$).

Table 13. Effects of different *Eimeria* infection doses on body composition parameters including total weight (kg), fat (kg), lean weight (kg), fat percentage (%), lean percentage (%), and lean:fat in broilers [1].

	CON	ED1	ED2	ED3	ED4	SEM [2]	*p* Value	Linear	Quadratic
D 21									
Total weight	0.825 [a]	0.737 [ab]	0.719 [ab]	0.709 [ab]	0.651 [b]	0.086	0.028	0.002	0.609
Fat	0.119	0.094	0.098	0.091	0.084	0.024	0.144	0.023	0.532
Lean weight	0.706 [a]	0.643 [ab]	0.620 [ab]	0.640 [ab]	0.567 [b]	0.071	0.038	0.005	0.838
Fat percentage	14.28	12.78	13.63	12.68	12.80	2.22	0.668	0.294	0.676
Lean percentage	85.71	87.22	86.37	87.34	87.20	2.345	0.657	0.287	0.661
Lean:fat	6.138	6.873	6.721	7.359	7.043	1.512	0.711	0.251	0.575
D 28									
Total weight	1.309	1.224	1.259	1.295	1.272	0.13	0.823	0.988	0.538
Fat	0.176	0.162	0.147	0.160	0.153	0.04	0.788	0.379	0.494
Lean weight	1.134	1.062	1.113	1.135	1.119	0.114	0.806	0.767	0.641
Fat percentage	13.37	13.02	11.70	12.32	12.08	2.73	0.825	0.364	0.608
Lean percentage	86.64	86.99	88.29	87.67	87.89	2.74	0.834	0.377	0.609
Lean:fat	6.553	7.289	8.063	7.670	7.462	2.174	0.81	0.441	0.366
D 35									
Total weight	2.19	1.70	2.04	2.03	1.90	0.37	0.238	0.603	0.497
Fat	0.382 [a]	0.346 [ab]	0.338 [ab]	0.342 [ab]	0.297 [b]	0.04	0.055	0.006	0.925
Lean weight	1.810	1.616	1.699	1.684	1.606	0.147	0.107	0.068	0.532
Fat percentage	17.40	17.67	16.60	16.85	15.60	1.72	0.293	0.060	0.52
Lean percentage	82.61	82.30	83.41	83.15	84.39	1.72	0.291	0.059	0.513
Lean:fat	4.772	4.671	5.051	5.048	5.540	0.695	0.258	0.043	0.455

[1] CON (unchallenged control); ED1 (*Eimeria* dose 1): *E. acervulina*: 31,250/*E. maxima*: 6250/*E. tenella*: 6250; ED2 (*Eimeria* dose 2): *E. acervulina*: 62,500/*E. maxima*: 12,500/*E. tenella*: 12,500; ED3 (*Eimeria* dose 3): *E. acervulina*: 125,000/*E. maxima*: 25,000/*E. tenella*: 25,000; ED4 (*Eimeria* dose 4): *E. acervulina*: 250,000/*E. maxima*: 50,000/*E. tenella*: 50,000. Oral gavage of *Eimeria* spp. inoculum (1 mL of PBS) was performed on D 15. Experimental groups were compared using PROC MIXED followed by the Tukey's HSD (honestly significant difference) test. Different letters in the same row means significant differences ($p < 0.05$) among the experimental groups. Orthogonal polynomial contrasts were conducted to see linear and quadratic patterns among the experimental groups. [2] Standard errors of the means.

4. Discussion

The purpose of the study was to investigate the effects of different *Eimeria* inoculation doses on growth performance, litter moisture content, nutrient digestion and absorption, incidence and severity of FPD, gut microbiota, oocyst shedding, *Eimeria* gene expression, and body composition in broilers raised in floor pens for 35 days. Cross contamination among pens in the *Eimeria* infection studies in floor pens would be problematic [27]. To minimize cross contamination between the unchallenged group and *Eimeria* challenged groups, an anti-coccidial drug (Coban 90) was supplemented in the unchallenged group (CON group) in the current study. The dose of Coban 90 (500 mg/kg) was determined based on our previous study (unpublished). Broilers challenged with *Eimeria* spp. and fed 500 mg/kg of Coban 90 showed improved body weight compared to the challenged group without Coban 90, and their BW was similar to that of the non-challenged group. Furthermore, the supplementation of 500 mg/kg of Coban 90 did not adversely affect the growth performance of broiler chickens without an *Eimeria* challenge. However, although workers were extra careful during the entire experimental period, there was still *Eimeria* cross contamination (e.g., oocyst shedding and lesion score) to the unchallenged group in the current study. These results indicate that supplementation of Coban 90 at 500 mg/kg did not completely inhibit the colonization of *Eimeria* spp. in broilers even with indirect infection [28]. However, significant statistical differences in body weight (BW) were observed between the control group and the *Eimeria* challenged groups on Day 35. Additionally, throughout the current study, the control group consistently had lesion scores for *Eimeria* spp. lower than 0.5 and a minimum gut permeability. These findings suggest that the level of Eimeria infection in the CON group did not have a significant impact on serving as the negative control for the CON group. The DFI was measured during the entire experiment period to check the severity of *Eimeria* infection and whether *Eimeria* re-infection occurred in floor pen conditions. The DFI could be one of the powerful and non-invasive parameters to indicate incidence and severity of *Eimeria* infection in broilers. Measuring DFI has benefits in easiness and time over other non-invasive methods including measuring core body temperature or fecal moisture content. Our previous study [29] showed that *Eimeria* spp. Infection reduced ADFI in broilers in the acute phase (0 to 6 days post infection (dpi)). The time points 5 dpi and 6 dpi were considered as the peak time points for *Eimeria* infection according to our previous studies based on the results of gut permeability and daily feed intake [14,29]. In the current study, DFI was dramatically decreased on 5 and 6 dpi and continued to be linearly reduced by increased *Eimeria* inoculation doses from 11 to 14 dpi (D 26 to 29), which potentially indicates that re-infection of *Eimeria* has occurred in the current study. This was supported by the results of *Eimeria* lesion and oocyst shedding of *Eimeria* spp. on D 28 and 35, while the average values for *Eimeria* lesion scores on D 35 were below 1 out of 4 in the current study. Once exposed to *Eimeria* infection, chickens develop strong humoral and cellular immunity against re-infection of *Eimeria* [30], which may demonstrate no statistical differences in gut permeability and intestinal morphology in broilers after the acute phase in the current study. Still, our current study showed that reinfection of *Eimeria* was able to decrease feed intake of broilers in the floor pen conditions. Reduced feed intake in broilers infected with *Eimeria* spp. might be due to alternation in immune response and endocrine system in broilers [31,32].

In the acute phase (0 to 6 dpi; D 15 to 21), reduced feed intake with impaired feed efficiency dramatically decreased BW and ADG in broilers infected with *Eimeria* spp. in the current study, which was in agreement with Teng et al. [14]. On D 21 to 28, BW and ADG were reduced, along with decreased ADFI, without affecting feed efficiency in *Eimeria*-infected broilers. These results indicate that severe *Eimeria* infection reduced growth rate of broilers by damaging the capacity of nutrient digestion and absorption and reducing the feed intake in broilers, but re-infection of *Eimeria* decreased growth rate via decreasing only feed intake in broilers. These indicate that decreasing feed intake could be a sensitive sign in *Eimeria*-infected broilers. On D 35, *Eimeria* infection reduced only BW without affecting ADG, ADFI, and FCR in the *Eimeria* challenged groups compared to the CON group on D 35. These results were in consistent with a study reported that *Eimeria* inoculation on D 15 decreased BW in broilers on D 42 in floor pens [33]. In previous studies, compensatory growth happened in broilers infected with *E. maxima* in the recovery phase (6 to 13 dpi) [12] and in broilers challenged with *E. acervulina* and *E. maxima* (14 to 21 dpi) [34] by improving feed intake or feed efficiency after the acute phase of *Eimeria* infection. However, potentially, infection of *E. acervulina*, *E. maxima*, and *E. tenella* would not induce compensatory growth in the current study because they induced severe damage in the gastrointestinal tract of broilers. More studies are required to explain the potential mode of actions of *Eimeria* infection on reduced feed intake and to elucidate compensatory growth after *Eimeria* infection in broilers.

Higher *Eimeria* inoculation doses above the threshold where the maximal reproductive potential reached may result in reduced oocyst shedding and downregulated *Eimeria* genes relating to viability and sexual reproduction due to crowding effects in the gastrointestinal tract [35]. In our previous study, while higher doses of *E. tenella* increased oocyst shedding 5 to 6 dpi, oocyst shedding 6 to 8 dpi was not affected by challenging doses [36]. This potentially designates that *Eimeria* can control themselves to determine maximal oocyst production in broilers, which is called crowding effect [35,37]. In the current study, the ED4 group (the highest dose group) had numerically similar oocyst shedding of E. maxima and *E. tenella* compared to the ED2 and ED3 groups in the cloaca content on D 22. *Eimeria* oocysts were counted in the cloaca content and litter instead of feces because it is not feasible to collect fresh fecal samples in the litter condition. To elucidate crowding effects of *Eimeria* spp., whole duodenal, jejunal, and cecal tissue samples were collected on 6 dpi, one day prior to peak date (7 dpi) for oocyst production of *E. maxima* and *E. tenella* [38], and gene expression of *APN*, *EF2*, *GAM56*, and *GAM82* of *E. maxima* were modulated by increased *Eimeria* inoculation doses along with quadratically modulated oocyst production of *E. maxima* in the current study. The *APN*, *EF*, and *GAM* of *Eimeria* spp. play important roles in *Eimeria* viability and sexual reproduction to produce oocysts [24,39,40]. The inconsistency between linearly increased gene expression of *APN*, *EF*, and *GAM* and quadratically modulated oocyst shedding of *E. maxima* is still in question because only transcriptional level was analyzed in the current study. However, our current study showed that different inoculation doses of *Eimeria* can alter *Eimeria* gene expression. Increased inoculation doses only affected *E. maxima* mRNA expression in the current study potentially because *E. maxima* is more sensitive to crowding effects due to their largest size among chicken *Eimeria* spp. [41].

In the current study, ileal digesta and litter moisture contents were increased due to *Eimeria* infection in the acute phase. In our previous study, *E. tenella* infection did not increase ileal digesta moisture content on 5 to 7 dpi and even decreased ileal moisture content on 6 dpi [36]. Increased ileal moisture content is mainly due to infection of *E. acervulina* and *E. maxima* in broilers. Increased digesta moisture content may imply shortened digesta transit time, which can reduce the capacity of nutrient digestion and absorption [42] and is associated with reduced nutrient digestibility in the current study. Litter moisture content was linearly increased in the acute phase and was quadratically increased by *Eimeria* infection on D 35 in the current study. Many studies demonstrated that litter moisture is closely associated with FPD in broiler chickens [6,43,44]. FPD is a type of skin inflammation that induces necrotic lesions on the plantar surface of foot pad in broilers [45]. FPD can reduce the marketability of chicken feet, be an entry route for pathogenic bacteria, cause lameness, and reduce the growth performance of broiler chickens [46]. Increased litter moisture content due to *Eimeria* infection can increase the incidence and severity of FPD in broilers by increasing the litter ammonia concentration [47,48]. The litter moisture content on D 35 and severity/incidence of FPD on D 35 showed similar trends among the experimental groups in the current study, while the severity and incidence of FPD was mild potentially because of the dry conditions of the room and early slaughter age (D 35) in the current study. A previous study by El-Wahab et al. [43] showed that litter moisture should be above 35% to induce FPD in poultry. *Eimeria* infection still has the potential to increase the incidence and severity of FPD by increasing litter moisture content in broilers.

Gut permeability measured by FITC-D4 is an important indicator to represent functionality and integrity of gut in broilers [49]. Increased gut permeability indicates that more pathogens and toxins can permeate into the blood stream across the epithelial layer, which can cause systemic infection in broilers [50]. In the current study, *Eimeria* infection increased gut permeability on 5 dpi in broilers, which is consistent with our previous study [14]. Our previous study demonstrated that *E. maxima* infection had a significant impact on increasing gut permeability [51], whereas *E. tenella* infection did not affect gut permeability in broilers [36]. Potentially, the infection of *E. acervulina* and *E. maxima* disrupts tight junction proteins and the mucus layer, which play an important role in maintaining gut barrier integrity, and this would make the intestinal wall thinner (more permeable) in broilers [52,53]. In contrast, Vicuña et al. [49] showed that FITC-D4 can be deposited in the cecal tissue, and *E. tenella* infection did not increase gut permeability potentially because *E. tenella* infection thickened the intestinal wall in the ceca [36]. No differences were observed in the gut permeability on D 27 in the current study, and this suggests that the re-infection of *Eimeria* did not severely damage gut functionality and integrity, while it reduced feed intake in broilers in the floor pen conditions.

Reduced growth performance and feed efficiency might be mainly attributed to the reduced capacity of nutrient digestion and absorption in *Eimeria*-infected broilers. In the current study, the AID of CP was linearly reduced by *Eimeria* infection in broilers in the acute phase (0 to 6 dpi), which is consistent with our previous study by Teng et al. [14], which reported that *Eimeria* infection reduced the AID of CP in *Eimeria*-infected broilers. *E. acervulina* and *E. maxima* are the main *Eimeria* spp. that directly reduce the AID of CP because they inhabit in the duodenum and jejunum, respectively [54,55]. Our previous study by Choi et al. [36] reported that *E. tenella* infection in the ceca did not directly affect AID in the acute (6 dpi) and recovery phase (9 dpi) in broilers. The AID method does not account for the endogenous loss of nutrients in the gastrointestinal tract of broilers [54]. Both nutrient disappearance and increased endogenous loss may have affected the AID of CP values in the *Eimeria*-infected broilers because *Eimeria* infection impaired the intestinal morphology and activities of jejunal brush border digestive enzymes in the current study. *Eimeria* infection is known to increase endogenous losses of proteins (e.g., plasma proteins, mucin, and cell debris) in the gastrointestinal tract during *Eimeria* colonization and reproduction activities [56,57]. The AID of CF was dramatically reduced by *Eimeria* infection, and AID values of CF were negative in the ED2, ED3, and ED4 groups in the current study. Consistently, Ghareeb et al. [58] reported that *E. maxima* infection dramatically reduced the AID of CF in broilers. The negative values for AID of CF indicate that there were high endogenous losses in *Eimeria*-infected broilers. Endogenous fat from the gastrointestinal tract includes bile, cell debris, intestinal secretions, and microbial lipids [59]. Adams et al. [60] demonstrated that infection of *E. acervulina* decreased the secretion of bile, which has an essential role in fat digestion by emulsifying fat in the gastrointestinal tract. Potentially, reduced bile secretion and increased cell debris from the gastrointestinal tract may have decreased fat digestibility and increased endogenous loss of fat, which resulted in negative values for the AID of CP in *Eimeria*-infected broilers in the current study. However, more studies are needed to specify the factors such as bile, cell debris, intestinal secretions, and microbial lipids that increased endogenous losses of fat in *Eimeria*-infected broilers.

Intestinal morphology is an important indicator to represent capacity of nutrient digestion and absorption in the gastrointestinal tract of chickens [61]. In the current study, infection of *E. acervulina* and *E. maxima* reduced VH and increased CD in both duodenum and jejunum on D 21. These results agree with several previous studies [62,63]. The reduced VH and increased CD implies there was high tissue turnover in the intestine by *Eimeria* infection because bigger crypts, which are reservoirs for enterocytes, indicates a high demand for new tissue in the villus [62,64]. Impaired jejunal morphology is highly associated with reduced activities of jejunal maltase in the current study. This is because mature enterocytes in the villus, which mainly express brush border digestive enzymes, would be quickly deceased due to a higher turnover rate in *Eimeria* infection conditions [65]. Furthermore, negatively modulated duodenal and jejunal morphology would explain decreased AID of CP and EE in the current study. While no differences were observed in the jejunal morphology on D 35, duodenal VH:CD were quadratically reduced with increased duodenal CD in *Eimeria*-infected broilers in the current study. Reduced VH:CD due to increased CD indicates more energy and nutrients are required for gut maintenance, which can result in growth retardation in broilers [62]. However, while statistical differences were not observed, the ED3 and ED4 groups exhibited numerically higher duodenal VH and numerically lower VH:CD compared to the control group on D 35. This numerical trend suggests that gut maintenance and new tissue generation may have occurred on D 35. This result also indicates that negative effects of *E. acervulina* infection lasted through to D 35 in broilers in the current study.

Chicken ceca have only crypts without villus, similarly to the colon of mammals (humans and pigs) [66]. In the current study, *E. tenella* infection deepened the cecal crypts as we observed in our previous study [36]. However, it is still uncertain whether thickened cecal wall due to *E. tenella* infection decreases the permeation of microbial metabolites (e.g., endotoxins and short chain fatty acids) across epithelium. In the current study, *Eimeria* infection negatively affected cecal microbiota in the acute and chronic phases. In the current study, *Eimeria* infection increased the phylum Proteobacteria and the family Enterobacteriaceae in the acute phase, which includes diverse pathogenic bacteria such as *Escherichia coli*, *Salmonella* spp., *Vibrio* spp., and *Pseudomonas* spp. [67]. Moreover, in the acute phase, *Eimeria* infection reduced the relative abundance of the family Christensenellaceae and Peptostreptococcaceae, which play important roles in fiber degradation and short chain fatty acid production in broilers [68]. Potentially, negatively altered cecal microbiota may have reduced volatile fatty acids (VFA) production in the ceca, and this would explain the reduced activities of SAP, which needs VFA production [36,69]. Whereas alpha diversity indices in the cecal microbial communities on D 21 were not affected by *Eimeria* infection, all alpha diversity indices including pielou evenness (evenness), faith phylogenetic diversity (biodiversity based on phylogeny), shannon entropy (richness and evenness), and observed features (richness) were linearly decreased on D 35 in the current study. Reduced alpha diversity indices suggest the presence of unhealthier and immature gut microbiota in chickens [70]. Furthermore, the relative abundance of the family Enterobacteriaceae was quadratically increased by *Eimeria* infection, and the relative abundance of the family Ruminococcaceae, which play an important role in fiber degradation and short-chain fatty acid production in the ceca of chickens [71], was linearly reduced by *Eimeria* infection on D 35 in the current study. These results imply that *Eimeria* infection on D 15 still negatively influenced cecal microbiota in broilers on D 35. Our previous study showed that *E. tenella* infection negatively influenced cecal microbiota toward increasing the abundance of pathogenic bacteria and reducing microbial VFA (e.g., an important energy source for chickens) production mainly by impairing the mucosal immune system of broilers and increasing the protein concentration in the cecal content [72]. While *E. tenella* infection would be the main factor to alter cecal microbiota, the infection of *E. acervulina* and *E. maxima* potentially can affect cecal microbiota by increasing endogenous loss of proteins and undigested dietary proteins in the gastrointestinal tract because the phylum Proteobacteria mainly ferment protein sources for their growth and reproduction [73–75]. More studies are required to investigate whether the infection of *E. acervulina* and *E. maxima* itself can alter cecal microbiota in broilers.

Body composition is a crucial parameter in broiler production because body composition is closely associated with meat yield and quality [76]. In the current study, body composition of broilers was altered by *Eimeria* infection. On D 35, the lean:fat ratio was linearly enhanced, and fat accumulation was reduced by increased *Eimeria* infection doses in the current study. Potentially, *Eimeria* infection stimulated the immune system, resulting in the excessive usage of energy sources (e.g., fat and glycogen) in broilers [77,78]. Furthermore, reduced fat digestibility and decreased cecal VFA production due to negatively altered microbiota on D 21 could result in decreased fat accumulation as a chronic effect on D 35. These results suggest that *Eimeria* infection can influence the body composition and meat quality of broilers.

5. Conclusions

In conclusion, *Eimeria* infection negatively affected the growth performance, gut health, gut barrier integrity, nutrient digestion and absorption, gut microbiota, and body composition of broilers in the acute phase. Increased *Eimeria* inoculation doses modulated the relative mRNA expression of *Eimeria* genes relating to viability and sexual reproduction. *Eimeria* infection negatively affected the growth performance, gut microbiota, FPD, and body composition in broilers, and the negative effects were prolonged to D 35 in the floor pen conditions.

Author Contributions: Conceptualization, J.C. and W.K.K.; methodology, J.C. and W.K.K.; validation, J.C. and W.K.K.; investigation, J.C., D.G., M.K.S., H.K., G.L., D.P., V.S.R.C. and J.L.; writing—original draft preparation, J.C.; writing—review and editing, J.C., D.G., M.K.S., H.K., G.L., D.P., V.S.R.C., J.L. and W.K.K.; visualization, J.C.; supervision, W.K.K.; project administration, W.K.K. All authors have read and agreed to the published version of the manuscript.

Funding: This study was financed in part by a cooperative agreement 6040-32000-080-000D from United States Department of Agriculture-Agricultural Research Service.

Institutional Review Board Statement: The Institutional Animal Care and Use Committee at the University of Georgia approved animal care and use protocols of this current study.

Informed Consent Statement: Not applicable.

Data Availability Statement: Data are unavailable due to privacy restrictions.

Acknowledgments: We appreciate Thiago Belem and Manuel Arango for their assistance at the farm.

Conflicts of Interest: The authors declare no conflict of interest.

References

1. Blake, D.P.; Knox, J.; Dehaeck, B.; Huntington, B.; Rathinam, T.; Ravipati, V.; Ayoade, S.; Gilbert, W.; Adebambo, A.O.; Jatau, I.D. Re-calculating the cost of coccidiosis in chickens. *Vet. Res.* **2020**, *51*, 115. [CrossRef]
2. Mesa, C.; Gómez-Osorio, L.; López-Osorio, S.; Williams, S.; Chaparro-Gutiérrez, J. Survey of coccidia on commercial broiler farms in Colombia: Frequency of *Eimeria* species, anticoccidial sensitivity, and histopathology. *Poult. Sci.* **2021**, *100*, 101239. [CrossRef] [PubMed]
3. Allen, P.C.; Fetterer, R. Recent advances in biology and immunobiology of *Eimeria* species and in diagnosis and control of infection with these coccidian parasites of poultry. *Clin. Microbiol. Rev.* **2002**, *15*, 58–65. [CrossRef] [PubMed]
4. Madlala, T.; Okpeku, M.; Adeleke, M.A. Understanding the interactions between *Eimeria* infection and gut microbiota, towards the control of chicken coccidiosis: A review. *Parasite* **2021**, *28*, 48. [CrossRef] [PubMed]
5. Zhang, J.J.; Wang, L.X.; Ruan, W.K.; An, J. Investigation into the prevalence of coccidiosis and maduramycin drug resistance in chickens in China. *Vet. Parasitol.* **2013**, *191*, 29–34. [CrossRef]
6. Taira, K.; Nagai, T.; Obi, T.; Takase, K. Effect of litter moisture on the development of footpad dermatitis in broiler chickens. *J. Vet. Med. Sci.* **2014**, *76*, 583–586. [CrossRef]
7. Sizmaz, O.; Barrett, N.W.; Lewis, J.; Yakout, H.; Persia, M.E. Effect of various concentration of butyric acid on growth performance, intestinal lesion scores, and body composition of broilers raised on used litter. *J. Appl. Poult. Res.* **2022**, *31*, 100296. [CrossRef]
8. Antoszczak, M.; Steverding, D.; Huczyński, A. Anti-parasitic activity of polyether ionophores. *Eur. J. Med. Chem.* **2019**, *166*, 32–47. [CrossRef] [PubMed]
9. Parker, C.D.; Lister, S.A.; Gittins, J. Impact assessment of the reduction or removal of ionophores used for controlling coccidiosis in the UK broiler industry. *Vet. Rec.* **2021**, *189*, e513. [CrossRef] [PubMed]
10. Teng, P.-Y.; Choi, J.; Yadav, S.; Tompkins, Y.; Kim, W.K. Effects of low-crude protein diets supplemented with arginine, glutamine, threonine, and methionine on regulating nutrient absorption, intestinal health, and growth performance of *Eimeria*-infected chickens. *Poul. Sci.* **2021**, *100*, 101427. [CrossRef]
11. Lee, J.-W.; Kim, D.-H.; Kim, Y.-B.; Jeong, S.-B.; Oh, S.-T.; Cho, S.-Y.; Lee, K.-W. Dietary encapsulated essential oils improve production performance of coccidiosis-vaccine-challenged broiler chickens. *Animals* **2020**, *10*, 481. [CrossRef] [PubMed]
12. Choi, J.; Tompkins, Y.H.; Teng, P.-Y.; Gogal, R.M., Jr.; Kim, W.K. Effects of tannic acid supplementation on growth performance, oocyst shedding, and gut health of in broilers infected with *Eimeria maxima*. *Animals* **2022**, *12*, 1378. [CrossRef]
13. Shields, S.; Greger, M. Animal welfare and food safety aspects of confining broiler chickens to cages. *Animals* **2013**, *3*, 386–400. [CrossRef]
14. Teng, P.-Y.; Yadav, S.; de Souza Castro, F.L.; Tompkins, Y.H.; Fuller, A.L.; Kim, W.K. Graded *Eimeria* challenge linearly regulated growth performance, dynamic change of gastrointestinal permeability, apparent ileal digestibility, intestinal morphology, and tight junctions of broiler chickens. *Poul. Sci.* **2020**, *99*, 4203–4216. [CrossRef]
15. Johnson, J.; Reid, W.M. Anticoccidial drugs: Lesion scoring techniques in battery and floor-pen experiments with chickens. *Exp. Parasitol.* **1970**, *28*, 30–36. [CrossRef] [PubMed]
16. Koo, B.; Choi, J.; Yang, C.; Nyachoti, C.M. Diet complexity and l-threonine supplementation: Effects on growth performance, immune response, intestinal barrier function, and microbial metabolites in nursery pigs. *J. Anim. Sci.* **2020**, *98*, skaa125. [CrossRef] [PubMed]
17. Sorin, M.C.; Ilie, V.; Ciurescu, G.; Anca, G.; Teodor, M. Influence of the dietary protein level on the incidence of footpad dermatitis in broiler chickens. *Indian J. Anim. Sci.* **2013**, *83*, 1084–1089.
18. Short, F.; Gorton, P.; Wiseman, J.; Boorman, K. Determination of titanium dioxide added as an inert marker in chicken digestibility studies. *Anim. Feed Sci. Technol.* **1996**, *59*, 215–221. [CrossRef]
19. Lin, Y.; Olukosi, O.A. Qualitative and quantitative profiles of jejunal oligosaccharides and cecal short-chain fatty acids in broiler chickens receiving different dietary levels of fiber, protein and exogenous enzymes. *J. Sci. Food Agric.* **2021**, *101*, 5190–5201. [CrossRef]
20. Fan, M.Z.; Matthews, J.C.; Etienne, N.M.; Stoll, B.; Lackeyram, D.; Burrin, D.G. Expression of apical membrane L-glutamate transporters in neonatal porcine epithelial cells along the small intestinal crypt-villus axis. *Am. J. Physiol. Gastrointest.* **2004**, *287*, G385–G398. [CrossRef]
21. Lackeyram, D.; Yang, C.; Archbold, T.; Swanson, K.C.; Fan, M.Z. Early weaning reduces small intestinal alkaline phosphatase expression in pigs. *J. Nutr.* **2010**, *140*, 461–468. [CrossRef]
22. Maroux, S.; Louvard, D.; Barath, J. The aminopeptidase from hog intestinal brush border. *Biochim. Biophys. Acta* **1973**, *321*, 282–295. [CrossRef] [PubMed]
23. Elgharbawy, A.A.; Hayyan, A.; Hayyan, M.; Rashid, S.N.; Nor, M.R.M.; Zulkifli, M.Y.; Alias, Y.; Mirghani, M.E.S. Shedding light on lipase stability in natural deep eutectic solvents. *Chem. Biochem. Eng. Q.* **2018**, *32*, 359–370. [CrossRef]

24. Schneiders, G.H.; Foutz, J.C.; Milfort, M.C.; Ghareeb, A.F.; Fuller, A.L.; Rekaya, R.; Williams, S.M.; Aggrey, S.E. Heat stress reduces sexual development and affects pathogenesis of *Eimeria maxima* in meat-type chickens. *Sci. Rep.* **2020**, *10*, 10736. [CrossRef] [PubMed]
25. Livak, K.J.; Schmittgen, T.D. Analysis of relative gene expression data using real-time quantitative PCR and the $2^{-\Delta\Delta CT}$ method. *Methods* **2001**, *25*, 402–408. [CrossRef] [PubMed]
26. Bolyen, E.; Rideout, J.R.; Dillon, M.R.; Bokulich, N.A.; Abnet, C.C.; Al-Ghalith, G.A.; Alexander, H.; Alm, E.J.; Arumugam, M.; Asnicar, F. Reproducible, interactive, scalable and extensible microbiome data science using QIIME 2. *Nat. Biotechnol.* **2019**, *37*, 852–857. [CrossRef]
27. Oviedo-Rondón, E.; Hume, M.; Hernández, C.; Clemente-Hernández, S. Intestinal microbial ecology of broilers vaccinated and challenged with mixed *Eimeria* species, and supplemented with essential oil blends. *Poult. Sci.* **2006**, *85*, 854–860. [CrossRef]
28. Yang, W.-C.; Yang, C.-Y.; Liang, Y.-C.; Yang, C.-W.; Li, W.-Q.; Chung, C.-Y.; Yang, M.-T.; Kuo, T.-F.; Lin, C.-F.; Liang, C.-L. Anti-coccidial properties and mechanisms of an edible herb, Bidens pilosa, and its active compounds for coccidiosis. *Sci. Rep.* **2019**, *9*, 2896. [CrossRef]
29. Tompkins, Y.H.; Choi, J.; Teng, P.-Y.; Yamada, M.; Sugiyama, T.; Kim, W.K. Reduced bone formation and increased bone resorption drive bone loss in *Eimeria* infected broilers. *Sci. Rep.* **2023**, *13*, 616. [CrossRef]
30. Wattrang, E.; Thebo, P.; Ibrahim, O.; Dalgaard, T.S.; Lundén, A. Parasite-specific proliferative responses of chicken spleen cells upon in vitro stimulation with *Eimeria tenella* antigen. *Parasitology* **2019**, *146*, 625–633. [CrossRef]
31. Humphrey, B.; Klasing, K. Modulation of nutrient metabolism and homeostasis by the immune system. *Worlds Poult. Sci. J.* **2004**, *60*, 90–100. [CrossRef]
32. Johnson, R. Immune and endocrine regulation of food intake in sick animals. *Domest. Anim. Endocrinol.* **1998**, *15*, 309–319. [CrossRef]
33. Ritzi, M.M.; Abdelrahman, W.; Mohnl, M.; Dalloul, R.A. Effects of probiotics and application methods on performance and response of broiler chickens to an *Eimeria* challenge. *Poult. Sci.* **2014**, *93*, 2772–2778. [CrossRef]
34. Voeten, A.; Braunius, W.; Orthel, F.; van Rijen, M.A. Influence of coccidiosis on growth rate and feed conversion in broilers after experimental infections with *Eimeria acervulina* and *Eimeria maxima*. *Vet. Q.* **1988**, *10*, 256–264. [CrossRef]
35. Williams, R. Quantification of the crowding effect during infections with the seven *Eimeria* species of the domesticated fowl: Its importance for experimental designs and the production of oocyst stocks. *Int. J. Parasitol.* **2001**, *31*, 1056–1069. [CrossRef]
36. Choi, J.; Ko, H.; Tompkins, Y.H.; Teng, P.-Y.; Lourenco, J.M.; Callaway, T.R.; Kim, W.K. Effects of *Eimeria tenella* infection on key parameters for feed efficiency in broiler chickens. *Animals* **2021**, *11*, 3428. [CrossRef]
37. Williams, R. Relative virulences of a drug-resistant and a drug-sensitive strain of *Eimeria acervulina*, a coccidium of chickens. *Vet. Parasitol.* **2006**, *135*, 15–23. [CrossRef]
38. Cha, J.O.; Zhao, J.; Yang, M.S.; Kim, W.I.; Cho, H.S.; Lim, C.W.; Kim, B. Oocyst-shedding patterns of three *Eimeria* species in chickens and shedding pattern variation depending on the storage period of *Eimeria tenella* oocysts. *J. Parasitol.* **2018**, *104*, 18–22. [CrossRef]
39. Gras, S.; Byzia, A.; Gilbert, F.B.; McGowan, S.; Drag, M.; Silvestre, A.; Niepceron, A.; Lecaille, F.; Lalmanach, G.; Brossier, F. Aminopeptidase N1 (EtAPN1), an M1 metalloprotease of the apicomplexan parasite *Eimeria tenella*, participates in parasite development. *Eukaryot. Cell* **2014**, *13*, 884–895. [CrossRef]
40. Walker, R.A.; Sharman, P.A.; Miller, C.M.; Lippuner, C.; Okoniewski, M.; Eichenberger, R.M.; Ramakrishnan, C.; Brossier, F.; Deplazes, P.; Hehl, A.B. RNA Seq analysis of the *Eimeria tenella* gametocyte transcriptome reveals clues about the molecular basis for sexual reproduction and oocyst biogenesis. *BMC Genom.* **2015**, *16*, 94. [CrossRef]
41. Dubey, J.; Jenkins, M. Re-evaluation of the life cycle of *Eimeria maxima* Tyzzer, 1929 in chickens (*Gallus domesticus*). *Parasitology* **2018**, *145*, 1051–1058. [CrossRef]
42. Noy, Y.; Sklan, D. Digestion and absorption in the young chick. *Poult. Sci.* **1995**, *74*, 366–373. [CrossRef]
43. El-Wahab, A.; Visscher, C.; Beineke, A.; Beyerbach, M.; Kamphues, J. Experimental studies on the effects of different litter moisture contents and exposure time to wet litter on development and severity of foot pad dermatitis in young fattening turkeys. *Arch. Geflugelkd.* **2012**, *76*, 55–62.
44. De Jong, I.C.; Gunnink, H.; Van Harn, J. Wet litter not only induces footpad dermatitis but also reduces overall welfare, technical performance, and carcass yield in broiler chickens. *J. Appl. Poult. Res.* **2014**, *23*, 51–58. [CrossRef]
45. Shepherd, E.; Fairchild, B. Footpad dermatitis in poultry. *Poul. Sci.* **2010**, *89*, 2043–2051. [CrossRef] [PubMed]
46. Mahmoud, U.T.; Darwish, M.H.; Ali, F.A.Z.; Amen, O.A.; Mahmoud, M.A.; Ahmed, O.B.; Abd El-Reda, G.; Osman, M.A.; Othman, A.A.; Abushahba, M.F. Zinc oxide nanoparticles prevent multidrug resistant *Staphylococcus*-induced footpad dermatitis in broilers. *Avian Pathol.* **2021**, *50*, 214–226. [CrossRef]
47. Mayne, R.; Else, R.; Hocking, P. High litter moisture alone is sufficient to cause footpad dermatitis in growing turkeys. *Br. Poult. Sci.* **2007**, *48*, 538–545. [CrossRef]
48. Cabrera, M.; Chiang, S. Water content effect on denitrification and ammonia volatilization in poultry litter. *Soil Sci. Soc. Am. J.* **1994**, *58*, 811–816. [CrossRef]
49. Vicuña, E.; Kuttappan, V.; Tellez, G.; Hernandez-Velasco, X.; Seeber-Galarza, R.; Latorre, J.; Faulkner, O.; Wolfenden, A.; Hargis, B.; Bielke, L. Dose titration of FITC-D for optimal measurement of enteric inflammation in broiler chicks. *Poult. Sci.* **2015**, *94*, 1353–1359. [CrossRef]

50. Grenier, B.; Applegate, T.J. Modulation of intestinal functions following mycotoxin ingestion: Meta-analysis of published experiments in animals. *Toxins* **2013**, *5*, 396–430. [CrossRef]
51. Teng, P.-Y.; Choi, J.; Tompkins, Y.; Lillehoj, H.; Kim, W. Impacts of increasing challenge with *Eimeria maxima* on the growth performance and gene expression of biomarkers associated with intestinal integrity and nutrient transporters. *Vet. Res.* **2021**, *52*, 81. [CrossRef] [PubMed]
52. Major, P.; Revajová, V.; Levkut, M.; Ševčíková, Z.; Spišáková, V.; Faixová, Z.; Levkutová, M.; Kožárová, I.; Goldová, M.; Levkut, M. Intestinal mucin dynamic and leukocytic responses of chickens infected with *Eimeria acervulina* and fed oregano supplemented diet. *Acta Vet. Brno* **2011**, *80*, 147–156. [CrossRef]
53. Choi, J.; Wang, L.; Liu, S.; Lu, P.; Zhao, X.; Liu, H.; Lahaye, L.; Santin, E.; Liu, S.; Nyachoti, M. Effects of a microencapsulated formula of organic acids and essential oils on nutrient absorption, immunity, gut barrier function, and abundance of enterotoxigenic *Escherichia coli* F4 in weaned piglets challenged with *E. coli* F4. *J. Anim. Sci.* **2020**, *98*, skaa259. [CrossRef]
54. Teng, P.-Y.; Yadav, S.; Shi, H.; Kim, W.K. Evaluating endogenous loss and standard ileal digestibility of amino acids in response to the graded severity levels of *E. maxima* infection. *Poult. Sci.* **2021**, *100*, 101426. [CrossRef]
55. Rochell, S.; Parsons, C.; Dilger, R. Effects of *Eimeria acervulina* infection severity on growth performance, apparent ileal amino acid digestibility, and plasma concentrations of amino acids, carotenoids, and α1-acid glycoprotein in broilers. *Poult. Sci.* **2016**, *95*, 1573–1581. [CrossRef]
56. Kiarie, E.G.; Leung, H.; Akbari Moghaddam Kakhki, R.; Patterson, R.; Barta, J.R. Utility of feed enzymes and yeast derivatives in ameliorating deleterious effects of coccidiosis on intestinal health and function in broiler chickens. *Front. Vet. Sci.* **2019**, *6*, 473. [CrossRef]
57. Gregory, M.; Catchpole, J. Ovine coccidiosis: Pathology of *Eimeria ovinoidalis* infection. *Int. J. Parasitol.* **1987**, *17*, 1099–1111. [CrossRef] [PubMed]
58. Ghareeb, A.F.; Schneiders, G.H.; Richter, J.N.; Foutz, J.C.; Milfort, M.C.; Fuller, A.L.; Yuan, J.; Rekaya, R.; Aggrey, S.E. Heat stress modulates the disruptive effects of *Eimeria maxima* infection on the ileum nutrient digestibility, molecular transporters, and tissue morphology in meat-type chickens. *PLoS ONE* **2022**, *17*, e0269131. [CrossRef]
59. Wang, L.; Wang, L.; Lyu, Z.; Huang, B.; Hu, Q.; Lai, C. Endogenous losses of fat and fatty acids in growing pigs are not affected by vegetable oil sources but by the method of estimation. *Animals* **2019**, *10*, 48. [CrossRef]
60. Adams, C.; Vahl, H.; Veldman, A. Interaction between nutrition and *Eimeria acervulina* infection in broiler chickens: Development of an experimental infection model. *Br. J. Nutr.* **1996**, *75*, 867–873. [CrossRef]
61. Song, J.; Lei, X.; Luo, J.; Everaert, N.; Zhao, G.; Wen, J.; Yang, Y. The effect of Epigallocatechin-3-gallate on small intestinal morphology, antioxidant capacity and anti-inflammatory effect in heat-stressed broilers. *J. Anim. Physiol. Anim. Nutr.* **2019**, *103*, 1030–1038. [CrossRef] [PubMed]
62. Giannenas, I.; Tsalie, E.; Triantafillou, E.; Hessenberger, S.; Teichmann, K.; Mohnl, M.; Tontis, D. Assessment of probiotics supplementation via feed or water on the growth performance, intestinal morphology and microflora of chickens after experimental infection with *Eimeria acervulina*, *Eimeria maxima* and *Eimeria tenella*. *Avian Pathol.* **2014**, *43*, 209–216. [CrossRef] [PubMed]
63. Morris, B.C.; Danforth, H.; Caldwell, D.; Pierson, F.; McElroy, A. Intestinal mucosal mast cell immune response and pathogenesis of two *Eimeria acervulina* isolates in broiler chickens. *Poult. Sci.* **2004**, *83*, 1667–1674. [CrossRef] [PubMed]
64. Miles, R.; Butcher, G.; Henry, P.; Littell, R. Effect of antibiotic growth promoters on broiler performance, intestinal growth parameters, and quantitative morphology. *Poult. Sci.* **2006**, *85*, 476–485. [CrossRef] [PubMed]
65. Jung, K.; Ahn, K.; Chae, C. Decreased activity of brush border membrane-bound digestive enzymes in small intestines from pigs experimentally infected with porcine epidemic diarrhea virus. *Res. Vet. Sci.* **2006**, *81*, 310–315. [CrossRef]
66. Liu, P.; Pieper, R.; Rieger, J.; Vahjen, W.; Davin, R.; Plendl, J.; Meyer, W.; Zentek, J. Effect of dietary zinc oxide on morphological characteristics, mucin composition and gene expression in the colon of weaned piglets. *PLoS ONE* **2014**, *9*, e91091. [CrossRef]
67. Ma, X.; Wang, Q.; Li, H.; Xu, C.; Cui, N.; Zhao, X. 16S rRNA genes Illumina sequencing revealed differential cecal microbiome in specific pathogen free chickens infected with different subgroup of avian leukosis viruses. *Vet. Microbiol.* **2017**, *207*, 195–204. [CrossRef]
68. Tavella, T.; Rampelli, S.; Guidarelli, G.; Bazzocchi, A.; Gasperini, C.; Pujos-Guillot, E.; Comte, B.; Barone, M.; Biagi, E.; Candela, M. Elevated gut microbiome abundance of *Christensenellaceae*, *Porphyromonadaceae* and *Rikenellaceae* is associated with reduced visceral adipose tissue and healthier metabolic profile in Italian elderly. *Gut Microbes* **2021**, *13*, 1880221. [CrossRef]
69. Koyama, H.; Ono, T. Induction by short-chain fatty acids of alkaline phosphatase activity in cultured mammalian cells. *J. Cell. Physiol.* **1976**, *88*, 49–56. [CrossRef]
70. Ty, M.; Taha-Abdelaziz, K.; Demey, V.; Castex, M.; Sharif, S.; Parkinson, J. Performance of distinct microbial based solutions in a Campylobacter infection challenge model in poultry. *Anim. Microbiome* **2022**, *4*, 2. [CrossRef]
71. Liu, J.; Stewart, S.N.; Robinson, K.; Yang, Q.; Lyu, W.; Whitmore, M.A.; Zhang, G. Linkage between the intestinal microbiota and residual feed intake in broiler chickens. *J. Anim. Sci. Biotechnol.* **2021**, *12*, 22. [CrossRef] [PubMed]
72. Choi, J.; Kim, W. Interactions of Microbiota and Mucosal Immunity in the Ceca of Broiler Chickens Infected with *Eimeria tenella*. *Vaccines* **2022**, *10*, 1941. [CrossRef] [PubMed]
73. Diether, N.E.; Willing, B.P. Microbial fermentation of dietary protein: An important factor in diet–microbe–host interaction. *Microorganisms* **2019**, *7*, 19. [CrossRef]

74. Williams, R. Intercurrent coccidiosis and necrotic enteritis of chickens: Rational, integrated disease management by maintenance of gut integrity. *Avian Pathol.* **2005**, *34*, 159–180. [CrossRef]
75. Zhao, J.; Zhang, X.; Liu, H.; Brown, M.A.; Qiao, S. Dietary protein and gut microbiota composition and function. *Curr. Protein Pept. Sci.* **2019**, *20*, 145–154. [CrossRef]
76. Choi, J.; Kong, B.; Bowker, B.C.; Zhuang, H.; Kim, W.K. Nutritional Strategies to Improve Meat Quality and Composition in the Challenging Conditions of Broiler Production: A Review. *Animals* **2023**, *13*, 1386. [CrossRef] [PubMed]
77. Yang, J.; Liu, L.; Sheikhahmadi, A.; Wang, Y.; Li, C.; Jiao, H.; Lin, H.; Song, Z. Effects of corticosterone and dietary energy on immune function of broiler chickens. *PLoS ONE* **2015**, *10*, e0119750. [CrossRef]
78. Kim, W.H.; Chaudhari, A.A.; Lillehoj, H.S. Involvement of T cell immunity in avian coccidiosis. *Front. Immunol.* **2019**, *10*, 2732. [CrossRef]

animals MDPI

Article

Effects of Combinations of Toxin Binders with or without Natural Components on Broiler Breeders Exposed to Ochratoxin A

Jihwan Lee [1], Hyunah Cho [2], Dongcheol Song [2], Seyeon Chang [2], Jaewoo An [2], Jeonghun Nam [3], Byoungkon Lee [3], Sowoong Kim [4], Woo Kyun Kim [1,*] and Jinho Cho [2,*]

1 Department of Poultry Science, University of Georgia, Athens, GA 30602, USA; jl26112@uga.edu
2 Department of Animal Science, Chungbuk National University, Cheongju 28644, Republic of Korea; hannah0928@naver.com (H.C.); paul741@hanmail.net (D.S.); angella2425@naver.com (S.C.); blueswing547@naver.com (J.A.)
3 Cherrybro Co., Ltd., Jincheon 27820, Republic of Korea; nam0353@cau.ac.kr (J.N.); bklee0418@cherrybro.com (B.L.)
4 Provimi Co., Ltd., Seoul 06158, Republic of Korea; bill_so-woong_kim@provimi.com
* Correspondence: wkkim@uga.edu (W.K.K.); jinhcho@chungbuk.ac.kr (J.C.); Tel.: +1-706-542-1346 (W.K.K.); +82-043-261-2544 (J.C.)

Simple Summary: Ochratoxin A (OTA) is known to be a highly toxic mycotoxin, and it is a secondary metabolite which is primarily produced by several fungi, including *Aspergillus* and *Penicillium genera*. In the poultry industry, many researchers have been searching for alternative ways to control the toxic effects of mycotoxins to safeguard the health and growth performance of birds. The clay minerals and natural components used in this study are considered useful for the mitigation of the negative effects of mycotoxins. However, only a limited number of studies have investigated the effects of supplementation with these feed additives on broiler breeders exposed to OTA. Our results indicate that toxin binders, with or without natural components, can be effective tools in the mitigation of OTA-induced problems due to their synergistic effects.

Citation: Lee, J.; Cho, H.; Song, D.; Chang, S.; An, J.; Nam, J.; Lee, B.; Kim, S.; Kim, W.K.; Cho, J. Effects of Combinations of Toxin Binders with or without Natural Components on Broiler Breeders Exposed to Ochratoxin A. *Animals* **2023**, *13*, 2266. https://doi.org/10.3390/ani13142266

Academic Editor: Sabine Gebhardt-Henrich

Received: 7 June 2023
Revised: 27 June 2023
Accepted: 3 July 2023
Published: 11 July 2023

Abstract: The objective of this study was to investigate the effects of toxin binders on broiler breeders fed an ochratoxin A (OTA)-contaminated diet. A total of 60 45-week-old female Arbor Acres broiler breeder birds with an initial body weight of 3.65 ± 0.35 kg were randomly divided into 6 treatment groups, with 10 replicates per group and 1 bird per replicate. The trial was conducted for 9 weeks (including 1 week of adaptation). Feed additive 1 (FA1) was composed of clay minerals (85% bentonite and 12% clinoptilolite) with 3% charcoal. FA2 was composed of clay minerals (66.1% aluminosilicates) with natural components (0.8% artichoke and rosemary plant extracts), 7% yeast extract, 0.5% beta-glucans, and 25.6% carriers. The dietary treatment groups were as follows: (1) birds fed an OTA-free basal diet (Negative Control; NC); (2) lipopolysaccharide (LPS)-challenged birds fed a diet including OTA (4 mg/kg) (Positive Control, PC); (3) the PC with 0.05% FA1 (Treatment 1, T1); (4) the PC with 0.10% FA1 (Treatment 2, T2); (5) the PC with 0.10% FA2 (Treatment 3, T3); and (6) the PC with 0.20% FA2 (Treatment 4, T4). The LPS challenge (an intramuscular injection of 1 mg *E. coli* O55:B5 LPS per kg of body weight) was performed on the first day of the experiment. The results of this experiment show that the PC treatment negatively affected ($p < 0.05$) egg production, hatchability, Haugh unit, bone mineralization, relative organ weight (abdominal fat, liver), the levels of glutamic oxaloacetic transaminase (GOT), high-density lipoprotein (HDL), and total cholesterol in the blood, and OTA accumulation in the liver compared with the NC. However, supplementation with toxin binders mitigated ($p < 0.05$) the negative effects of the OTA. Specifically, supplementation with 0.10% FA1 and 0.10% FA2 increased ($p < 0.05$) eggshell strength by week 4, and the Haugh unit and bone mineralization (phosphorous) by week 8, while decreasing ($p < 0.05$) the relative weight of the liver and the levels of GOT and HDL in the blood. Supplementation with 0.10% FA2 led to greater improvements in various parameters, including laying performance and bone mineralization, than the other treatments. In conclusion, toxin binders with or without natural components can be effective tools in the mitigation of OTA-induced problems due to their synergistic effects.

Keywords: broiler breeder; ochratoxin A; toxin binder; clay mineral; natural component

1. Introduction

Mycotoxins are well-known toxic secondary metabolites produced by certain fungi that grow on feed ingredients. Among them, ochratoxin A (OTA) is considered a highly toxic mycotoxin, and it is primarily produced by several fungi, including the Aspergillus and Penicillium genera [1,2]. OTA, which is the most common mycotoxin contaminant in poultry diets, is not destroyed by temperatures below 180 °C [3]. Moreover, OTA is chemically stable, so even if the fungi which produce OTA are killed, the residual toxins can still produce adverse effects, such as the contamination of an animal's tissues, milk, and eggs, and thereby threaten human consumers [4]. Furthermore, OTA induces nephron toxicity, hepatotoxicity, teratogenicity, and immune toxicity in poultry, which are more sensitive to this toxin than mammals [5–8]. In addition to these direct and harmful effects on poultry, OTA exposure can result in the presence of OTA in the tissues and eggs of poultry. In previous studies, OTA was detected in 35% of egg samples and 41% of meat samples, and the continuous consumption of OTA was found to further increase the amount of OTA residue in eggs, and this could potentially be harmful to human health [9,10]. The supplementation of natural substances which bind mycotoxins in contaminated feed can mitigate mycotoxin-induced adverse effects. This is currently one of the most promising approaches to solving the problems outlined above [11]. Among these natural substances, clay minerals are known to have the ability to adsorb mycotoxins [12]. They also convert mycotoxins into less toxic metabolites, prevent their absorption in the intestines, and help facilitate the direct excretion of toxins in the feces [13]. Charcoal is also considered an effective toxin binder. Numerous in vitro studies have reported that charcoal can effectively absorb mycotoxins and alleviate mycotoxin-induced immune suppression, consequently improving the growth performance of broilers in vivo [14,15]. In addition to the use of these toxin binders, several additional strategies have been proposed to minimize the harmful effects of mycotoxins [16]. Interest in the effects of plant extracts, yeast extracts, and prebiotics on mycotoxins has recently increased because of their ability to minimize fungal and bacteria growth [17–20]. Moreover, lipopolysaccharide (LPS) has been widely used to induce immune stress in animal models [21]. Many studies have shown that LPS-induced immune stress can cause various negative effects in poultry, such as reduced growth performance, intestinal mucosal damage, and immunosuppression [22]. In rodents, oral challenges with LPS disturbed the subjects' response to other xenobiotic agents and thus increased the toxicity of trichothecenes such as deoxynivalenol [23]. Therefore, we simultaneously conducted an OTA challenge via feed and intramuscular LPS injections to maximize the negative effects of OTA in broiler breeders. However, only a limited number of studies have investigated the synergistic effects achieved by using other natural substances (i.e., plant extracts, yeast extracts and prebiotics) along with toxin binders (i.e., clay minerals and charcoals). Therefore, this study was conducted to investigate the effects of toxin binders (i.e., clay minerals and charcoals), with or without natural substances (i.e., plant extracts, yeast extracts, and prebiotics), on the laying performance, egg quality, relative organ weight, bone mineralization, blood traits, and toxin concentrations of broiler breeders challenged with ochratoxin A and LPS.

2. Materials and Methods

2.1. Ethics

The experimental protocols describing the management and care of animals were reviewed and approved by the Animal Care and Use Committee of Chungbuk National University (Cheongju, Republic of Korea; CBNUA-2045-22-01).

2.2. Preparation of Ochratoxin A

OTA was cultured according to the method of Ruan et al. [24]. In moldy corn, *Aspergillus ochraceus* was identified using Czapek Dox Agar (MB cell, Seoul, Republic of Korea), and then was enriched to 8×10^{10} CFU/mL using Czapek Dox Media (CDM) (MB cell, South Korea). Corn was inoculated with this *A. ochraceus* suspension twice a day (08:00 and 20:00) and cultured at 29 °C. On day 14 of incubation, contaminated corn was terminated by autoclaving at 121 °C for 15 min. OTA content was measured using ELISA Kits (Romer AgraQuant, Romer Labs, Jalan Bukit Merah, Singapore).

2.3. Experimental Animal and Design

A total of sixty 45-week-old Arbor Acres female broiler breeders (initial body weight of 3.65 ± 0.35 kg) were used in this experiment. All birds were randomly allocated to six dietary treatments in a randomized complete block design. The experiment was conducted for 9 weeks including a 1 week adaptation period. Feed additive 1 (FA1) was composed of 85% clinoptinolite, 12% bentonite and 3% charcoal. FA2 was composed of 66.1% aluminosilicates, 0.8% plant extracts (artichoke and rosemary), 7% yeast extracts, 0.5% beta-glucans and 25.6% carriers. The dietary treatments were as follows: (1) Negative Control (NC): OTA free diets; (2) Positive Control (PC): OTA (4 mg/kg) with LPS challenge; (3) Treatment 1 (T1): PC with 0.05% FA1 in the diet; (4) Treatment 2 (T2): PC with 0.10% FA1; (5) Treatment 3 (T3): PC with 0.10% FA2; and (6) Treatment 4 (T4): PC with 0.20% FA2. LPS challenge was performed on the first day of the experiment, and 1 mg of *E. coli* O55:B5 LPS per kg of body weight was intramuscularly injected according to the method of Xu et al. [25]. Experimental feed was provided by Cherry Buro Co., Ltd. (Jincheon, Republic of Korea; Table 1). The broiler breeders were raised in steel cages (W: 59 cm, D: 43 cm, H: 51 cm). All birds were given limited access to feed and water throughout the experiment. During the experiment, all chickens were housed in a room with 16 h of light and 8 h of darkness.

Table 1. Ingredients and chemical composition of the basal experimental diets (as fed basis).

Items	0–8 Weeks
Ingredients (%)	
Corn	54.12
Soybean meal, 45%	15.00
DDGS 28%	10.00
Corn Gluten Feed	6.27
Wheat Pollards	2.50
Rice pollards	1.50
Animal Fats	0.50
L-Lys-SO_4	0.08
DL-Methionine	0.12
L-Threonine	0.02
L-Tryptophan	0.13
Salt	0.21
Limestone	8.97
Mineral premix [1]	0.22
Vitamin premix [2]	0.11
Choline	0.25
Total	100
Calculated Value	
ME, Kcal/Kg	2710
Crude protein, %	15.42
Crude Fat, %	3.86
Crude Fiber, %	3.85

Table 1. *Cont.*

Items	0–8 Weeks
Total Lys, %	0.749
Total TSAA, %	0.684
Calcium, %	3.50
Available P, %	0.41

[1] Provided per kg of diet: 37.5 mg Zn (as $ZnSO_4$), 37.5 mg of Mn (MnO_2), 37.5 mg of Fe (as $FeSO_4{\bullet}7H_2O$), 3.75 mg of Cu (as $CuSO_4{\bullet}5H_2O$), 0.83 mg of I (as KI), and 0.23 mg of Se (as $Na_2SeO_3{\bullet}5H2O$). [2] Provided per kg of diet: 15,000 IU of vitamin A, 3750 IU of vitamin D3, 37.5 mg of vitamin E, 2.55 mg of vitamin K3, 3 mg of thiamin, 7.5 mg of riboflavin, 4.5 mg of vitamin B6, 24 µg of vitamin B12, 51 mg of niacin, 1.5 mg of folic acid, 0.2 mg of biotin and 13.5 mg of pantothenic acid.

2.4. Laying Performance

During the experiment period, eggs were collected at 3:00 p.m. every day. The egg production values were calculated for the total number of eggs (normal eggs, broken eggs, and shell less eggs) and for the number of normal eggs specifically. Egg weight was calculated by dividing the total normal egg weight by the total number of normal eggs.

2.5. Fertility and Hatchability

During the entire period, broken, contaminated, deformed, and small eggs (under 48 g) were excluded from fertility and hatchability evaluations. Fertility was calculated on the 8th day of the incubation by dividing the number of fertilized eggs by the total number of eggs incubated. Hatchability was calculated by dividing the number of new hatched chicks by the number of eggs in the incubator (MX-1000CD, Gimhae, Republic of Korea).

2.6. Egg Quality

Eggshell strength was measured using a texture analyzer (model 081002, FHK, Fujihara Ltd., Tokyo, Japan) and in units of compressive force loaded onto a unit of eggshell surface area. Eggshell thickness was measured at three random locations (top, middle, and bottom) using a micrometer dial pipe gauge (model 7360, Mitutoyo Co., Ltd., Kawasaki, Japan), and the average value was used to calculate the eggshell strength. Egg shell color was measured at the bottom of the egg using a colorimeter (model CM-25cG, konica minolta, Osaka, Japan) for L* (lightness), a* (redness), and b* (yellowness). Egg yolk color was measured using a colorimetric Roche color fan (Hoffman-La Roche Ltd., Basel, Switzerland).

$$1\ (\textit{pale yellow}) \sim 15\ (\textit{deep orange})$$

The Haugh unit (HU) was measured using a tripod micrometer (AMES, Waltham, MA, USA) for albumin height after breaking the egg on a flat glass surface, and the HU value was calculated based on the equation below, where H represents albumin length (mm) and W represents egg weight (g).

$$HU = 100 \log (H + 7.57 - 1.7\ W^{0.37})$$

2.7. Bone Mineralization

At the end of the experiment, the right tibiae of the slaughtered broiler breeders were removed for bone mineralization evaluation. To establish the ash content, the constant weight of the crucible was obtained using the direct incineration method. An approximately 1 g bone sample was taken, incinerated in a furnace (Ward, South Australia, Australia) at 200 °C for 1 h, 300 °C for 1 h, and 600 °C for 3 h, and then cooled in a desiccator for 30 min. The ash sample was cooled and weighed, and the amount of ash was calculated from the difference in weight. The ash samples were dissolved in a hydrochloric acid solution (1:1, v/v). Calcium (Ca) levels were measured with an atomic absorption spectrophotometer (AAS-3300, PERKIN ELMER, Norwalk, CT, USA), and phosphorus (P) levels were measured at 470 nm with a spectrophotometer (UV-2450, Shimadzu Co., Tokyo, Japan) according to the molybdenum blue colorimetric method [26].

2.8. Relative Organ Weight

At the end of the experiment, the weights of the liver, spleen and abdominal fat were measured. These values were expressed in terms of relative weight to live weight.

$$\text{Relative organ weight (g/kg)} = \text{organ weight (g)/body weight (kg)}$$

2.9. Blood Profile

At the end of the experiment, blood samples were collected with tubes (Becton Dickinson, Franklin Lakes, NJ, USA) from the wing vein in all birds. The serum was separated by means of centrifugation at 3000 rpm and 4 °C for 20 min. High-density lipoprotein (HDL) and total cholesterol (Total-C) concentrations in the blood were analyzed using an automatic blood analyzer (ADVIA 1650, Bayer, Tokyo, Japan). Glutamic oxaloacetic transaminase (GOT) and glutamic pyruvic transaminase (GPT) activities were analyzed using an automatic biochemistry analyzer (Hitachi 747, Hitachi, Tokyo, Japan).

2.10. OTA Accumulation in the Feed and Liver

After the experiment was completed, the feed samples were collected before starting the experiment and stored at −20 °C until analysis. Liver samples from all individual birds were collected and weighted. The samples were ground using liquid nitrogen and stored at −20 °C until analysis. OTA analysis was performed by extracting 2.5 g of the liver with 20 mL of 1% $NaHCO_3$/methanol (30:70, *v*/*v*). After centrifugation at 3000 rpm for 10 min, 12 mL of the supernatant was mixed with an equal volume of PBS and applied to an OchraTest WB immunocolumn (Vicam, Fleurus, Belgium). The eluate was filtered through a 0.45 μL syringe filter and analyzed using high-pressure liquid chromatography (HPLC Prominence, Shimadzu, Kyoto, Japan) equipped with a fluorescence detector. HPLC separations were performed using a Phenomenex Luna C18(2) 3 μm, 150 4.60 mm column equipped with a Gemini C18, 4.3 mm Security Guard pre-column (Phenomenex Inc., Torrance, CA, USA). A flow rate of 0.7 mL/min, an injection volume of 40 L, and a column oven temperature of 30 °C were set. Fluorescence detection was performed at an excitation wavelength of 327 nm and an emission wavelength of 462 nm. Conformity with the retention time of the standard product measured at the same time was the criterion for the qualitative test. A calibration curve based on the analytical concentration of the standard product was generated, and the obtained peak height was used for the quantitative test.

2.11. Statistical Analysis

All data were analyzed by means of one-way ANOVA using SPSS software (ver. 25.0; IBM, Armonk, NY, USA). Tukey's multiple range test examined differences among treatment groups, which were considered significant at $p < 0.05$.

3. Results

3.1. Laying Performance

The OTA-contaminated diet (PC) significantly decreased ($p < 0.05$) egg production compared to the negative control (NC) (Table 2). However, supplementation with 0.10% FA2 in the OTA-contaminated diet (T3) significantly increased egg production compared to the PC. There were no significant differences in egg production among the natural-additive-supplemented groups (T1, T2, T3 and T4).

Table 2. Effect of combinations of natural feed additives on egg production and egg weight in broiler breeders fed with an ochratoxin A diet [1].

Items	NC	PC	T1	T2	T3	T4	SEM	*p*-Value
Egg production, %	95.00 a	79.91 c	83.13 bc	86.11 bc	88.49 b	82.54 bc	1.763	<0.001
Egg weight, g/egg	64.59	62.58	63.40	64.86	63.74	63.68	0.761	0.401

[1] Abbreviations: NC = Negative Control, basal diet; PC = Positive Control, basal diet with + 4 ppm ochratoxin A (OTA) + lipopolysaccharide (LPS) challenge; T1 = basal diet with FA1 including 0.05% clay mineral and charcoal + 4 ppm OTA + LPS challenge; T2 = basal diet with FA1 including 0.10% clay mineral and charcoal + 4 ppm OTA + LPS challenge; T3 = basal diet with FA2 including 0.10% clay mineral, plant extract, yeast extract and beta-glucan + 4 ppm OTA + LPS challenge; T4 = basal diet with FA2 including 0.20% clay mineral, plant extract, yeast extract and beta-glucan + 4 ppm OTA + LPS challenge; SEM, standard error of mean. a–c = Means with different letters are significantly different ($p < 0.05$). N = 10.

3.2. Fertility and Hatchability

The OTA-contaminated diet (PC) significantly decreased ($p < 0.05$) hatchability compared to the NC group (Table 3). Although there was no significant difference ($p > 0.05$), the natural-feed-additive-supplemented groups showed numerically increased hatchability compared to the PC group.

Table 3. Effect of combinations of natural feed additives on fertility and hatchability in broiler breeders fed with an ochratoxin A diet [1].

Items	NC	PC	T1	T2	T3	T4	SEM	*p*-Value
Fertility, %	91.67	85.94	85.65	87.96	86.11	86.57	1.730	0.119
Hatchability, %	86.25 a	75.26 b	77.08 ab	78.24 ab	78.70 ab	77.31 ab	2.282	0.024

[1] Abbreviations: NC = Negative Control, basal diet; PC = Positive Control, basal diet with + 4 ppm ochratoxin A (OTA) + lipopolysaccharide (LPS) challenge; T1 = basal diet with FA1 including 0.05% clay mineral and charcoal + 4 ppm OTA + LPS challenge; T2 = basal diet with FA1 including 0.10% clay mineral and charcoal + 4 ppm OTA + LPS challenge; T3 = basal diet with FA2 including 0.10% clay mineral, plant extract, yeast extract and beta-glucan + 4 ppm OTA + LPS challenge; T4 = basal diet with FA2 including 0.20% clay mineral, plant extract, yeast extract and beta-glucan + 4 ppm OTA + LPS challenge; SEM, standard error of mean. a–b = Means with different letters are significantly different ($p < 0.05$). N = 10.

3.3. Egg Quality

There was no significant difference ($p > 0.05$) in egg quality at 4 and 8 weeks, except for the Haugh unit (HU) and eggshell strength, among the treatments (Tables 4 and 5). The OTA contaminated diet (PC) significantly decreased ($p < 0.05$) HU at 4 and 8 weeks compared to the NC group. At 4 weeks, supplementation with FA1 and FA2 increased HU to similar levels to the NC group. At 8 weeks, supplementation with 0.10% FA1 (T2) and 0.10% FA2 (T3) significantly improved HU compared to the PC group. Regarding eggshell strength, the PC demonstrated significantly reduced eggshell strength compared to the NC group at 4 weeks, whereas supplementation with 0.10% FA1 (T2) and 0.10% FA2 (T3) significantly increased eggshell strength compared to the PC group.

Table 4. Effect of combinations of natural feed additives on egg quality in broiler breeders fed with an ochratoxin A diet after 4 weeks of the experiment [1].

Items	NC	PC	T1	T2	T3	T4	SEM	*p*-Value
Eggshell color (Hunter color)								
L*	72.39	77.37	77.47	80.73	78.07	78.49	1.769	0.091
a*	8.11	6.98	7.67	6.10	8.59	7.48	0.692	0.377
b*	21.25	22.09	21.82	20.92	23.26	21.33	0.904	0.634

Table 4. *Cont.*

Items	NC	PC	T1	T2	T3	T4	SEM	*p*-Value
Egg yolk color	8.00	7.25	7.22	7.56	7.44	7.44	0.358	0.662
Haugh unit	86.24 a	78.51 b	83.28 ab	84.18 ab	80.99 ab	84.24 ab	1.561	0.030
Eggshell strength (kg/cm^2)	4.81 a	4.18 c	4.38 bc	4.54 b	4.55 b	4.29 c	0.053	<0.001
Eggshell Thickness (μm)	411.27	404.83	404.56	407.33	408.48	405.67	6.211	0.972

[1] Abbreviations: NC = Negative Control, basal diet; PC = Positive Control, basal diet with + 4 ppm ochratoxin A (OTA) + lipopolysaccharide (LPS) challenge; T1 = basal diet with FA1 including 0.05% clay mineral and charcoal + 4 ppm OTA + LPS challenge; T2 = basal diet with FA1 including 0.10% clay mineral and charcoal + 4 ppm OTA + LPS challenge; T3 = basal diet with FA2 including 0.10% clay mineral, plant extract, yeast extract and beta-glucan + 4 ppm OTA + LPS challenge; T4 = basal diet with FA2 including 0.20% clay mineral, plant extract, yeast extract and beta-glucan + 4 ppm OTA + LPS challenge; L*, lightness; a*, redness; b*, yellowness; SEM, standard error of mean. a–c = Means with different letters are significantly different ($p < 0.05$). N = 10.

Table 5. Effect of combinations of natural feed additives on egg quality in broiler breeders fed with and ochratoxin A diet after 8 weeks of the experiment [1].

Items	NC	PC	T1	T2	T3	T4	SEM	*p*-Value
Eggshell color (Hunter color)								
L*	73.81	78.79	78.89	82.15	79.49	79.91	1.788	0.094
a*	9.53	8.40	9.09	7.51	10.01	8.90	0.692	0.382
b*	22.67	23.51	23.24	22.03	24.68	22.75	0.908	0.560
Egg yolk color	8.30	7.50	7.67	8.11	8.22	7.78	0.291	0.327
Haugh unit	87.19 a	70.69 c	78.84 abc	82.20 ab	85.95 ab	77.91 bc	2.115	<0.001
Eggshell strength (kg/cm^2)	4.42	3.49	3.92	3.40	3.36	3.87	0.358	0.266
Eggshell Thickness (μm)	415.50	396.08	397.52	416.52	406.00	402.19	10.972	0.686

[1] Abbreviations: NC = Negative Control, basal diet; PC = Positive Control, basal diet with + 4 ppm ochratoxin A (OTA) + lipopolysaccharide (LPS) challenge; T1 = basal diet with FA1 including 0.05% clay mineral and charcoal + 4 ppm OTA + LPS challenge; T2 = basal diet with FA1 including 0.10% clay mineral and charcoal + 4 ppm OTA + LPS challenge; T3 = basal diet with FA2 including 0.10% clay mineral, plant extract, yeast extract and beta-glucan + 4 ppm OTA + LPS challenge; T4 = basal diet with FA2 including 0.20% clay mineral, plant extract, yeast extract and beta-glucan + 4 ppm OTA + LPS challenge; L*, lightness; a*, redness; b*, yellowness; SEM, standard error of mean. a–c = Means with different letters are significantly different ($p < 0.05$). N = 10.

3.4. Bone Mineralization

The OTA-contaminated diet (PC) significantly decreased ($p < 0.05$) ash, Ca and *p* content in the tibia compared to the NC group (Table 6). Supplementation with 0.10% FA2 (T3) significantly increased ($p < 0.05$) ash content in the tibia in comparison to the PC group. The FA2-supplemented groups (T3 and T4) had a higher Ca content in the tibia than the PC group. Moreover, supplementation with 0.10% FA1 (T2) and two levels of FA2 significantly increased P content in the tibia compared to the PC group.

Table 6. Effect of combinations of natural feed additives on bone mineralization in broiler breeders fed with an ochratoxin A diet [1].

Items	NC	PC	T1	T2	T3	T4	SEM	*p*-Value
Ash, %	31.75 a	22.27 c	23.25 c	25.12 bc	27.40 b	26.19 bc	0.980	<0.001
Ca, %	10.73 a	5.87 c	6.21 c	6.75 c	8.80 b	7.97 b	0.317	<0.001
P, %	5.04 a	3.29 c	3.72 bc	4.04 b	4.31 b	4.12 b	0.189	<0.001

[1] Abbreviations: NC = Negative Control, basal diet; PC = Positive Control, basal diet with + 4 ppm ochratoxin A (OTA) + lipopolysaccharide (LPS) challenge; T1 = basal diet with FA1 including 0.05% clay mineral and charcoal + 4 ppm OTA + LPS challenge; T2 = basal diet with FA1 including 0.10% clay mineral and charcoal + 4 ppm OTA + LPS challenge; T3 = basal diet with FA2 including 0.10% clay mineral, plant extract, yeast extract and beta-glucan + 4 ppm OTA + LPS challenge; T4 = basal diet with FA2 including 0.20% clay mineral, plant extract, yeast extract and beta-glucan + 4 ppm OTA + LPS challenge; Ca, calcium; P, phosphorus; SEM, standard error of mean. a–c = Means with different letters are significantly different ($p < 0.05$). N = 10.

3.5. Relative Organ Weight

The OTA-contaminated diet (PC) significantly increased ($p < 0.05$) the relative weight of the liver and abdominal fat in comparison to the NC group (Table 7). However, all of the natural-feed-additive-supplemented groups (T1–T4) displayed a lower relative liver weight than the PC group. In particular, supplementation with 0.10% FA1 (T2) and 0.10% FA2 (T3) significantly decreased ($p < 0.05$) the relative weight of the liver to similar values to the NC group. Supplementation with 0.10% FA2 (T3) alleviated ($p < 0.05$) the increased relative weight of abdominal fat caused by the OTA-contaminated diet (PC) and even decreased ($p < 0.05$) it to similar values to the NC group. There was no significant difference ($p > 0.05$) in the relative weight of the spleen among the treatment groups.

Table 7. Effect of combinations of natural feed additives on the relative weights of the liver, spleen and abdominal fat in broiler breeders fed with an ochratoxin A diet [1].

Items	NC	PC	T1	T2	T3	T4	SEM	*p*-Value
Liver, g/kg	19.55 c	24.56 a	22.38 b	21.27 bc	20.81 bc	21.73 b	0.498	<0.001
Spleen, g/kg	0.73	0.67	0.70	0.68	0.71	0.68	0.021	0.337
Abdominal fat, g/kg	28.57 b	31.74 a	30.70 ab	29.51 ab	28.60 b	30.02 ab	0.654	0.012

[1] Abbreviations: NC = Negative Control, basal diet; PC = Positive Control, basal diet with + 4 ppm ochratoxin A (OTA) + lipopolysaccharide (LPS) challenge; T1 = basal diet with FA1 including 0.05% clay mineral and charcoal + 4 ppm OTA + LPS challenge; T2 = basal diet with FA1 including 0.10% clay mineral and charcoal + 4 ppm OTA + LPS challenge; T3 = basal diet with FA2 including 0.10% clay mineral, plant extract, yeast extract and beta-glucan + 4 ppm OTA + LPS challenge; T4 = basal diet with FA2 including 0.20% clay mineral, plant extract, yeast extract and beta-glucan + 4 ppm OTA + LPS challenge; SEM, standard error of mean. a–c = Means with different letters are significantly different ($p < 0.05$). N = 10.

3.6. Blood Profiles

The OTA-contaminated diet (PC) significantly increased ($p < 0.05$) GOT and HDL levels in blood, whereas it decreased ($p < 0.05$) total cholesterol levels in the blood in comparison to the NC group (Table 8). In particular, supplementation with two levels of FA1 and 0.10% FA2 significantly mitigated ($p < 0.05$) the increased GOT levels in the blood caused by the OTA-contaminated diet (PC). The natural-feed-additive-supplemented groups demonstrated significantly increased ($p < 0.05$) total cholesterol levels in the blood, displaying similar values to the NC group. Supplementation with 0.10% FA1 and two levels of FA2 remarkably decreased ($p < 0.05$) HDL levels in the blood to similar values to the NC group. There was no significant difference in the GPT levels in the blood among the treatment groups.

Table 8. Effect of combinations of natural feed additives on the blood profile in broiler breeders fed with an ochratoxin A diet [1].

Items	NC	PC	T1	T2	T3	T4	SEM	*p*-Value
GOT, U/L	137.00 d	229.88 a	198.44 bc	172.89 c	191.56 bc	213.56 ab	5.280	<0.001
GPT, U/L	3.70	4.75	4.33	4.11	4.33	4.22	0.342	0.442
Total-C, mg/dL	119.70 a	95.25 b	119.22 a	116.89 a	119.33 a	117.44 a	4.373	0.005
HDL, mg/dL	49.00 c	78.25 a	70.11 ab	52.44 c	53.33 c	60.33 bc	3.948	<0.001

[1] Abbreviations: NC = Negative Control, basal diet; PC = Positive Control, basal diet with + 4 ppm ochratoxin A (OTA) + lipopolysaccharide (LPS) challenge; T1 = basal diet with FA1 including 0.05% clay mineral and charcoal + 4 ppm OTA + LPS challenge; T2 = basal diet with FA1 including 0.10% clay mineral and charcoal + 4 ppm OTA + LPS challenge; T3 = basal diet with FA2 including 0.10% clay mineral, plant extract, yeast extract and beta-glucan + 4 ppm OTA + LPS challenge; T4 = basal diet with FA2 including 0.20% clay mineral, plant extract, yeast extract and beta-glucan + 4 ppm OTA + LPS challenge; GOT, glutamic oxaloacetic transaminase; GPT, glutamic pyruvic transaminase; Total-C, total cholesterol; HDL, high-density lipoprotein; SEM, standard error of mean. a–d = Means with different letters are significantly different ($p < 0.05$). N = 10.

3.7. OTA Accumulation

The OTA-contaminated diet (PC) significantly increased ($p < 0.05$) OTA accumulation in the feed and liver in comparison to the NC group (Table 9). Supplementation with natural feed additives significantly decreased ($p < 0.05$) OTA accumulation in the liver compared to the PC group.

Table 9. Effect of combinations of natural feed additives on OTA deposition in the liver and feed of broiler breeders fed with an ochratoxin A diet [1].

Items	NC	PC	T1	T2	T3	T4	SEM	*p*-Value
Feed, ppm	ND	4.02	4.01	4.00	4.01	4.01	0.003	0.173
Liver, ppb	ND	48.07 a	33.18 b	30.79 bc	29.13 c	33.19 b	0.760	<0.001

[1] Abbreviation: NC = Negative Control, basal diet; PC = Positive Control, basal diet with + 4 ppm ochratoxin A (OTA) + lipopolysaccharide (LPS) challenge; T1 = basal diet with FA1 including 0.05% clay mineral and charcoal + 4 ppm OTA + LPS challenge; T2 = basal diet with FA1 including 0.10% clay mineral and charcoal + 4 ppm OTA + LPS challenge; T3 = basal diet with FA2 including 0.10% clay mineral, plant extract, yeast extract and beta-glucan + 4 ppm OTA + LPS challenge; T4 = basal diet with FA2 including 0.20% clay mineral, plant extract, yeast extract and beta-glucan + 4 ppm OTA + LPS challenge; ND, non-detect; SEM, standard error of mean. a–c = Means with different letters are significantly different ($p < 0.05$). N = 10.

4. Discussion

In the current study, ochratoxin A (OTA)-contaminated diets caused significantly lower egg production and hatchability and numerically decreased egg weight compared to the negative control. This result is in agreement with the studies of Zahoor-ul-Hassan et al. [9] and Eid et al. [27], who reported that ochratoxin caused lower egg mass, egg production, average egg weight, and hatchability along with reduced feed consumption. According to a previous study, OTA has the highest embryotoxicity among all mycotoxins, and thus can increase embryo mortality and decrease fertilized egg hatchability [5]. Moreover, Konrad and Röschenthaler [28] reported that OTA can negatively influence the synthesis of deoxyribonucleic acid (DNA), ribonucleic acid (RNA), and proteins. These authors suggested that the reduction in egg performance may be attributed to these disorders in metabolic pathways caused by OTA.

The supplementation of FA1, which mainly consisted of toxin binders (i.e., clay mineral and charcoal), and FA2, which consisted of toxin binders with natural components (i.e., clay mineral, plant extract, yeast extract and beta-glucan), used in this study mitigated the reduced egg production and hatchability caused by OTA, respectively. Toxin binders such as clay mineral and charcoal, which are the main components in FA1, bind and immobilize the mycotoxin in the gastrointestinal tract (GIT), thereby reducing the absorption of toxins into the GIT in poultry [29–31].

In this study, the improved egg production and hatchability caused by supplementing toxin binders (i.e., clay mineral and charcoal) may be attributed to the abovementioned mechanism and the beneficial roles of binders in minimizing the harmful effects of mycotoxins. Likewise, the addition of natural feed additives (i.e., plant extract, yeast extract and beta-glucan) improved egg production and hatchability; the group supplemented with 0.10% natural additives (T3) had the highest egg production and hatchability among the treatments. Thanissery et al. [32] and Esper et al. [33] reported that plant and yeast extracts are effective against molds such as *Aspergillus* and *A. fumigatus* and bacteria such as *Clostridium perfringens*. Moreover, previous studies suggested that plant extracts and prebiotics stimulated the release of endogenous digestive enzymes, which enhance nutrient digestion [34,35]. As mentioned above, mycotoxins can interfere with metabolic pathways such as carbohydrate, protein and lipid metabolisms. However, natural components such as plant extracts, yeast extracts, and beta-glucan could support these metabolisms by secreting endogenous enzymes and inhibiting the growth of molds as a fundamental cause. Thus, the improved egg production and hatchability observed in the current study may be due to the beneficial effects of toxin binders and natural components. Moreover, the

Haugh units at 8 weeks were enhanced by supplementing FA1 and FA2 in the current study. Our results regarding Haugh unit improvement provide clear evidence supporting the hypothesis that the addition of natural feed additives can improve digestibility by enhancing enzyme activity.

It has been reported that egg quality, bone mineralization, and the relative weight of the liver and abdominal fat can be negatively affected by mycotoxins. Egg internal and shell quality are important parameters for the worldwide egg industry [36]. In the current study, eggshell strength at 4 weeks and Haugh units at 8 weeks were decreased, while the OTA contaminated diet increased the relative weight of the liver and abdominal fat. Jia et al. [36] observed the lowest eggshell strength in the mycotoxin-affected groups. According to Zaghini et al. [37], eggshell strength is affected by shell thickness, and they observed reduced eggshell thickness along with eggshell strength. This is consistent with the current study which indicated that shell thickness in the OTA-contaminated group (PC) was noticeably thinner than in the negative control. This poor eggshell quality may be due to poor calcium and phosphorus absorption, supported by bone mineralization in the current study. We found that the OTA-contaminated diet caused a reduction in ash, Ca and P contents in the tibia. Yildirim et al. [38] reported that high levels of mycotoxin in diets can reduce the circulation level of parathyroid hormone (PTH), which is an essential regulator of extracellular calcium and phosphate metabolism. PTH increases Ca re-absorption and P excretion in the kidney and enhances Ca absorption in the intestine by stimulating the synthesis of 1,25-dihydroxy vitamin D in the kidney [39]. The dysregulation of the above-mentioned mechanisms may negatively affect egg quality and bone mineralization.

In the current study, the addition of FA1 and FA2 alleviated poor eggshell strength at 4 weeks and bone mineralization in the tibia. These results are in agreement with the results of Darmawan and Ozturk [40] showing that eggshell strength linearly increased with increasing clay mineral levels. According to the study by Elliott et al. [41], clay minerals have a high mineral content, ion exchange capacity, and calcium affinity; these functions positively affect eggshell quality. Kriseldi et al. [42] suggested that clay minerals increased alkaline phosphatase, influencing bone mineralization and phytate degradation in the small intestine.

Plant extracts, yeast extracts, and prebiotics have been reported to affect eggshell formation. Lokaewmanee et al. [43] reported that plant extracts such as red clover and garlic enhanced eggshell strength, and this beneficial result could be attributed to the improvement in the morphology of the small intestine. Moreover, according to a review by Miranda et al. [44], plant extracts mitigated pro-inflammatory response and improved antioxidant enzyme release, thereby increasing bone repair, osteogenesis, formation and mineralization of bone. Thus, their beneficial functions may cause the improvement in Ca and P availability for eggshell formation. However, numerous studies reported that yeast and prebiotics did not affect eggshell formation [45,46]. The current study indicated that improved eggshell and bone mineralization may be due to clay minerals and plant extracts.

In the current study, the OTA-contaminated diet increased the relative weight of the liver and abdominal fat while decreasing total cholesterols in the blood compared to the negative control. Our results are in agreement with previous studies showing that mycotoxins, including deoxynivalenol (DON), zearalenone (ZEA), and OTA, caused fat deposition in the liver and reductions in cholesterol and triglyceride contents in the blood [37,47–49]. These authors suggested that triglyceride and cholesterol in chickens fed mycotoxins-contaminated diets were transported into the liver, and thus fat accumulation in the liver was increased, while cholesterol and triglyceride levels in the blood were decreased. Our results agree with the abovementioned mechanisms. Furthermore, in the current study, we found that the OTA-contaminated diet caused high OTA accumulation in the liver and GOT concentration in the blood. These findings are in agreement with previous studies reporting that chickens exposed to mycotoxins had high mycotoxin accumulation in the liver and activated GOT in the serum [11,50,51]. Mycotoxins are mainly metabolized in the liver and transformed into their reactive metabolites, and thus cause hepatocyte cancerization and liver damage [52]. According to Rashidi et al. [49], mycotoxins induced increased hepatocyte permeability, and then transaminases could shift from the damaged hepatocyte to the bloodstream, consequently causing increased activity of transaminases like GOT and GPT. The increased levels of GOT in the blood and OTA accumulation in the liver observed in the current study may be explained by the abovementioned mechanisms. Clay minerals have been known to have high mycotoxin adsorption capacity due to their high surface area and ion exchange capacity [17,53]. Moreover, plant extracts that have antimicrobial and antifungal properties can inhibit bacteria and mold [54]. In the current study, supplementation with natural feed additives (FA1 and FA2) mitigated the relative weight of the liver and abdominal fat, the GOT level in the blood, and OTA accumulation in the liver. These findings may be attributed to the inhibition of OTA by clay minerals and plant extracts.

5. Conclusions

In the current study, we found that supplementation with toxin binders (FA1) and toxin binders with natural components (FA2) mitigated the negative effects of OTA and even alleviated Haugh units, the relative weight of the liver and abdominal fat, total cholesterol, and HDL, bringing them to similar levels to those observed in non-challenged treatments (NC group). In particular, the group supplemented with 0.10% FA2 demonstrated improved parameters compared to the other treatment groups. The results of our study indicate that toxin binders with or without natural components can be very useful in the mitigation of OTA-contamination-induced problems by inhibiting OTA and improving nutrient absorption and host health. Moreover, we observed tenuous synergistic effects between clay minerals and natural components.

Author Contributions: Conceptualization, J.L., B.L; S.K., W.K.K. and J.C.; investigation, J.L., H.C., D.S., S.C., J.A., J.N., B.L. and S.K.; writing—original draft preparation, J.L.; writing—review and editing, J.L., W.K.K. and J.C.; supervision, W.K.K. and J.C. All authors have read and agreed to the published version of the manuscript.

Funding: This research was funded by Provimi Singapore Pte Ltd. (Singapore).

Institutional Review Board Statement: Not applicable.

Informed Consent Statement: Not applicable.

Data Availability Statement: No new data were generated in this manuscript.

Conflicts of Interest: Cherrybro Company and Provimi Company, who were involved in the provision of data for this study, did not interfere with the co-authors' access to all of the study's data, analyzing and interpreting the data, or preparing and publishing manuscripts independently. All authors declare no conflict of interest.

References

1. Amézqueta, S.; González-Peñas, E.; Murillo-Arbizu, M.; de Cerain, A.L. Ochratoxin A decontamination: A review. *Food Control* **2009**, *20*, 326–333. [CrossRef]
2. El-Sayed, R.A.; Jebur, A.B.; Kang, W.; El-Esawi, M.A.; El-Demerdash, F.M. An overview on the major mycotoxins in food products: Characteristics, toxicity, and analysis. *J. Future Foods* **2022**, *2*, 91–102. [CrossRef]
3. Raters, M.; Matissek, R. Thermal stability of aflatoxin B 1 and ochratoxin A. *Mycotoxin Res.* **2008**, *24*, 130–134. [CrossRef] [PubMed]
4. Al Shap, N.F.; El-Sherbeny, E.M.E.; El Masry, D. The efficacy of metal nanocomposite (Fe_3O_4/CuO/ZnO) to ameliorate the toxic effects of ochratoxin in broilers. *BMC Vet. Res.* **2022**, *18*, 312. [CrossRef] [PubMed]
5. Hassan, Z.U.; Khan, M.Z.; Khan, A.; Javed, I.; Hussain, Z. Effects of individual and combined administration of ochratoxin A and aflatoxin B1 in tissues and eggs of White Leghorn breeder hens. *J. Sci. Food Agr.* **2012**, *92*, 1540–1544. [CrossRef]
6. Hameed, M.R.; Khan, M.Z.; Saleemi, M.K.; Khan, A.; Akhtar, M.; Hassan, Z.U.; Hussain, Z. Study of ochratoxin A (OTA)-induced oxidative stress markers in broiler chicks. *Toxin. Rev.* **2017**, *36*, 270–274. [CrossRef]
7. Bhatti, S.A.; Khan, M.Z.; Saleemi, M.K.; Hassan, Z.U. Impact of dietary Trichosporon mycotoxinivorans on ochratoxin A induced immunotoxicity; in vivo study. *Food Chem. Toxicol.* **2019**, *132*, 110696. [CrossRef]
8. Wang, W.; Zhai, S.; Xia, Y.; Wang, H.; Ruan, D.; Zhou, T.; Zhu, Y.; Zhang, H.; Zhang, M.; Ye, H.; et al. Ochratoxin A induces liver inflammation: Involvement of intestinal microbiota. *Microbiome* **2019**, *7*, 151. [CrossRef]

9. Zahoor-ul-Hassan, K.M.; Khan, A.; Javed, I. Pathological responses of White Leghorn breeder hens kept on ochratoxin A contaminated feed. *Pak. Vet. J.* **2010**, *30*, 118–123.
10. Iqbal, S.Z.; Nisar, S.; Asi, M.R.; Jinap, S. Natural incidence of aflatoxins, ochratoxin A and zearalenone in chicken meat and eggs. *Food Control* **2014**, *43*, 98–103. [CrossRef]
11. Alharthi, A.S.; Al Sulaiman, A.R.; Aljumaah, R.S.; Alabdullatif, A.A.; Ferronato, G.; Alqhtani, A.H.; Al-Garadi, M.A.; Al-sornokh, H.; Abudabos, A.M. The efficacy of bentonite and zeolite in reducing aflatoxin B1 toxicity on production performance and intestinal and hepatic health of broiler chickens. *Ital. J. Anim. Sci.* **2022**, *21*, 1181–1189. [CrossRef]
12. Nazarizadeh, H.; Pourreza, J. Evaluation of three mycotoxin binders to prevent the adverse effects of aflatoxin B1 in growing broilers. *J. Appl. Anim. Res.* **2019**, *47*, 135–139. [CrossRef]
13. Nadziakiewicza, M.; Kehoe, S.; Micek, P. Physico-chemical properties of clay minerals and their use as a health promoting feed additive. *Animals* **2019**, *9*, 714. [CrossRef] [PubMed]
14. Burchacka, E.; Łukaszewicz, M.; Kułażyński, M. Determination of mechanisms of action of active carbons as a feed additive. *Bioorg. Chem.* **2019**, *93*, 102804. [CrossRef] [PubMed]
15. Bhatti, S.A.; Khan, M.Z.; Saleemi, M.K.; Hassan, Z.U. Combating immunotoxicity of aflatoxin B1 by dietary carbon supplementation in broiler chickens. *Environ. Sci. Pollut. Res.* **2021**, *28*, 49089–49101. [CrossRef]
16. Abidin, Z.; Khatoon, A.; Numan, M. Mycotoxins in broilers: Pathological alterations induced by aflatoxins and ochratoxins, diagnosis and determination, treatment and control of mycotoxicosis. *Worlds. Poult. Sci. J.* **2011**, *67*, 485–496. [CrossRef]
17. Manafi, M.; Hedayati, M.; Yari, M. Aflatoxicosis and herbal detoxification: The effectiveness of thyme essence on performance parameters and antibody titers of commercial broilers fed aflatoxin B1. *Res. Zool.* **2014**, *4*, 43–50.
18. El Khoury, R.; Atoui, A.; Mathieu, F.; Kawtharani, H.; El Khoury, A.; Maroun, R.G.; El Khoury, A. Antifungal and antiochratoxigenic activities of essential oils and total phenolic extracts: A comparative study. *Antioxidants* **2017**, *6*, 44. [CrossRef]
19. Adeogun, J.B.; Abu, O.A.; Ewuola, E.O. Effect of yeast beta-glucans on aflatoxin absorption in the gut of broiler chickens. *Niger. J. Anim. Prod.* **2021**, *48*, 198–208.
20. Kudupoje, M.B.; Malathi, V.; Yiannikouris, A. Impact of a Natural Fusarial Multi-Mycotoxin Challenge on Broiler Chickens and Mitigation Properties Provided by a Yeast Cell Wall Extract and a Postbiotic Yeast Cell Wall-Based Blend. *Toxins* **2022**, *14*, 315. [CrossRef]
21. Yang, X.J.; Li, W.L.; Feng, Y.; Yao, J.H. Effects of immune stress on growth performance, immunity, and cecal microflora in chickens. *Poult. Sci.* **2011**, *90*, 2740–2746. [CrossRef]
22. Wang, Y.; Wang, F.; Yang, D.; Tang, X.; Li, H.; Lv, X.; Lu, D.; Wang, H. Berberine in combination with yohimbine attenuates sepsis-induced neutrophil tissue infiltration and multiorgan dysfunction partly via IL-10-mediated inhibition of CCR2 expression in neutrophils. *Int. Immunopharmacol.* **2016**, *35*, 217–225. [CrossRef] [PubMed]
23. Islam, Z.; Pestka, J.J. LPS priming potentiates and prolongs proinflammatory cytokine response to the trichothecene deoxynivalenol in the mouse. *Toxicol. Appl. Pharmacol.* **2006**, *211*, 53–63. [CrossRef]
24. Ruan, D.; Wang, W.C.; Lin, C.X.; Fouad, A.M.; Chen, W.; Xia, W.G.; Wang, S.; Luo, X.; Zhang, W.H.; Yan, S.J.; et al. Effects of curcumin on performance, antioxidation, intestinal barrier and mitochondrial function in ducks fed corn contaminated with ochratoxin A. *Animal* **2019**, *13*, 42–52. [CrossRef] [PubMed]
25. Xu, L.; Eicher, S.D.; Applegate, T.J. Effects of increasing dietary concentrations of corn naturally contaminated with deoxynivalenol on broiler and turkey poult performance and response to lipopolysaccharide. *Poult. Sci.* **2011**, *90*, 2766–2774. [CrossRef]
26. Zinzadze, C. Colorimetric methods for the determination of phosphorus. *Ind. Eng. Chem. Anal. Ed.* **1935**, *7*, 227–230. [CrossRef]
27. Eid, Y.Z.; Hassan, R.A.; El-soud, S.A.; Eldebani, N. The Protective Role of Silymarin to Ameliorate the Adverse Effects of Ochratoxin-A in Laying Hens on Productive Performance, Blood Biochemistry, Hematological and Antioxidants Status. *Braz. J. Poult. Sci.* **2022**, *24*, eRBCA-2021. [CrossRef]
28. Konrad, I.; Röschenthaler, R. Inhibition of phenylalanine tRNA synthetase from Bacillus subtilis by ochratoxin A. *Febs Lett.* **1977**, *83*, 341–347. [CrossRef]
29. Miazzo, R.; Rosa, C.A.R.; Carvalho, E.D.Q.; Magnoli, C.; Chiacchiera, S.M.; Palacio, G.; Saenz, M.; Kikot, A.; Basaldella, E.; Dalcero, A. Efficacy of synthetic zeolite to reduce the toxicity of aflatoxin in broiler chicks. *Poult. Sci.* **2000**, *79*, 1–6. [CrossRef]
30. Bhatti, S.A.; Khan, M.Z.; Hassan, Z.U.; Saleemi, M.K.; Saqib, M.; Khatoon, A.; Akhter, M. Comparative efficacy of Bentonite clay, activated charcoal and Trichosporon mycotoxinivorans in regulating the feed-to-tissue transfer of mycotoxins. *J. Sci. Food Agric.* **2018**, *98*, 884–890. [CrossRef]
31. Saminathan, M.; Selamat, J.; Abbasi Pirouz, A.; Abdullah, N.; Zulkifli, I. Effects of nano-composite adsorbents on the growth performance, serum biochemistry, and organ weights of broilers fed with aflatoxin-contaminated feed. *Toxins* **2018**, *10*, 345. [CrossRef] [PubMed]
32. Thanissery, R.; McReynolds, J.L.; Conner, D.E.; Macklin, K.S.; Curtis, P.A.; Fasina, Y.O. Evaluation of the efficacy of yeast extract in reducing intestinal *Clostridium perfringens* levels in broiler chickens. *Poult. Sci.* **2010**, *89*, 2380–2388. [CrossRef] [PubMed]
33. Esper, R.H.; Gonçalez, E.; Marques, M.O.; Felicio, R.C.; Felicio, J.D. Potential of essential oils for protection of grains contaminated by aflatoxin produced by *Aspergillus flavus*. *Front. Microbiol.* **2014**, *5*, 269. [CrossRef] [PubMed]
34. Teng, P.Y.; Kim, W.K. Roles of prebiotics in intestinal ecosystem of broilers. *Front. Vet. Sci.* **2018**, *5*, 245. [CrossRef]

35. Nazarizadeh, H.; Mohammad Hosseini, S.; Pourreza, J. Effect of plant extracts derived from thyme and chamomile on the growth performance, gut morphology and immune system of broilers fed aflatoxin B1 and ochratoxin A contaminated diets. *Ital. J. Anim. Sci.* **2019**, *18*, 1073–1081. [CrossRef]
36. Jia, R.; Ma, Q.; Fan, Y.; Ji, C.; Zhang, J.; Liu, T.; Zhao, L. The toxic effects of combined aflatoxins and zearalenone in naturally contaminated diets on laying performance, egg quality and mycotoxins residues in eggs of layers and the protective effect of Bacillus subtilis biodegradation product. *Food Chem. Toxicol.* **2016**, *90*, 142–150. [CrossRef] [PubMed]
37. Zaghini, A.; Martelli, G.; Roncada, P.; Simioli, M.; Rizzi, L. Mannanoligosaccharides and aflatoxin B1 in feed for laying hens: Effects on egg quality, aflatoxins B1 and M1 residues in eggs, and aflatoxin B1 levels in liver. *Poult. Sci.* **2005**, *84*, 825–832. [CrossRef]
38. Yildirim, E.; Yalcinkaya, I.; Kanbur, M.U.R.A.T.; Cinar, M.; Oruc, E. Effects of yeast glucomannan on performance, some biochemical parameters and pathological changes in experimental aflatoxicosis in broiler chickens. *Rev. Med. Vet.* **2011**, *162*, 413–420.
39. Leung, E.K.Y. Parathyroid hormone. In *Advances in Clinical Chemistry*; Elsevier: Amsterdam, The Netherlands, 2021; Volume 101, pp. 41–93.
40. Darmawan, A.; Ozturk, E. The Impact of Bentonite Feed Additives on Laying Hens Performance and Egg Quality: A Meta Analysis. *Iran. J. Appl. Anim. Sci.* **2022**, *12*, 647–653.
41. Elliott, C.T.; Connolly, L.; Kolawole, O. Potential adverse effects on animal health and performance caused by the addition of mineral adsorbents to feeds to reduce mycotoxin exposure. *Mycotoxin Res.* **2020**, *36*, 115–126. [CrossRef]
42. Kriseldi, R.; Johnson, J.A.; Walk, C.L.; Bedford, M.R.; Dozier III, W.A. Influence of exogenous phytase supplementation on phytate degradation, plasma inositol, alkaline phosphatase, and glucose concentrations of broilers at 28 days of age. *Poult. Sci.* **2021**, *100*, 224–234. [CrossRef] [PubMed]
43. Lokaewmanee, K.; Yamauchi, K.E.; Komori, T.; Saito, K. Eggshell quality, eggshell structure and small intestinal histology in laying hens fed dietary Pantoea-6® and plant extracts. *Ital. J. Anim. Sci.* **2014**, *13*, 3163. [CrossRef]
44. Miranda, L.L.; Guimaraes-Lopes, V.D.P.; Altoe, L.S.; Sarandy, M.M.; Melo, F.C.S.A.; Novaes, R.D.; Goncalves, R.V. Plant extracts in the bone repair process: A systematic review. *Mediat. Inflamm.* **2019**, *2019*, 1296153. [CrossRef] [PubMed]
45. Sacakli, P.; Ergun, A.; Koksal, B.H.; Ozsoy, B.; Cantekin, Z. Effects of inactivated brewer's yeast (*Saccharomyces cereviciae*) on egg production, serum antibody titres and cholesterol levels in laying hens. *Vet. Ir. Zootech.* **2013**, *61*, 53–60.
46. Sarwar, F.; Akhtar, R.; Akram, Q.; Rizwan, H.M.; Naeem, M.A.; Azad, A.; Rahman, M.S.; Saleem, M.M.; Anwar, M.N.; Zahid, B.; et al. Effects of Saccharomyces Cerevisiae Supplemented Diet on Production Performance, Egg Quality and Humoral Immunity in Black Australorp and Fayoumi Layers. *Braz. J. Poult. Sci.* **2023**, *25*, eRBCA-2021.
47. Andretta, I.; Kipper, M.; Lehnen, C.R.; Lovatto, P.A. Meta-analysis of the relationship of mycotoxins with biochemical and hematological parameters in broilers. *Poult. Sci.* **2012**, *91*, 376–382. [CrossRef]
48. Ghareeb, K.; Awad, W.A.; Böhm, J. Ameliorative effect of a microbial feed additive on infectious bronchitis virus antibody titer and stress index in broiler chicks fed deoxynivalenol. *Poult. Sci.* **2012**, *91*, 800–807. [CrossRef]
49. Riahi, I.; Marquis, V.; Ramos, A.J.; Brufau, J.; Esteve-Garcia, E.; Pérez-Vendrell, A.M. Effects of Deoxynivalenol-contaminated diets on productive, morphological, and physiological indicators in broiler Chickens. *Animals* **2020**, *10*, 1795. [CrossRef]
50. Salem, R.; El-Habashi, N.; Fadl, S.E.; Sakr, O.A.; Elbialy, Z.I. Effect of probiotic supplement on aflatoxicosis and gene expression in the liver of broiler chicken. *Environ. Toxicol. Pharmacol.* **2018**, *60*, 118–127. [CrossRef]
51. Śliżewska, K.; Cukrowska, B.; Smulikowska, S.; Cielecka-Kuszyk, J. The effect of probiotic supplementation on performance and the histopathological changes in liver and kidneys in broiler chickens fed diets with aflatoxin B1. *Toxins* **2019**, *11*, 112. [CrossRef]
52. Ismail, I.E.; Farag, M.R.; Alagawany, M.; Mahmoud, H.K.; Reda, F.M. Efficacy of some feed additives to attenuate the hepato-renal damage induced by aflatoxin B1 in rabbits. *J. Anim. Physiol. Anim. Nutr.* **2020**, *104*, 1343–1350. [CrossRef] [PubMed]
53. Murray, H.H. Traditional and new applications for kaolin, smectite, and palygorskite: A general overview. *Appl. Clay Sci.* **2000**, *17*, 207–221. [CrossRef]
54. Vaičiulienė, G.; Bakutis, B.; Jovaišienė, J.; Falkauskas, R.; Gerulis, G.; Bartkienė, E.; Klupšaitė, D.; Klementavičiūtė, J.; Baliukonienė, V. Effects of Ethanol Extracts of Origanum vulgare and Thymus vulgaris on the Mycotoxin Concentrations and the Hygienic Quality of Maize (*Zea mays* L.) Silage. *Toxins* **2022**, *14*, 298. [CrossRef] [PubMed]

animals

Review

Osteoimmunology: A Link between Gastrointestinal Diseases and Skeletal Health in Chickens

Milan Kumar Sharma, Prafulla Regmi, Todd Applegate, Lilong Chai and Woo Kyun Kim *

Department of Poultry Science, University of Georgia, Athens, GA 30602, USA; milan.sharma@uga.edu (M.K.S.); pregmi@uga.edu (P.R.); applegt@uga.edu (T.A.); lchai@uga.edu (L.C.)
* Correspondence: wkkim@uga.edu

Simple Summary: Gastrointestinal diseases in poultry are more significant because of high economic loss due to production loss, reduced feed efficiency, mortality, compromising bird welfare, and sometimes zoonotic importance. With the removal of antibiotic growth promoters in the diet, gastrointestinal disease incidences have grown in recent years. Gastrointestinal diseases have been shown to affect bone growth negatively; however, possible mechanisms have yet to be confirmed in chickens.

Abstract: Bone serves as a multifunctional organ in avian species, giving structural integrity to the body, aiding locomotion and flight, regulating mineral homeostasis, and supplementing calcium for eggshell formation. Furthermore, immune cells originate and reside in the bone marrow, sharing a milieu with bone cells, indicating a potential interaction in functions. In avian species, the prevalence of gastrointestinal diseases can alter the growth and the immune response, which costs a great fortune to the poultry industry. Previous studies have shown that coccidiosis and necrotic enteritis can dramatically reduce bone quality as well. However, possible mechanisms on how bone quality is influenced by these disease conditions have not yet been completely understood, other than the reduced feed intake. On the other hand, several mediators of the immune response, such as chemokines and cytokines, play a vital role in the differentiation and activation of osteoclasts responsible for bone resorption and osteoblasts for bone formation. In the case of *Eimeria* spp./*Clostridium perfringens* coinfection, these mediators are upregulated. One possible mechanism for accelerated bone loss after gastrointestinal illnesses might be immune-mediated osteoclastogenesis via cytokines-RANKL-mediated pathways. This review article thus focuses on osteoimmunological pathways and the interaction between host immune responses and bone biology in gastrointestinal diseases like coccidiosis and necrotic enteritis affecting skeletal health.

Keywords: osteoimmunology; coccidiosis; necrotic enteritis; skeletal health

Citation: Sharma, M.K.; Regmi, P.; Applegate, T.; Chai, L.; Kim, W.K. Osteoimmunology: A Link between Gastrointestinal Diseases and Skeletal Health in Chickens. *Animals* **2023**, *13*, 1816. https://doi.org/10.3390/ani13111816

Academic Editor: Velmurugu Ravindran

Received: 8 May 2023
Revised: 22 May 2023
Accepted: 23 May 2023
Published: 30 May 2023

1. Introduction

The avian skeletal system is a complex organ with multiple functions providing structural integrity to the body and maintaining homeostasis of minerals such as calcium and phosphorus [1–3]. Furthermore, bone marrow is the primary site for hematopoiesis, harboring hematopoietic stem cells, and mature immune cells (B cells, neutrophils, macrophages, and T cells), defending the body against invading pathogens [4–7]. Immune cells originate and some of them reside within the bone marrow and share the same microenvironment as bone cells, indicating a potential for interactions. The interaction between the bone and immune cells has already been described wherein osteoblasts are a regulator of the hematopoietic stem cells and osteoclast having the same origin as the immune cells, i.e., macrophages and dendritic cells [4,6,7]. Moreover, several mediators of the immune response, such as chemokines and cytokines, play a vital role in the differentiation and activation of osteoclasts and osteoblasts [4–7]. Since skeletal and immune systems are closely associated with each other, immune cells, chemokines, and cytokines may play a critical role in regulating bone remodeling during disease conditions in avian species. Osteoimmunology, which studies possible interactions between the immune system and bone biology and how they cross-communicate to modulate their respective functions under one umbrella, might be a link in describing bone loss in chickens when infected with gastrointestinal diseases [4,6,8].

Recently, skeletal disorders in broilers and laying hens have resurfaced with advancements in genetics and modern production systems. The selective breeding for higher growth rates in broilers has led to the development of immature skeletal systems unable to withstand the rapidly increasing body mass, resulting in lameness and bone disorders [9,10]. Similarly, the selection of laying hens for early sexual maturity, small body conformation, and maximum egg production has put significant strain on their skeletal systems, as medullary bone serves as a labile source of calcium for eggshell production, potentially contributing to bone disorders [1,2,11–13]. Moreover, the intensive production systems utilized in modern poultry farming involve high stocking densities, which provide ideal conditions for transmitting gastrointestinal diseases such as coccidiosis, necrotic enteritis, salmonellosis, etc. [14–16]. These gastrointestinal diseases have a substantial economic impact on the poultry industry primarily due to a reduction in body weight and worsened feed conversion ratio [17–20]. The estimated global loss due to coccidiosis is around 15 billion dollars, whereas that of necrotic enteritis is 6 billion dollars to the poultry industry [16,21]. These gastrointestinal diseases also cause severe inflammation, affecting birds' bone growth and remodeling. The negative impacts of gastrointestinal diseases and their effect on bone in poultry have already been observed in broilers and laying hens [8,22–28]. The economic losses associated with bone disorders resulting from gastrointestinal diseases and the possible mechanism other than reduced feed intake for bone loss in these conditions have not been fully understood. Therefore, it is crucial to understand the complex relationship between gastrointestinal diseases, immune response, and skeletal disorders in poultry to develop effective intervention strategies for controlling and preventing these conditions, ultimately promoting the health and welfare of poultry in modern production systems.

In this review, we connect gastrointestinal diseases with skeletal disorders, focusing on the immune responses during and after coccidiosis and necrotic enteritis infections. Understanding this link might lead to a better understanding of the underlying mechanisms and potential therapeutic strategies to control gastrointestinal and skeletal disorders in poultry. Ultimately, understanding the complex relationship between the immune and skeletal systems in avian species might be of great importance for promoting their health and welfare while maintaining their maximum genetic potential.

2. Bone Biology of Broilers and Layers

2.1. Long Bone Growth

Avian skeletal development begins as early as the embryonic stage and reaches maturity during rearing [1,3,29]. During post-hatch growth, the cartilaginous template from the embryonic stage is replaced by the process of endochondral ossification [30,31]. In the early phase, a thin bony collar is gradually formed around the diaphysis of the cartilaginous bone template through perichondral ossification, allowing lengthwise growth. Lengthwise growth takes place at the primary ossification site through endochondral ossification [1,30]. At the end of the cartilage template or the epiphyseal growth plate, chondrocytes transform into proliferative chondrocytes and are arranged in columns within an extracellular matrix containing type II collagen. Following proliferation, these chondrocytes differentiate into hypertrophic chondrocytes, which secrete type X collagen and later undergo apoptosis and resorbed. Other than type II and type X collagen, chondrocytes also secrete organic matrix containing proteoglycans, glycoproteins, and growth factors, which regulate the further development of chondrocytes [1]. The lengthwise growth of the bone starts at the primary ossification site where chondrocytes lay down the cartilage. In contrast, chondroclast reabsorbs the extracellular matrix, while alkaline phosphatase secreted by chondrocytes and osteoblasts initiates bone calcification [1].

Osteoblasts, derived from mesenchymal stem cells, secrete type I collagen and raise the concentration of Ca^{++} and PO_4^{3-}, forming calcium hydroxyapatite crystals around the matrix [1]. During bone elongation, cartilage is resorbed, resulting in the formation of marrow spaces. The bone is resorbed and remodeled by a collaborative effort of osteoclasts and osteoblasts ultimately resulting in the formation of the trabecular bone in the marrow space [1,2]. Bone modeling allows for longitudinal and periosteal growth during early growth stages. However, hormonal changes after sexual maturity in egg-laying hens lead to the cessation of cortical and trabecular bone formation as well as longitudinal growth of long bones [1,2,12,32]. As hens sexually mature, circulating levels of estrogen significantly increase, which stimulates the function of osteoblasts for the formation of medullary bone at the expense of cortical and trabecular bones, while inhibiting osteoclastic bone resorption. Towards the end of the lay, when circulating estrogen declines, the function of osteoblast is reversed towards the structural bone formation [1,2,32]. Nevertheless, flat bones such as keel bones and ribs continue to ossify even after hens enter laying to provide structural support during egg-laying and maintain their strength. Before the first oviposition, the inner diameter of the long bones increases by approximately 20% through intramembranous ossification and enables the deposition of medullary bone along the marrow cavity. Osteoblasts create a network of immature osteoid along the outer surface of the periosteum, which eventually undergoes calcification to form cortical bones, while osteoclasts resorb bone on the endosteal surface, resulting in the widening of the bone [1]. At sexual maturity, the estrogen surge alters osteoblasts' function toward medullary bone formation [1,2,32]. Medullary bone is a secondary bone structure with a highly woven texture deposited in the marrow cavities and on the trabecular surfaces. During the pre-lay stage of the growth, medullary bone in marrow space accumulates rapidly and continues to deposit for the rest of the laying period [1,2,32,33]. In the meantime, blood supply plays a crucial role in the development and growth of bones in chickens, and avian vasculature penetrates deeper into the growth plate to support their fast growth rate [34]. The avian blood vasculature not only triggers the calcification of the cartilage matrix and formation of bone marrow, but also recruits osteoblast and osteoclast precursor cells to bone remodeling sites. It also supplies nutrients for collagen and bone formation, removes the metabolites from bone resorption, delivers systemic hormones and precursor cells, and provides angiogenic and angiocrine signals regulating bone formation, resorption, and remodeling [34,35].

2.2. Bone Remodeling

Bone remodeling is a dynamic and continuous process to maintain skeletal integrity and functionality under various stressors including physiological, hormonal, or mechanical [1,6,32,33]. This dynamic process involves the resorption and formation of bones by osteoclast and osteoblast, respectively. Osteoclasts are multinucleated bone-resorbing cells derived from monocyte/macrophage cells of hematopoietic origin [5,36]. They are more abundant in bone surfaces and are engaged in bone degradation. Whereas osteoblasts are bone-forming cells that originate from mesenchymal stem cells residing in the bone matrix of the bone marrow. These metabolically active cells synthesize the collagenous and non-collagenous bone matrix and facilitate calcification [1,5,6]. During this process, osteoblasts are trapped in the bone matrix and are converted to osteocytes, which regulate the pace of bone modeling and remodeling [5,6,37]. In fast-growing broilers, bone remodeling occurs rapidly due to the fast growth rate and high metabolic demand of bone tissue [29]. Bone remodeling occurs in both types of structural bone tissue, cortical and trabecular. Cortical bone is denser and provides structural support to the skeleton, while the trabecular bone is more porous and plays a role in mineral metabolism and load-bearing function [1,2]. Trabecular bone, due to its larger surface area and higher metabolic activity compared to cortical bone, is more susceptible to higher rates of bone remodeling [8,22].

However, medullary bone in laying hens is highly vascularized and plays a crucial role in maintaining calcium homeostasis [1,2,13,32]. Furthermore, it serves as a labile source of calcium for eggshell formation, and the process of its deposition and resorption occurs continuously throughout the laying period. Therefore, medullary bone remodeling in egg-laying chickens is a significant physiological process that allows a hen to meet the increased demand for calcium during egg production.

The dynamic process of bone remodeling in chickens is tightly regulated by the osteoclastogenesis, and osteoblastogenesis at the cellular level and is initiated by the change in the secretion of hormones for calcium homeostasis (parathyroid, calcitriol, calcitonin, or estrogen hormones), the status of lay, mechanical stress, or immune responses. Bone remodeling is completed in three different phases: (i) initiation of resorption by osteoclasts, (ii) transitioning from resorption to the formation of new bone (reversal), and (iii) formation of new bones [6,7,38]. A change in calcium homeostasis, mechanical stress, or damage to the bone stimulates osteocytes and osteoblasts to recruit osteoclast precursor cells in bone surfaces. An increase in the expression of receptor activator of nuclear factor kappa-B ligand (RANKL), decreased in osteoprotegerin (OPG) from osteoblast and increased in sclerostin from osteocytes, stimulates RANKL-mediated osteoclastogenesis [37,39–43]. These multinucleated osteoclasts then degrade and resorb the mineralized matrix by producing collagenolytic enzymes (Cathepsin K, Tartrate resistant acid phosphatase) and hydrochloric acids and undergo apoptosis [44]. The resorption of mineralized bones is then followed by the reversal phase, where the mononuclear cells remove the remaining collagen [45]. Furthermore, the increased concentration of insulin-like growth factors (I, II) and transforming growth factors B recruit mesenchymal stem cells to bone resorption sites and differentiate into osteoblasts [46]. The newly differentiated osteoblasts lay down the collagenous bone matrix in the resorbed sites called osteoid, which later calcifies. The bone remodeling concludes once the equilibrium is achieved, and then osteoblasts either undergo apoptosis or embed themselves in the bone matrix and differentiate into osteocytes [43,47]. The detailed process of bone remodeling is visually illustrated in Figure 1.

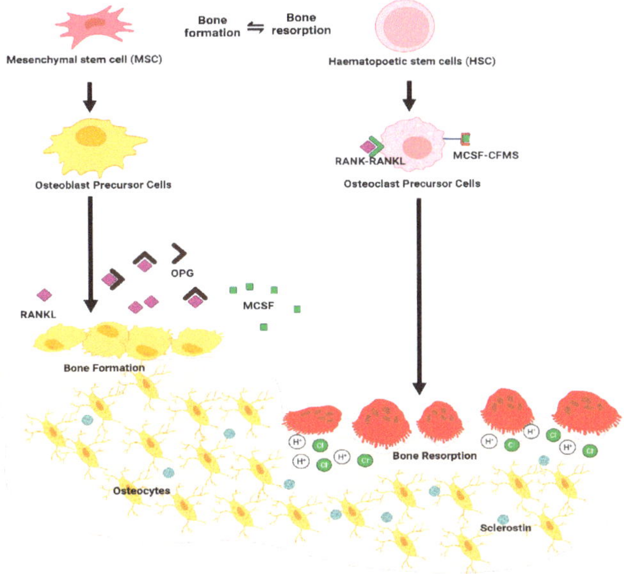

Figure 1. The mechanism of mesenchymal and hematopoietic cell differentiation into osteoblast and osteoclast, along with the bone remodeling process. RANKL: Receptor activator of nuclear factor κB ligand, RANK: Receptor activator of nuclear factor κB, OPG: Osteoprotegerin, MCSF: Macrophage colony-stimulating factor, CFMS: Colony stimulating factor receptor, H^+: Hydrogen ion, Cl^-: chloride ion.

3. Host-Immune Response in Coccidiosis and Necrotic Enteritis

Coccidiosis and necrotic enteritis have the most significant impact on the poultry industry, with more than 20 billion dollars in loss globally per annum [16,21]. Coccidiosis is caused by several species of *Eimeria* (*Eimeria acervulina, Eimeria brunetti, Eimeria maxima, Eimeria mitvai, Eimeria necatrix, Eimeria mitis, Eimeria praecox, Eimeria tenella, Eimeria hagani*) either alone or in combination [15,19,48,49]. Necrotic enteritis is a highly prevalent gastrointestinal disease in poultry, caused by *Clostridium perfringens* strains expressing NetB toxins and, most of the time, it is secondary to other predisposing factors [50–52]. Coccidiosis is one of the major predisposing factors other than high dietary proteins, dysbiosis, or enteric infections altering the immune status of birds and diets, increasing the passage rate of intestinal content [50,52,53]. Coccidiosis and necrotic enteritis affect most of the gastrointestinal tract by disrupting the gastrointestinal barrier, causing severe acute and chronic inflammation of the gastrointestinal lining, tissue necrosis, and reducing the development of immune organs [17–19,54–56].

The gut-associated lymphoid tissues (GALT) serve as the body's first line of defense against infection, disease development, and other threats by eradicating infectious pathogens at an early stage [57–59]. Pattern recognition receptors, such as Toll-like receptors, are well adapted to the intestinal epithelial cell barrier for microbial sensing through a microbial-associated molecular pattern. GALT functions as antigen recognition and presentation, followed by the release of intestinal antibodies and activation of the cell-mediated immune response in chickens when infected with gastrointestinal diseases [50,52,59,60]. Following this, activated B and T cells move to the lamina propria, where immune responses are put into action [58,59,61–63]. Adaptive immunity depends on the T and B cells found in the lamina propria. The immune system reacts swiftly to infection as early as 3 h, mainly indicated by an increased concentration of polymorphonuclear leukocytes (primarily heterophils) in the intestinal villi, the antigen's point of invasion [63].

The gastrointestinal mucosal lining is a part of the intestinal defense mechanism and helps recognize the pathogen and activate innate immune responses. Toll-like receptors recognize the microbial-associated pattern factor in and on microbes and initiate the signal transduction cascade that leads to the transcription of numerous genes involved in the immune response through RANK-mediated pathways. Activation of the antigen-presenting cells (APCs), such as dendritic cells and macrophages, by the pathogens and upregulation of pro-inflammatory cytokines and chemokines by innate immune cells are essential for developing adaptive immunity [50,52,58,59,61,64]. Upon activation, APCs migrate to regional lymphoid organs, where they upregulate the production of pro-inflammatory cytokines, such as IL-1β, IL-6, IL-8, and IL-12, facilitating the activation of naive CD4+ T-cells. Subsequently, these activated T-cells differentiate into various subsets, including Th1, Th2, Treg, and Th17 [62,65–69]. It has also been observed that necrotic enteritis and coccidiosis significantly increase the number of naive CD4+ cells, which is supposed to differentiate into different subsets and cytotoxic T-reg cell (CD8+) of a cell-mediated immune response [51,70].

In *Clostridium perfringens*/*Eimeria* spp. co-infected necrotic enteritis, Th1 cells are activated as part of the adaptive immune response. The cytokines produced by Th1 cells (IFN-γ, IL-2, IL-12, and TNF-α) can help clear the infection by promoting the activation of macrophages and other immune cells that can eliminate the intracellular pathogens [50,58,62,65,69]. Studies have demonstrated that cytokines associated with cell-mediated immunity, such as IFN-γ, IL-2, and TNF-alpha, are upregulated during coccidiosis and necrotic enteritis [17,19,65]. However, IFN-γ and IL-2 were reported to be downregulated in necrotic enteritis to prevent the activation of innate immune system cells and suppress antigen presentation by APCs [71]. These cytokines can promote the activation and proliferation of other immune cells, such as cytotoxic T cells and natural killer cells. Furthermore, Th2 cytokines such as IL-3, IL-4, IL-9, IL-10, and IL-13 are also upregulated during necrotic enteritis as a part of the humoral defense system [50]. Previous studies have suggested that Th2 cytokines such as IL-4, IL-9, IL-10, and IL-13 may play a role in limiting the severity of the disease by promoting tissue repair and limiting inflammation and promoting antigen-specific B cell activation and antibody production. Th17 cells play a crucial role in microbial defense at mucosal surfaces, as they upregulate the production of cytokines IL-17 and IL-22, secretion of antimicrobial peptides, IgA, and the induction of inflammation, all of which contribute to defense against invading pathogens. Furthermore, Th17 cells also aid in the regeneration of intestinal epithelial cells [72,73]. The upregulation of immune response is critical in protecting chickens against various infectious diseases including coccidiosis and necrotic enteritis. These immune responses are not limited to the intestinal mucosa but might also have an influence on skeletal homeostasis altering the skeletal integrity.

4. Shift in Gut Microbiome and Immune Response

Diverse microbial communities naturally colonize the GIT of chickens and are an integral part of the intestinal ecosystem alongside the intestinal epithelial lining and immune system [74–77]. Microbiomes and immune cells residing in the GIT coexist in equilibrium while maintaining the functionality to respond against harmful pathogens. The crosstalk between the intestinal microbiome and immune system, and epithelial cells strengthens their symbiotic relationship. The host immune system regulates the colonization and microbial composition, whereas the interaction between the microbiome helps in the development and functionality of the immune system [74,75,77,78]. It has been observed that gut microbiome plays a significant role in goblet cell development [79], B cell development and activation in lamina propria and Bursa of Fabricius [80], T cell differentiation [81,82] and development of GALT [75,80]. Furthermore, metabolites produced by the gut microbiota such as short-chain fatty acids, indole, tryptamine, vitamins, and bacteriocins were reported to increase the production of anti-inflammatory cytokines by altering inflammasomes thus reducing the harmful inflammation [74,82].

However, a shift in the commensal microbial population was observed after coccidiosis and necrotic enteritis infection. A decrease in segmented filamentous bacterial population was observed in the intestine and ceca of infected birds [83]. These bacterial populations are involved in the differentiation of Th17 cells from naïve CD4+ T cells along with the IgA production [84]. Likewise, a shift in lactic acid-forming bacterial populations (either increased or decreased) was also observed following the infection [83,85–90]. These bacterial populations were involved in maintaining tight junction integrity and immunomodulation by stimulating TNF-α, INF-γ, and IL-12 [91]. Similarly, short-chain fatty acids-producing bacterial populations were decreased in chickens following the infection [83,85–90].

5. Interaction between Immune Response against Gastrointestinal Disorders and Bone Biology

Since mature immune cells (B cells, macrophages, and T cells), hematopoietic stem cells, and bone cells (osteoblasts, osteoclasts, and osteocytes) all coexist in the same milieu (bone marrow), they interact with one another to perform their respective functions [4,6,7]. Osteoclasts originate from hematopoietic stem cells, from which blood cells and other immune cells originate too. Osteoclasts are derived from monocytes/macrophages, which are circulating white blood cells that play a significant role in immune functions. In contrast, osteoblasts are derived from mesenchymal stem cells. Human and animal studies have revealed that several immune cytokines released from activated immune cells influence bone cell activity and remodeling through RANKL during inflammatory conditions [51–53,71,72].

The nuclear factor kappa-ligand (RANK) is a common link between the immune and skeletal systems. RANKL, along with macrophage colony-stimulating factor (MCSF), signals the monocytes and macrophages to differentiate into osteoclasts, known as osteoclastogenesis [5,6,41]. Osteoblasts and osteocytes express RANKL under the influence of hormones and physiological factors [37,41,43,47]. Pro-inflammatory cytokines from activated immune T cells, TNF-α, IL1β, and IL-17 intensely expressed the RANKL [42,92,93], which is key to the differentiation of osteoclasts after binding with their receptor RANK from monocyte-osteoclast precursors, and their activity. The OPG extensively regulates the interaction of RANKL with RANK, a decoy receptor for RANKL expressed by osteocytes and osteoblasts [94]. The interaction between the immunological pathways connecting to the bone remodeling is visually illustrated in Figure 2.

The process of bone remodeling is continuous, and the balance between bone resorption and bone formation is maintained through the coordinated activities of osteoclasts and osteoblasts. Depending on the severity of the infection, coccidiosis and necrotic enteritis can induce both acute and chronic inflammation of the gastrointestinal system, which negatively affects the growth and might alter the bone modeling and remodeling of birds [50,69,70]. Previously, it has been observed that during coccidiosis (broilers and laying hen pullets), the number and activity of osteoclasts increased, reducing cortical and trabecular bone volume [8,22,95]. The microstructural architecture of the long bones was measured using X-ray microtomography. The authors found that coccidiosis decreased the bone volume as a fraction of tissue volume (up to 5%), bone mineral content (BMC) (up to 61%), and bone mineral density (BMD) (up to 56%) of both cortical and trabecular bone in infected pullets on 6 and 14 days post-inoculation (DPI) [8,22]. Comparable effects were observed, including decreased BMC, breaking strength, and ash% in broilers infected with *Eimeria* spp. [25,27,28]. Likewise, Sharma et al. (2022) observed a decrease in BMD (up to 35%), BMC (up to 41%), and bone volume (up to 35%) of femurs when laying hen pullets were infected with *Eimeria* spp. The one possible mechanism for this loss in bone properties might be a drastic reduction in feed intake, as observed a couple of days after infection [19,22,95]. However, restriction of feed intake in pair feeding control as that of infected broilers did not have as adverse an effect as that of the infected birds, indicating immunological responses play a significant factor in bone loss. Feed restriction led to a reduction of 10% in BMC, while inflammation resulting from *Eimeria* spp. infection caused an additional 11% loss [8]. It is also interesting to note that while inflammation subsided by 14 DPI, it took longer for the skeletal health of pullets infected with *Eimeria* spp. to recover, requiring up to 28 days post-inoculation, as observed by Sharma et al. 2022.

Eimeria infection, alone or combined with the *Clostridium perfringens*, upregulates several cytokines that play a critical role in RANKL expression and osteoclastogenesis. During the infection, naive CD4+ cells differentiated into several subsets, including Th1, Th2, Treg, and Th17 [50,52]. These activated T cells produced several pro-inflammatory and anti-inflammatory cytokines with stimulatory or inhibitory effects on osteoclastogenesis based on the expression of RANKL [4–6,92,93,96–99]. Either alone or together with the *Clostridium perfringens*, *Eimeria* spp. infection upregulates the expression of the pro-inflammatory cytokines (TNF-α, IL-1β, and IL-6) of the innate immune system and APCs, which express the secretion of RANKL [19,50,52]. Furthermore, TNF- α is also a potent inhibitor of osteoblast differentiation by several transcription factors and decreases the phenotype selection of precursor cells to the osteoblast pathway [100]. Among the pro-inflammatory cytokines, IL-17 from Th17 expresses the highest level of RANKL and, has a strong osteoclastogenic effect, was upregulated during the infection. IL-17 further increases local inflammation, and the production of TNF-α and IL-6 further increases the production of RANKL [92]. The production of RANKL by these cytokines activated the RANK signaling pathways in osteoclast precursor cells. It increased the number of osteoclasts (up to 300%), their activity, and elevated bone resorption, as observed previously in chickens by Tompkins et al. 2023 and Sharma et al. 2022.

Furthermore, peripheral APCs (dendritic cells and macrophages) are actively involved in tissue inflammation, and their number rapidly increases during infection. These cells have the capacity to differentiate into osteoclasts, which is governed by cytokine RANKL-RANK expression [101]. Although CD4+ inhibits osteoclastogenesis by increasing the secretion of transforming growth factor-β, while cytokines INF-γ, IL-4, IL-10, and IL-12 by inhibiting the RANK pathways, their functionality is masked by the upregulation of the IL-17, which has strong osteogenic properties [7,92]. On the other hand, IL-17A has been shown to inhibit the wnt signaling pathway, resulting in reduced expression of osteoblast differentiation and early osteocyte markers, thereby inhibiting the osteoblastic differentiation of bone marrow mesenchymal stem cells [102]. The interaction between immune cells and inflammatory cytokines produced during coccidiosis and necrotic enteritis and bone remodeling is illustrated in Table 1.

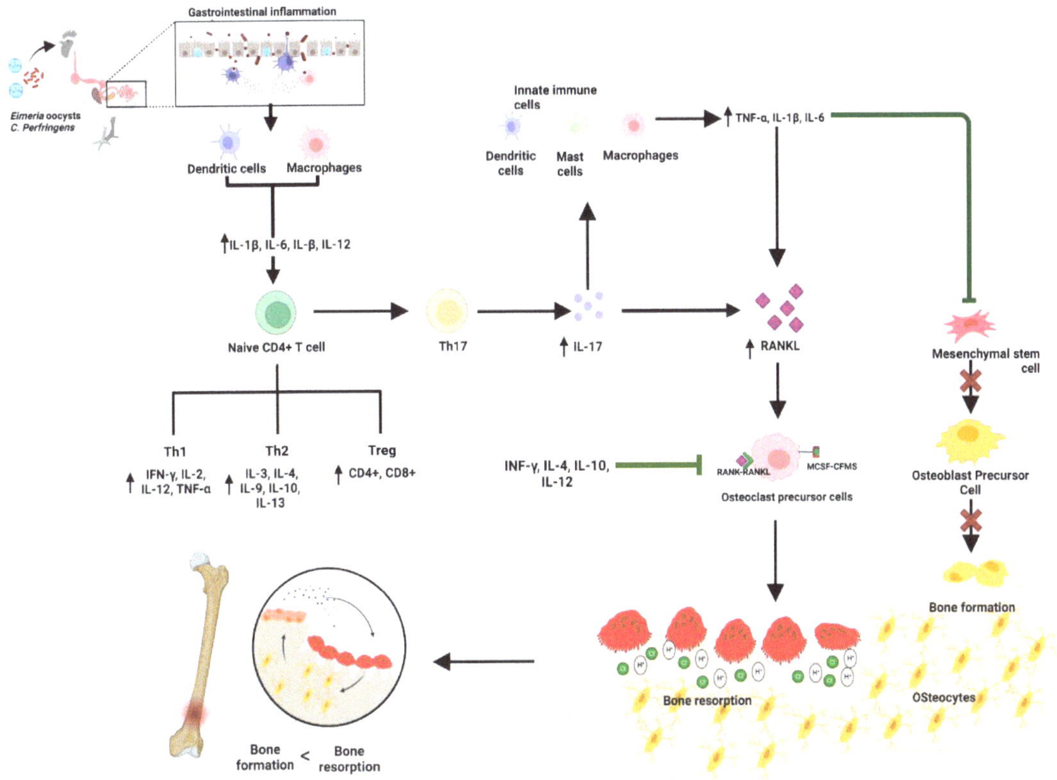

Figure 2. Schematic representation of the interaction between immune responses and bone biology following gastrointestinal disorder (coccidiosis and necrotic enteritis) in chickens. IL-1β: Interleukin 1 β, IL-6: Interleukin 6, IL-12: Interleukin 12, Th1: T helper cells 1 (Cell-mediated immunity), Th2: T helper cells 2 (Humoral immunity), Treg: Regulatory T cells, Th17: T helper cells 17, IFN-γ: Interferon γ, IL-2: Interleukin 2, TNF-α: Tumor necrosis factor α, IL-3: Interleukin 3; IL-4: Interleukin 4, IL-9: Interleukin 9, IL-10: Interleukin 10; IL-13: Interleukin 13; IL-17: Interleukin 17, RANKL: Receptor activator of nuclear factor κB ligand, RANK: Receptor activator of nuclear factor κB, OPG: Osteoprotegerin, MCSF: Macrophage colony-stimulating factor, CFMS: Colony stimulating factor receptor, H^+: Hydrogen ion, Cl^-: chloride ion.

Table 1. Role of innate immune cells and inflammatory cytokines produced during coccidiosis and necrotic enteritis in osteoimmune system.

Disease	Immune Responses	Cytokines	Effect on Immunity	Bone Resorption	Bone Formation	Possible Mechanism
Coccidiosis Necrotic enteritis	Innate Immunity	Natural killer cells	Antigen recognition and phagocytosis	↑	↓	Osteoclast differentiation, maturation, and activation
	Macrophages Dendritic cells	IL-1β	Proinflammation	↑	↓	Direct activates the RANK signaling to promote osteoclastogenesis [92,93]
		IL-6	Th17 induction	↑	↓	Activation of osteoclastogenesis [99,101,102]
		IL-8	Proinflammation	↑	↓	Osteoclastic activation through RANKL [103]
		IL-12	Proinflammation	↓	↑	Inhibition of RANKL-mediated osteoclast formation [92]
		TNF-α	Proinflammation	↑	↓	Indirect osteoclastic activation through RANKL [97,98,100]
Coccidiosis Necrotic enteritis	Adaptive Immune Response					
	Th1	IFN-γ	Cellular immunity	↓	↑	Inhibit osteoclastogenesis [92]
		IL-2	Proinflammation	↓	↑	Inhibition of RANKL [104]
		IL-12	Proinflammation	↓	↑	Inhibition of RANKL-initiated osteoclastogenesis [92]
		TNF-α	Proinflammation	↑	↓	Indirect osteoclastic activation through RANKL [97,98,100]
	Th2	IL-3	Proinflammation	↓	↑	Blocks RANKL-induced osteoclastogenesis [105]
		IL-4	Humoral immunity	↓	↑	Inhibit osteoclastogenesis, Osteoprotegerin [106]
		IL-9	Antiinflammation	↑	↓	Unknown
		IL-10	Antiinflammation	↓	↑	Suppress bone resorption [92]
		IL-13	Antiinflammation	↓	↑	Inhibit osteoclastogenesis, Osteoprotegerin [106]
	Th17	IL-17	Proinflammation	↑	↓	Induction of RANK and RANKL expression [99,101,102]
		IL-22	Proinflammation	↑	↓	Unknown [99,101,102]
	Tregs	CD4+	Helper T cells	↓	↑	TGF, IL-4, and IL10 [107]
		CD8+	Cytotoxic T cells	↓	↑	Production of OPG [107]

IL-1β: Interleukin 1 β, IL-6: Interleukin 6. IL-8: Interleukin 8, IL-12: Interleukin 12, TNF-α: Tumor necrosis factor α, IFN-γ: Interferon γ, IL-2: Interleukin 2, IL-3: Interleukin 3; IL-4: Interleukin 4, IL-9: Interleukin 9, IL-10: Interleukin 10; IL-13: Interleukin 13, IL-17: Interleukin 17, IL-22: Interleukin 22, RANKL: Receptor activator of nuclear factor κB ligand, RANK: Receptor activator of nuclear factor κB, OPG: Osteoprotegerin, Th1: T helper cells 1 (Cell-mediated immunity), Th2: T helper cells 2 (Humoral immunity), Tregs: Regulatory T cells, Th17: T helper cells 17, ↑ increase, ↓ decrease.

6. Gut Microbiome and Bone Homeostasis

It has been established that the microbiome helps in nutrient absorption and utilization, regulating the immune response, and reinforcement of the gastrointestinal barrier to prevent microbial translocation [76,108]. The active metabolites from the commensal microbial populations, such as short-chain fatty acids, can stimulate Treg cell activity, which suppresses inflammation and may help reduce bone loss during inflammatory conditions [82,108,109]. However, dysbiosis following coccidiosis or necrotic enteritis might lead to changes in microbial populations that compromise the microbiota's immunomodulatory functions. Similarly, lactic acid-forming bacteria can help maintain tight junctions in the intestinal epithelial cells, which prevent harmful bacteria from translocating into circulation and bones [110]. Bacterial translocation into bones, as seen in cases such as bacterial chondronecrosis and osteomyelitis in broilers, can initiate inflammatory responses in the femoral head and lead to necrosis. These bacteria further increase the inflammatory responses in the femoral head, leading to necrosis [111,112]. Furthermore, the Gut microbiome plays a significant role in Th17 differentiation. IL-17 from Th17 plays a significant role in RANKL-mediated osteoclastogenesis. Microbial dysbiosis disrupts the equilibrium of the pro-osteoclastogenic pathway and induces osteoclast-mediated bone loss in multiple ways, including the differentiation and inhibition of anti-osteoclastogenic Th1, Th2, and Treg subsets [113].

7. Conclusions

In conclusion, we summarized that coccidiosis and necrotic enteritis are two of the most significant gastrointestinal diseases in poultry, with a substantial economic impact on the poultry industry. These diseases affect not only birds' performance but also their bones' quality, thereby impacting their overall welfare. The delicate balance between bone-resorbing cells, osteoclasts, and bone-forming cells, osteoblasts, plays a crucial role in bone remodeling, modulated by the immune cells. However, during gastrointestinal infections in chickens, the increased immune response can disrupt this balance through the RANKL-RANK pathway mediated by inflammatory and pro-inflammatory cytokines, ultimately leading to bone loss. Furthermore, microbial dysbiosis further exaggerates bone loss during infection through immune-mediated response or through breaching the gastrointestinal barrier. Although several antibiotic alternatives have been proposed to enhance the immune response against these diseases, increasing the immune response might not always be helpful for maintaining bone quality as some of them might increase osteoclastic activities. Therefore, further studies are necessary to better comprehend the molecular mechanisms of bone loss and their relationship with bone homeostasis in poultry. This understanding will help develop novel therapeutic approaches that selectively improve the immune responses against these diseases without affecting bone homeostasis, ultimately improving chickens' health, welfare and performance.

Author Contributions: Conceptualization and writing, M.K.S.; writing—review and editing, P.R., T.A., L.C. and W.K.K.; supervision, W.K.K.; funding acquisition, W.K.K. All authors have read and agreed to the published version of the manuscript.

Funding: This research received no external funding.

Institutional Review Board Statement: Not applicable.

Informed Consent Statement: Not applicable.

Data Availability Statement: No new data were generated in this study.

Conflicts of Interest: The authors declare no conflict of interest.

References

1. Whitehead, C.C. Overview of Bone Biology in the Egg-Laying Hen. *Poult. Sci.* **2004**, *83*, 193–199. [CrossRef] [PubMed]
2. Kim, W.K.; Bloomfield, S.A.; Sugiyama, T.; Ricke, S.C. Concepts and Methods for Understanding Bone Metabolism in Laying Hens. *Worlds Poult. Sci. J.* **2012**, *68*, 71–82. [CrossRef]
3. Regmi, P. *Influence of Housing Systems on Bone Properties of Laying Hens*; Michigan State University: Ann Arbor, MI, USA, 2015.
4. Tsukasaki, M.; Takayanagi, H. Osteoimmunology: Evolving Concepts in Bone-Immune Interactions in Health and Disease. *Nat. Rev. Immunol.* **2019**, *19*, 626–642. [CrossRef] [PubMed]
5. Lorenzo, J.; Horowitz, M.; Choi, Y. Osteoimmunology: Interactions of the Bone and Immune System. *Endocr. Rev.* **2008**, *29*, 403–440. [CrossRef]
6. Dar, H.Y.; Azam, Z.; Anupam, R.; Mondal, R.K.; Srivastava, R.K. Osteoimmunology: The Nexus between Bone and Immune System. *Front. Biosci. Landmark* **2018**, *23*, 464–492.
7. Okamoto, K.; Nakashima, T.; Shinohara, M.; Negishi-Koga, T.; Komatsu, N.; Terashima, A.; Sawa, S.; Nitta, T.; Takayanagi, H. Osteoimmunology: The Conceptual Framework Unifying the Immune and Skeletal Systems. *Physiol. Rev.* **2017**, *97*, 1295–1349. [CrossRef]
8. Tompkins, Y.H.; Choi, J.; Teng, P.Y.; Yamada, M.; Sugiyama, T.; Kim, W.K. Reduced Bone Formation and Increased Bone Resorption Drive Bone Loss in *Eimeria* Infected Broilers. *Sci. Rep.* **2023**, *13*, 616. [CrossRef]
9. Bradshaw, R.H.; Kirkden, R.D.; Broom, D.M. A Review of the Aetiology and Pathology of Leg Weakness in Broilers in Relation to Welfare. *Avian Poult. Biol. Rev.* **2002**, *13*, 45–104. [CrossRef]
10. Kierończyk, B.; Rawski, M.; Józefiak, D.; Świątkiewicz, S. Infectious and Non-Infectious Factors Associated with Leg Disorders in Poultry—A Review. *Ann. Anim. Sci.* **2017**, *17*, 645–669. [CrossRef]
11. Wilson, S.; Duff, S.R.I.; Whitehead, C.C. Effects of Age, Sex and Housing on the Trabecular Bone of Laying Strain Domestic Fowl. *Res. Vet. Sci.* **1992**, *53*, 52–58. [CrossRef]
12. Bishop, S.C.; Fleming, R.H.; Mccormagk, H.A.; Flock, D.K.; Whitehead, C.C. Inheritance of Bone Characteristics Affecting Osteoporosis in Laying Hens. *Br. Poult. Sci.* **2000**, *41*, 33–40. [CrossRef] [PubMed]
13. Sharma, M.K.; White, D.; Chen, C.; Kim, W.K.; Adhikari, P. Effects of the Housing Environment and Laying Hen Strain on Tibia and Femur Bone Properties of Different Laying Phases of Hy-Line Hens. *Poult. Sci.* **2021**, *100*, 100933. [CrossRef] [PubMed]
14. Chapman, H.D. Milestones in Avian Coccidiosis Research: A Review. *Poult. Sci.* **2014**, *93*, 501–511. [CrossRef]
15. Chapman, H.D. *Coccidiosis in Egg Laying Poultry*; Elsevier Inc.: Amsterdam, The Netherlands, 2017; ISBN 9780128011515.
16. Blake, D.P.; Knox, J.; Dehaeck, B.; Huntington, B.; Rathinam, T.; Ravipati, V.; Ayoade, S.; Gilbert, W.; Adebambo, A.O.; Jatau, I.D.; et al. Re-Calculating the Cost of Coccidiosis in Chickens. *Vet. Res.* **2020**, *51*, 115. [CrossRef]
17. Teng, P.Y.; Choi, J.; Tompkins, Y.; Lillehoj, H.; Kim, W. Impacts of Increasing Challenge with *Eimeria Maxima* on the Growth Performance and Gene Expression of Biomarkers Associated with Intestinal Integrity and Nutrient Transporters. *Vet. Res.* **2021**, *52*, 81. [CrossRef]
18. Teng, P.Y.; Yadav, S.; de Souza Castro, F.L.; Tompkins, Y.H.; Fuller, A.L.; Kim, W.K. Graded *Eimeria* Challenge Linearly Regulated Growth Performance, Dynamic Change of Gastrointestinal Permeability, Apparent Ileal Digestibility, Intestinal Morphology, and Tight Junctions of Broiler Chickens. *Poult. Sci.* **2020**, *99*, 4203–4216. [CrossRef]
19. Sharma, M.K.; Liu, G.; White, D.L.; Tompkins, Y.H.; Kim, W.K. Effects of Mixed *Eimeria* Challenge on Performance, Body Composition, Intestinal Health, and Expression of Nutrient Transporter Genes of Hy-Line W-36 Pullets (0–6 Wks of Age). *Poult. Sci.* **2022**, *101*, 102083. [CrossRef]
20. Poudel, S.; Tabler, G.T.; Lin, J.; Zhai, W.; Zhang, L. Riboflavin and Bacillus Subtilis Effects on Growth and Woody-Breast of Ross 708 Broilers with or without *Eimeria* Spp. Challenge. *J. Anim. Sci. Technol.* **2022**, *64*, 443. [CrossRef]
21. Wade, B.; Keyburn, A.L.; Haring, V.; Ford, M.; Rood, J.I.; Moore, R.J. The Adherent Abilities of *Clostridium Perfringens* Strains Are Critical for the Pathogenesis of Avian Necrotic Enteritis. *Vet. Microbiol.* **2016**, *197*, 53–61. [CrossRef] [PubMed]
22. Tompkins, Y.H.; Teng, P.; Pazdro, R.; Kim, W.K. Long Bone Mineral Loss, Bone Microstructural Changes and Oxidative Stress After *Eimeria* Challenge in Broilers. *Front. Physiol.* **2022**, *13*, 1344. [CrossRef]
23. Zanu, H.K.; Kheravii, S.K.; Morgan, N.K.; Bedford, M.R.; Swick, R.A. Interactive Effect of Dietary Calcium and Phytase on Broilers Challenged with Subclinical Necrotic Enteritis: 3. Serum Calcium and Phosphorus, and Bone Mineralization. *Poult. Sci.* **2020**, *99*, 3617–3627. [CrossRef] [PubMed]
24. Cheng, Y.H.; Zhang, N.; Han, J.C.; Chang, C.W.; Hsiao, F.S.H.; Yu, Y.H. Optimization of Surfactin Production from Bacillus Subtilis in Fermentation and Its Effects on *Clostridium Perfringens*-Induced Necrotic Enteritis and Growth Performance in Broilers. *J. Anim. Physiol. Anim. Nutr.* **2018**, *102*, 1232–1244. [CrossRef] [PubMed]
25. Shi, H.; Wang, J.; Teng, P.Y.; Tompkins, Y.H.; Jordan, B.; Kim, W.K. Effects of Phytase and Coccidial Vaccine on Growth Performance, Nutrient Digestibility, Bone Mineralization, and Intestinal Gene Expression of Broilers. *Poult. Sci.* **2022**, *101*, 102124. [CrossRef] [PubMed]
26. Oikeh, I.; Development, R. *Coccidiosis in Modern Broiler Chickens: Targeted Nutritional Modulations for Consequences on Bone Quality*; Newcastle University: Newcastle, UK, 2019.
27. Oikeh, I.; Sakkas, P.; Blake, D.P.; Kyriazakis, I. Interactions between Dietary Calcium and Phosphorus Level, and Vitamin D Source on Bone Mineralization, Performance, and Intestinal Morphology of Coccidia-Infected Broilers. *Poult. Sci.* **2019**, *98*, 5679–5690. [CrossRef] [PubMed]

28. Sakkas, P.; Oikeh, I.; Blake, D.P.; Smith, S.; Kyriazakis, I. Dietary Vitamin D Improves Performance and Bone Mineralisation, but Increases Parasite Replication and Compromises Gut Health in *Eimeria*-Infected Broilers. *Br. J. Nutr.* **2019**, *122*, 676–688. [CrossRef] [PubMed]
29. Applegate, T.J.; Lilburn, M.S. Growth of the Femur and Tibia of a Commercial Broiler Line. *Poult. Sci.* **2002**, *81*, 1289–1294. [CrossRef]
30. Gilbert, S.F. *Early Vertebrate Development: Mesoderm and Endoderm. Developmental Biology*, 5th ed.; Sinauer Assoc. Inc.: Sunderland, MA, USA, 1997; pp. 341–357.
31. Regmi, P.; Deland, T.S.; Steibel, J.P.; Robison, C.I.; Haut, R.C.; Orth, M.W.; Karcher, D.M. Effect of Rearing Environment on Bone Growth of Pullets. *Poult. Sci.* **2015**, *94*, 502–511. [CrossRef]
32. Beck, M.M.; Hansen, K.K. Role of Estrogen in Avian Osteoporosis. *Poult. Sci.* **2004**, *83*, 200–206. [CrossRef]
33. Webster, A.B. Welfare Implications of Avian Osteoporosis. *Poult. Sci.* **2004**, *83*, 184–192. [CrossRef]
34. Huang, S.; Kong, A.; Cao, Q.; Tong, Z.; Wang, X. The Role of Blood Vessels in Broiler Chickens with Tibial Dyschondroplasia. *Poult. Sci.* **2019**, *98*, 6527–6532. [CrossRef]
35. Yan, B.; Zhang, Z.; Jin, D.; Cai, C.; Jia, C.; Liu, W.; Wang, T.; Li, S.; Zhang, H.; Huang, B.; et al. MTORC1 Regulates PTHrP to Coordinate Chondrocyte Growth, Proliferation and Differentiation. *Nat. Commun.* **2016**, *7*, 11151. [CrossRef] [PubMed]
36. Kacena, M.A.; Nelson, T.; Clough, M.E.; Lee, S.K.; Lorenzo, J.A.; Gundberg, C.M.; Horowitz, M.C. Megakaryocyte-Mediated Inhibition of Osteoclast Development. *Bone* **2006**, *39*, 991–999. [CrossRef] [PubMed]
37. Aarden, E.M.; Nijweide, P.J.; Burger, E.H. Function of Osteocytes in Bone. *J. Cell Biochem.* **1994**, *55*, 287–299. [CrossRef]
38. Sims, N.A.; Gooi, J.H. Bone Remodeling: Multiple Cellular Interactions Required for Coupling of Bone Formation and Resorption. *Semin Cell Dev. Biol.* **2008**, *19*, 444–451. [CrossRef]
39. Takahashi, N.; Akatsu, T.; Udagawa, N.; Sasaki, T.; Yamaguchi, A.; Moseley, J.M.; Martin, T.J.; Suda, T. Osteoblastic Cells are Involved in Osteoclast Formation. *Endocrinology* **1988**, *123*, 2600–2602. [CrossRef] [PubMed]
40. Burger, E.H.; Van der Meer, J.W.M.; Nijweide, P.J. Osteoclast Formation from Mononuclear Phagocytes: Role of Bone-Forming Cells. *J. Cell Biol.* **1984**, *99*, 1901–1906. [CrossRef]
41. Nakashima, T.; Hayashi, M.; Fukunaga, T.; Kurata, K.; Oh-Hora, M.; Feng, J.Q.; Bonewald, L.F.; Kodama, T.; Wutz, A.; Wagner, E.F.; et al. Evidence for Osteocyte Regulation of Bone Homeostasis through RANKL Expression. *Nat. Med.* **2011**, *17*, 1231–1234. [CrossRef]
42. Danks, L.; Komatsu, N.; Guerrini, M.M.; Sawa, S.; Armaka, M.; Kollias, G.; Nakashima, T.; Takayanagi, H. RANKL Expressed on Synovial Fibroblasts Is Primarily Responsible for Bone Erosions during Joint Inflammation. *Ann. Rheum. Dis.* **2016**, *75*, 1187–1195. [CrossRef]
43. Knothe Tate, M.L.; Adamson, J.R.; Tami, A.E.; Bauer, T.W. The Osteocyte. *Int. J. Biochem. Cell Biol.* **2004**, *36*, 1–8. [CrossRef]
44. Teitelbaum, S.L. Bone Resorption by Osteoclasts. *Science* **2000**, *289*, 1504–1508. [CrossRef]
45. Newby, A.C. Metalloproteinase Expression in Monocytes and Macrophages and Its Relationship to Atherosclerotic Plaque Instability. *Arter. Thromb. Vasc. Biol.* **2008**, *28*, 2108–2114. [CrossRef] [PubMed]
46. Tang, Y.; Wu, X.; Lei, W.; Pang, L.; Wan, C.; Shi, Z.; Zhao, L.; Nagy, T.R.; Peng, X.; Hu, J.; et al. TGF-B1–Induced Migration of Bone Mesenchymal Stem Cells Couples Bone Resorption with Formation. *Nat. Med.* **2009**, *15*, 757–765. [CrossRef] [PubMed]
47. Klein-Nulend, J.; Nijweide, P.J.; Burger, E.H. Osteocyte and Bone Structure. *Curr. Osteoporos. Rep.* **2003**, *1*, 5–10. [CrossRef]
48. Conway, D.P.; Mckenzie, M.E.; Conway, D.P.; Mckenzie, M.E. *Poultry Coccidiosis*; John Wiley & Sons: Hoboken, NJ, USA, 2007; ISBN 9780813822020.
49. Abdisa, T.; Hasen, R.; Tagesu, T.; Regea, G.; Tadese, G. Poultry Coccidiosis and Its Prevention, Control. *J. Vet. Anim. Res. NF* **2019**, *2*, 1–6.
50. Fathima, S.; Al Hakeem, W.G.; Shanmugasundaram, R.; Selvaraj, R.K. Necrotic Enteritis in Broiler Chickens: A Review on the Pathogen, Pathogenesis, and Prevention. *Microorganisms* **2022**, *10*, 1958. [CrossRef]
51. Daneshmand, A.; Kermanshahi, H.; Mohammed, J.; Sekhavati, M.H.; Javadmanesh, A.; Ahmadian, M.; Alizadeh, M.; Razmyar, J.; Kulkarni, R.R. Intestinal Changes and Immune Responses during *Clostridium Perfringens*-Induced Necrotic Enteritis in Broiler Chickens. *Poult. Sci.* **2022**, *101*, 101652. [CrossRef]
52. Alizadeh, M.; Shojadoost, B.; Boodhoo, N.; Astill, J.; Taha-Abdelaziz, K.; Hodgins, D.C.; Kulkarni, R.R.; Sharif, S. Necrotic Enteritis in Chickens: A Review of Pathogenesis, Immune Responses and Prevention, Focusing on Probiotics and Vaccination. *Anim. Health Res. Rev.* **2021**, *22*, 147–162. [CrossRef]
53. Moore, R.J. Necrotic Enteritis Predisposing Factors in Broiler Chickens. *Avian Pathol.* **2016**, *45*, 275–281. [CrossRef]
54. Kogut, M.H.; Klasing, K. An Immunologist's Perspective on Nutrition, Immunity, and Infectious Diseases: Introduction and Overview. *J. Appl. Poult. Res.* **2009**, *18*, 103–110. [CrossRef]
55. Awad, W.A.; Hess, C.; Hess, M. Enteric Pathogens and Their Toxin-Induced Disruption of the Intestinal Barrier through Alteration of Tight Junctions in Chickens. *Toxins* **2017**, *9*, 60. [CrossRef]
56. Poudel, S.; Zhang, L.; Tabler, G.T.; Lin, J.; Zhai, W. Effects of Riboflavin and Bacillus Subtilis on Internal Organ Development and Intestinal Health of Ross 708 Male Broilers with or without Coccidial Challenge. *Poult. Sci.* **2021**, *100*, 100973. [CrossRef] [PubMed]
57. Yun, C.H.; Lillehoj, H.S.; Lillehoj, E.P. Intestinal Immune Responses to Coccidiosis. *Immunology* **2000**, *24*, 303–324. [CrossRef] [PubMed]

58. Lee, Y.; Lu, M.; Lillehoj, H.S. Coccidiosis: Recent Progress in Host Immunity and Alternatives to Antibiotic Strategies. *Vaccines* **2022**, *10*, 215.
59. Shivaramaiah, C.; Barta, J.R.; Hernandez-Velasco, X.; Téllez, G.; Hargis, B.M. Coccidiosis: Recent Advancements in the Immunobiology of *Eimeria* Species, Preventive Measures, and the Importance of Vaccination as a Control Tool against These Apicomplexan Parasites. *Vet. Med. Res. Rep.* **2014**, *5*, 23–24.
60. Lee, Y.; Lee, S.-H.; Deepthi Gadde, U.; Oh, S.-T.; Lee, S.-J.; Lillehoj, H.S. Dietary Allium Hookeri Reduces Inflammatory Response and Increases Expression of Intestinal Tight Junction Proteins in LPS-Induced Young Broiler Chicken. *Res. Vet. Sci.* **2017**, *112*, 149–155. [CrossRef]
61. Lillehoj, H.S.; Lillehoj, E.P. Avian Coccidiosis. A Review of Acquired Intestinal Immunity and Vaccination Strategies: American Association of Avian Pathologists Stable. *Avian Dis.* **2016**, *44*, 408–425.
62. Hong, Y.H.; Lillehoj, H.S.; Lee, S.H.; Dalloul, R.A.; Lillehoj, E.P. Analysis of Chicken Cytokine and Chemokine Gene Expression Following *Eimeria Acervulina* and *Eimeria Tenella* Infections. *Vet. Immunol. Immunopathol.* **2006**, *114*, 209–223. [CrossRef]
63. Lillehoj, H.S.; Trout, J.M. Avian Gut-Associated Lymphoid Tissues and Intestinal Immune Responses to *Eimeria* Parasites. *Clin. Microbiol. Rev.* **1996**, *9*, 349–360. [CrossRef]
64. Kim, W.H.; Chaudhari, A.A.; Lillehoj, H.S. Involvement of T Cell Immunity in Avian Coccidiosis. *Front. Immunol.* **2019**, *10*, 2732.
65. Hong, Y.H.; Lillehoj, H.S.; Lillehoj, E.P.; Lee, S.H. Changes in Immune-Related Gene Expression and Intestinal Lymphocyte Subpopulations Following *Eimeria Maxima* Infection of Chickens. *Vet. Immunol. Immunopathol.* **2006**, *114*, 259–272. [CrossRef]
66. Trinchieri, G.; Pflanz, S.; Kastelein, R.A. The IL-12 Family of Heterodimeric Cytokines: New Players in the Regulation of T Cell Responses. *Immunity* **2003**, *19*, 641–644. [CrossRef]
67. Scott, P. IL-12: Initiation Cytokine for Cell-Mediated Immunity. *Science* **1993**, *260*, 496–497. [CrossRef] [PubMed]
68. Collier, C.T.; Hofacre, C.L.; Payne, A.M.; Anderson, D.B.; Kaiser, P.; Mackie, R.I.; Gaskins, H.R. Coccidia-Induced Mucogenesis Promotes the Onset of Necrotic Enteritis by Supporting *Clostridium Perfringens* Growth. *Vet. Immunol. Immunopathol.* **2008**, *122*, 104–115. [CrossRef] [PubMed]
69. Lee, K.W.; Lillehoj, H.S.; Jeong, W.; Jeoung, H.Y.; An, D.J. Avian Necrotic Enteritis: Experimental Models, Host Immunity, Pathogenesis, Risk Factors, and Vaccine Development. *Poult. Sci.* **2011**, *90*, 1381–1390. [CrossRef]
70. Ruhnke, I.; Andronicos, N.M.; Swick, R.A.; Hine, B.; Sharma, N.; Kheravii, S.K.; Wu, S.B.; Hunt, P. Immune Responses Following Experimental Infection with *Ascaridia Galli* and Necrotic Enteritis in Broiler Chickens. *Avian Pathol.* **2017**, *46*, 602–609. [CrossRef] [PubMed]
71. Fasina, Y.O.; Lillehoj, H.S. Characterization of Intestinal Immune Response to *Clostridium Perfringens* Infection in Broiler Chickens. *Poult. Sci.* **2019**, *98*, 188–198. [CrossRef]
72. Kolls, J.K.; Khader, S.A. The Role of Th17 Cytokines in Primary Mucosal Immunity. *Cytokine Growth Factor. Rev.* **2010**, *21*, 443–448. [CrossRef]
73. Walliser, I.; Göbel, T.W. Chicken IL-17A Is Expressed in Aβ and Γδ T Cell Subsets and Binds to a Receptor Present on Macrophages, and T Cells. *Dev. Comp. Immunol.* **2018**, *81*, 44–53. [CrossRef]
74. Kogut, M.H. The Gut Microbiota and Host Innate Immunity: Regulators of Host Metabolism and Metabolic Diseases in Poultry? *J. Appl. Poult. Res.* **2013**, *22*, 637–646. [CrossRef]
75. Broom, L.J.; Kogut, M.H. The Role of the Gut Microbiome in Shaping the Immune System of Chickens. *Vet. Immunol. Immunopathol.* **2018**, *204*, 44–51. [CrossRef]
76. Kogut, M.H.; Lee, A.; Santin, E. Microbiome and Pathogen Interaction with the Immune System. *Poult. Sci.* **2020**, *99*, 1906–1913. [CrossRef] [PubMed]
77. Broom, L.J. Host–Microbe Interactions and Gut Health in Poultry—Focus on Innate Responses. *Microorganisms* **2019**, *7*, 139. [CrossRef] [PubMed]
78. Allaire, J.M.; Crowley, S.M.; Law, H.T.; Chang, S.Y.; Ko, H.J.; Vallance, B.A. The Intestinal Epithelium: Central Coordinator of Mucosal Immunity. *Trends Immunol.* **2018**, *39*, 677–696. [CrossRef]
79. Cheled-Shoval, S.L.; Gamage, N.S.W.; Amit-Romach, E.; Forder, R.; Marshal, J.; Van Kessel, A.; Uni, Z. Differences in Intestinal Mucin Dynamics between Germ-Free and Conventionally Reared Chickens after Mannan-Oligosaccharide Supplementation. *Poult. Sci.* **2014**, *93*, 636–644. [CrossRef]
80. Shira, E.B.; Sklan, D.; Friedman, A. Impaired Immune Responses in Broiler Hatchling Hindgut Following Delayed Access to Feed. *Vet. Immunol. Immunopathol.* **2005**, *105*, 33–45. [CrossRef]
81. Kaspers, B.; Lettmann, S.; Roell, S. Development of the Gut Associated Immune System. In Proceedings of the European Symposium on Poultry Nutrition, Prague, Czech Republic, 24–27 August 2015; Volume 20.
82. Levy, M.; Blacher, E.; Elinav, E. Microbiome, Metabolites and Host Immunity. *Curr. Opin. Microbiol.* **2017**, *35*, 8–15. [CrossRef]
83. Wu, S.B.; Stanley, D.; Rodgers, N.; Swick, R.A.; Moore, R.J. Two Necrotic Enteritis Predisposing Factors, Dietary Fishmeal and *Eimeria* Infection, Induce Large Changes in the Caecal Microbiota of Broiler Chickens. *Vet. Microbiol.* **2014**, *169*, 188–197. [CrossRef] [PubMed]
84. Caballero, S.; Pamer, E.G. Microbiota-Mediated Inflammation and Antimicrobial Defense in the Intestine. *Ann. Rev. Immunol.* **2015**, *33*, 227–256. [CrossRef] [PubMed]
85. Yang, Q.; Liu, J.; Wang, X.; Robinson, K.; Whitmore, M.A.; Stewart, S.N.; Zhao, J.; Zhang, G. Identification of an Intestinal Microbiota Signature Associated with the Severity of Necrotic Enteritis. *Front. Microbiol.* **2021**, *12*, 2377. [CrossRef]

86. Antonissen, G.; Eeckhaut, V.; Van Driessche, K.; Onrust, L.; Haesebrouck, F.; Ducatelle, R.; Moore, R.J.; Van Immerseel, F. Microbial Shifts Associated with Necrotic Enteritis. *Avian Pathol.* **2016**, *45*, 308–312. [CrossRef]
87. Xu, S.; Lin, Y.; Zeng, D.; Zhou, M.; Zeng, Y.; Wang, H.; Zhou, Y.; Zhu, H.; Pan, K.; Jing, B.; et al. Bacillus Licheniformis Normalize the Ileum Microbiota of Chickens Infected with Necrotic Enteritis. *Sci. Rep.* **2018**, *8*, 1744. [CrossRef]
88. Wang, W.; Gao, X.; Zheng, S.; Lancuo, Z.; Li, Y.; Zhu, L.; Hou, J.; Hai, J.; Long, X.; Chen, H.; et al. The Gut Microbiome and Metabolome of Himalayan Griffons (*Gyps Himalayensis*): Insights into the Adaptation to Carrion-Feeding Habits in Avian Scavengers. *Avian. Res.* **2021**, *12*, 52. [CrossRef]
89. Stanley, D.; Wu, S.-B.; Rodgers, N.; Swick, R.A.; Moore, R.J. Differential Responses of Cecal Microbiota to Fishmeal, *Eimeria* and *Clostridium Perfringens* in a Necrotic Enteritis Challenge Model in Chickens. *PLoS ONE* **2014**, *9*, e104739. [CrossRef]
90. Pietruska, A.; Bortoluzzi, C.; Hauck, R. A Meta-Analysis of the Effect of *Eimeria* Spp. and/or *Clostridium Perfringens* Infection on the Microbiota of Broiler Chickens. *Poult. Sci.* **2023**, *102*, 102652. [CrossRef]
91. Tsai, Y.T.; Cheng, P.C.; Pan, T.M. The Immunomodulatory Effects of Lactic Acid Bacteria for Improving Immune Functions and Benefits. *Appl. Microbiol. Biotechnol.* **2012**, *96*, 853–862. [CrossRef] [PubMed]
92. Sato, K.; Suematsu, A.; Okamoto, K.; Yamaguchi, A.; Morishita, Y.; Kadono, Y.; Tanaka, S.; Kodama, T.; Akira, S.; Iwakura, Y.; et al. Th17 Functions as an Osteoclastogenic Helper T Cell Subset That Links T Cell Activation and Bone Destruction. *J. Exp. Med.* **2006**, *203*, 2673–2682. [CrossRef] [PubMed]
93. Takayanagi, H.; Ogasawara, K.; Hida, S.; Chiba, T.; Murata, S.; Sato, K.; Takaoka, A.; Yokochi, T.; Oda, H.; Tanaka, K.; et al. T-Cell-Mediated Regulation of Osteoclastogenesis by Signaling Crosstalk between RANKL and IFN-γ. *Nature* **2000**, *408*, 600–605. [CrossRef]
94. O'Brien, C.A.; Nakashima, T.; Takayanagi, H. Osteocyte Control of Osteoclastogenesis. *Bone* **2013**, *54*, 258–263. [CrossRef]
95. Sharma, M.K.; White, D.L.; Tompkins, Y.H.; Kim, W.K. Effect of mixed *Eimeria* challenge on skeletal health of Hy-Line W-36 pullets. In Proceedings of the International Poultry Scientific Forum, Atlanta, GA, USA, 25–27 January 2022; p. 70.
96. Iwakura, Y.; Ishigame, H.; Saijo, S.; Nakae, S. Functional Specialization of Interleukin-17 Family Members. *Immunity* **2011**, *34*, 149–162. [CrossRef] [PubMed]
97. Kobayashi, K.; Takahashi, N.; Jimi, E.; Udagawa, N.; Takami, M.; Kotake, S.; Nakagawa, N.; Kinosaki, M.; Yamaguchi, K.; Shima, N.; et al. Tumor Necrosis Factor α Stimulates Osteoclast Differentiation by a Mechanism Independent of the Odf/Rankl–Rank Interaction. *J. Exp. Med.* **2000**, *191*, 275–286. [CrossRef]
98. Gilbert, L.; He, X.; Farmer, P.; Rubin, J.; Drissi, H.; Van Wijnen, A.J.; Lian, J.B.; Stein, G.S.; Nanes, M.S. Expression of the Osteoblast Differentiation Factor RUNX2 (Cbfa1/AML3/Pebp2αA) Is Inhibited by Tumor Necrosis Factor-α. *J. Biol. Chem.* **2002**, *277*, 2695–2701. [CrossRef] [PubMed]
99. Kotake, S.; Udagawa, N.; Takahashi, N.; Matsuzaki, K.; Itoh, K.; Ishiyama, S.; Saito, S.; Inoue, K.; Kamatani, N.; Gillespie, M.T.; et al. IL-17 in Synovial Fluids from Patients with Rheumatoid Arthritis Is a Potent Stimulator of Osteoclastogenesis. *J. Clin. Investig.* **1999**, *103*, 1345–1352. [CrossRef] [PubMed]
100. Kotake, S.; Nanke, Y. Effect of TNFα on Osteoblastogenesis from Mesenchymal Stem Cells. *Biochim. Biophys. Acta BBA Gen. Subj.* **2014**, *1840*, 1209–1213. [CrossRef] [PubMed]
101. Dhodapkar, K.M.; Barbuto, S.; Matthews, P.; Kukreja, A.; Mazumder, A.; Vesole, D.; Jagannath, S.; Dhodapkar, M.V. Dendritic Cells Mediate the Induction of Polyfunctional Human IL17-Producing Cells (Th17-1 Cells) Enriched in the Bone Marrow of Patients with Myeloma. *Blood* **2008**, *112*, 2878–2885. [CrossRef]
102. Uluçkan, Ö.; Jimenez, M.; Karbach, S.; Jeschke, A.; Graña, O.; Keller, J.; Busse, B.; Croxford, A.L.; Finzel, S.; Koenders, M.; et al. Chronic Skin Inflammation Leads to Bone Loss by IL-17-Mediated Inhibition of Wnt Signaling in Osteoblasts. *Sci. Transl. Med.* **2016**, *8*, 330ra37. [CrossRef] [PubMed]
103. Bendre, M.S.; Montague, D.C.; Peery, T.; Akel, N.S.; Gaddy, D.; Suva, L.J. Interleukin-8 Stimulation of Osteoclastogenesis and Bone Resorption Is a Mechanism for the Increased Osteolysis of Metastatic Bone Disease. *Bone* **2003**, *33*, 28–37. [CrossRef]
104. Sun, H.; Zhao, Y.; Wang, K.; Zhu, L.; Dong, J.; Zhao, J.; Wang, Y.; Li, H.; Sun, X.; Lu, Y. Low Dose IL-2 Suppress Osteoclastogenesis in Collagen-Induced Arthritis via JNK Dependent Pathway. *Immun. Inflamm. Dis* **2020**, *8*, 727–735. [CrossRef]
105. Srivastava, R.K.; Tomar, G.B.; Barhanpurkar, A.P.; Gupta, N.; Pote, S.T.; Mishra, G.C.; Wani, M.R. IL-3 Attenuates Collagen-Induced Arthritis by Modulating the Development of Foxp3+ Regulatory T Cells. *J. Immunol.* **2011**, *186*, 2262–2272. [CrossRef]
106. Stein, N.C.; Kreutzmann, C.; Zimmermann, S.P.; Niebergall, U.; Hellmeyer, L.; Goettsch, C.; Schoppet, M.; Hofbauer, L.C. Interleukin-4 and Interleukin-13 Stimulate the Osteoclast Inhibitor Osteoprotegerin by Human Endothelial Cells through the STAT6 Pathway. *J. Bone Miner. Res.* **2008**, *23*, 750–758. [CrossRef]
107. Kelchtermans, H.; Geboes, L.; Mitera, T.; Huskens, D.; Leclercq, G.; Matthys, P. Activated CD4+CD25+ Regulatory T Cells Inhibit Osteoclastogenesis and Collagen-Induced Arthritis. *Ann. Rheum. Dis.* **2009**, *68*, 744–750. [CrossRef]
108. Hernandez, C.J.; Guss, J.D.; Luna, M.; Goldring, S.R. Links Between the Microbiome and Bone. *J. Bone Miner. Res.* **2016**, *31*, 1638–1646. [CrossRef] [PubMed]
109. Ibáñez, L.; Rouleau, M.; Wakkach, A.; Blin-Wakkach, C. Gut Microbiome and Bone. *Jt. Bone Spine* **2019**, *86*, 43–47. [CrossRef]
110. Chen, P.; Xu, T.; Zhang, C.; Tong, X.; Shaukat, A.; He, Y.; Liu, K.; Huang, S. Effects of Probiotics and Gut Microbiota on Bone Metabolism in Chickens: A Review. *Metabolites* **2022**, *12*, 1000. [CrossRef]
111. Wideman, R.F. Bacterial Chondronecrosis with Osteomyelitis and Lameness in Broilers: A Review. *Poult. Sci.* **2016**, *95*, 325–344. [CrossRef]

112. Wideman, R.F.; Prisby, R.D. Bone Circulatory Disturbances in the Development of Spontaneous Bacterial Chondronecrosis with Osteomyelitis: A Translational Model for the Pathogenesis of Femoral Head Necrosis. *Front. Endocrinol.* **2013**, *3*, 183. [CrossRef] [PubMed]
113. Zhang, J.; Lu, Y.; Wang, Y.; Ren, X.; Han, J. The impact of the intestinal microbiome on bone health. *Intractable Rare Dis. Res.* **2018**, *7*, 148–155. [CrossRef] [PubMed]

animals MDPI

Article

Effects of Environmental Enrichments on Welfare and Hepatic Metabolic Regulation of Broiler Chickens

Seong W. Kang [1,*,†], Karen D. Christensen [2], Michael T. Kidd Jr. [1] and Sara K. Orlowski [1]

1 Department of Poultry Science, Center of Excellence for Poultry Science, University of Arkansas, Fayetteville, AR 72701, USA; michaelk@uark.edu (M.T.K.J.); orlowski@uark.edu (S.K.O.)
2 Tyson Foods, Inc., Springdale, AR 72762, USA; karen.christensen@tyson.com
* Correspondence: swkang@uark.edu
† Current address: Poultry Science Center, 1260 W. Maple Street, University of Arkansas, Fayetteville, AR 72701, USA.

Simple Summary: Environmental enrichment (EE) has been suggested to increase environmental complexity and can benefit broilers' welfare. In the previous lighting enrichment study in commercial broilers, a variable light intensity (VL) lighting program or gradient lighting program improved the voluntary natural behavior and activity for better health and production efficiency of broilers. In the present study, we observed the best engagement in enrichment huts (EHs) from the three types of EEs (board, ramp, and hut), implying that EHs provide a lower light intensity area which birds can rest inside for safety and mental health. The combined effect of two EEs, VL lighting and EHs, on mental health and hepatic metabolic function indicate that the enriched lighting and hut program improve the mental health and hepatic fatty acid/glucose metabolic functions, suggesting that the combined enriched lighting and hut programs make broilers sentient to light intensity in the broiler house and improve mental health and nutrient-use efficiency in the liver.

Abstract: The aims of this study were to find suitable environmental enrichment (EE) and evaluate the combined effect of two EEs, variable light intensity (VL) lighting program and EH, on mental health and hepatic metabolic regulation in commercial broilers. To find the advantageous EEs for broilers, three different EEs (board, hut, and ramp) were tested in trial 1. EEs were placed and the engagement of birds to EEs, dustbathing behavior, and daily physical activity were observed. Birds treated with huts showed higher engagement than the board- or ramp-treated birds ($p < 0.05$). The results of dustbathing behavior and daily physical activity indicated that the environmental hut (EH) is the most favorable enrichment for broilers. In the second trial, to test the effect of EHs on mental health and hepatic metabolic conditions, the brain and liver were sampled from the four treatment birds (20 lx_Con, 20 lx_Hut, VL_Con and VL_Hut) on day 42. The lower expression of TPH2 (tryptophan hydroxylase 2) of VL_Hut birds than those of VL_Con and 20 lx_Hut treated birds suggests the combining effect of EHs with the VL lighting program on the central serotonergic homeostasis of broilers. Reduced expressions of TH (tyrosine hydroxylase), GR (glucocorticoid receptor), BDNF (brain-derived neurotrophic factor) of VL_Hut treated birds compared to those of VL_Con and 20 lx_Hut birds suggest lower stress, stress susceptibility, and chronic social stress in VL_Hut treated birds. The expression of CPT1A (carnitine palmitoyl transferase 1) increased over three-fold in the liver of VL_Con birds compared to 20 lx_Con birds ($p < 0.05$). EHs treatment in VL birds (VL_Hut) significantly decreased CPT1A but not in 20 lx birds (20 lx_Hut). The expression of ACCα (acetyl-CoA carboxylase alpha) was significantly decreased in VL_Con birds compared to 20 lx_Con birds. There was no significant difference in the hepatic FBPase (fructose-1,6-bisphosphatase), GR, and 11β-HSD1 (11 β-hydroxysteroid dehydrogenease-1) expression between 20 lx_Con and VL_Con birds, but EHs significantly stimulated GR in 20 lx_Hut birds, and stimulated FBPase and 11β-HSD1 expression in the VL_Hut birds compared to 20 lx_Con birds, suggesting that the VL lighting program reduced fatty acid synthesis and increased fatty acid β-oxidation in the broilers' liver and VL_Hut improved the hepatic de novo glucose production. Taken together, the results suggest that the stimulated voluntary activity by EHs in the light-enriched broiler house improved mental health and hepatic metabolic

Citation: Kang, S.W.; Christensen, K.D.; Kidd, M.T., Jr.; Orlowski, S.K. Effects of Environmental Enrichments on Welfare and Hepatic Metabolic Regulation of Broiler Chickens. *Animals* **2024**, *14*, 557. https://doi.org/10.3390/ani14040557

Academic Editor: Jeffrey Downing

Received: 29 December 2023
Revised: 1 February 2024
Accepted: 5 February 2024
Published: 7 February 2024

function of broilers and may indicate that the improved hepatic metabolic function contributes to efficient nutritional support for broilers.

Keywords: enrichment hut; variable light intensity; hepatic fatty acid; welfare; serotonin; dopamine; BDNF

1. Introduction

Intensification systems in broiler production to produce more efficient meat products have many benefits including improved biosecurity and environmental control, but have a negative effect on animal stress and welfare by preventing animals from showing innate behaviors that they would perform in their natural environment [1,2]. Therefore, finding novel methods to reduce the stress, and improve welfare and natural behaviors associated with barren broiler houses on-farm are critically needed. Animal environmental enrichments (EEs) can be defined as any safe materials or objects added to the animal environment which engage the brain and encourage the expression of natural behavior and exercise [3,4].

Environmental light, especially light intensity, is an important component that can exert influence on behaviors, welfare, and production for intensively housed broilers [5–7]. In previous light preference studies, birds showed proclivity for the brighter light area when they are presenting active behaviors but for darker light areas when resting [8,9]. In an enriched light study in commercial broiler houses, which provided LED light over the feedlines and adjusted dimmer light areas on the side wall and middle of houses, birds showed enlarged natural behaviors and daily activity, which improved footpad condition and leg health of birds [10]. However, knowledge of the effects of other environmental enrichments on broilers' behavior and other welfare indicators including mental and hepatic functions is still limited [11].

Animal welfare is the outcome of physical and psychological well-being, which is not a continual state but rather the result of certain mental changes underlying innate motivated behaviors and learned responses [12]. The neurobiological system related to food seeking is perhaps one of the most critical components for survival and generates the attention and motivation that helps animals to find, detain, and acquire the object of their want [12,13]. The food seeking and reward system consists mostly of dopaminergic neurons in the midbrain ventral tegmental area (VTA) [14,15]. In recent studies, serotonin (5-HT), dopamine (DA), and brain-derived neurotropic factor (BDNF) were suggested in the assessment of animal welfare as positive indicators in the VTA [10,16,17]. BDNF is a stress- and physical-movement-dependent neurotrophic factor associated with emotional and cognitive function and was suggested as an animal health indicator of environmental enrichment (EE) in domestic swine [17,18]. Physical activity is essential to the augmented expression of BDNF in the brain and may enhance memory performance and reduce depressive symptoms by promoting neurogenesis and neuronal differentiation [19,20]. Glucocorticoids (GCs, corticosterone in birds) are glucoregulatory hormones that are synthesized in response to stimuli including stress and are regularly used in the assessment of animal welfare [21]. Their action depends on the GC receptor (GR) which translocate into the nucleus upon ligand binding and regulates transcription of responsive genes. VTA-GR is well known for regulating the stress and reward system regulating feeding behavior [22,23]. In recent studies of broiler chickens, avian VTA was suggested as the important region, consisting of a variety of neurons of the avian midbrain, that is involved in light perception by Opn4 (melanopsin) and might be involved in broilers' welfare [10,24].

Physical exercise and play have a significant beneficial impact on brain and liver health in mammalian species including humans [25–27]. The liver is the central metabolic organ in controlling metabolic homeostasis, acting as the primary site for lipid metabolism, where more than 90% of de novo fatty acids are synthesized in chickens [28,29]. Under healthy conditions, the lipid metabolism is regulated in the liver to meet systemic energy needs in the fed and fasted states by the tightly regulated processes of fatty acid uptake, synthesis, and oxidation pathways [30]. When one or more of these processes are abnormally regulated, excess lipid accumulation can occur [30]. Hepatic lipid metabolism closely interacts with glucose metabolism in liver diseases [31,32]. The liver stores glucose in the form of glycogen and releases glucose into circulation by either glycogenolysis or gluconeogenesis [33]. In the feed state, hepatic glucose production is suppressed by insulin secretion, and the glucose ingested is stored in part as glycogen [34]. Gluconeogenesis is an intricate process that requires several enzymatic steps, which are under the regulation of hormones, nutrient intake, stress conditions, and substrate concentrations. Fructose -1,6-bisphosphatase (FBPase) is a key enzyme of gluconeogenesis, and its deficiency causes hypoglycemia and fatty liver disease [35]. FBPase deficiency should be considered as an etiology of hepatic steatosis [35]. Increased levels of circulating GC levels have been associated with pathological conditions characterized by fatty liver disease [36,37].

Therefore, addressing the midbrain-VTA-welfare-related genes and hepatic metabolic regulating genes may provide critical data to understand possible adaptive physiological responses of broilers to environment enrichments which are involved in the mental and hepatic welfare of broilers. In the present study, we hypothesized that when broilers in commercial houses are provided appropriate environmental enrichments, it will stimulate birds' innate natural behavior, and consequently, cause voluntary movement for the consumption of feed and water, and improve physical activity and mental and hepatic welfare of birds.

2. Materials and Methods

2.1. Experimental Design and Animal Housing

Trial 1 was performed to test the engagement of birds to different enrichments (board, hut, and ramp), and the effects of the selected enrichment on the daily physical activity, natural behavior, and dustbathing. Day-old broilers (Cobb 700, mixed sex, 19,200 birds/house, stocking density_12.3 birds/m^2) were housed in four commercial broiler houses (Tyson Foods Broiler Welfare Research Farm (BWRF)). Two replicate trials were performed, and each house was composed of 4 quadrant sections. Each quadrant of the house was placed with 4800 chicks with all source flocks equally represented in each quadrant. Each house was equipped with standard feeders, waterers, and brooders (12.8 m $\times$ 122 m, wood shavings). Two different light intensity lighting programs (20 lux (lx) and variable light intensity (VL)) were installed, and the light intensity (LED) was measured in 9 different areas of the house. The averages of light intensity in the 20 lx and VL house were 26.16 $\pm$ 0.70 lx, and 2.07/40.4 $\pm$ 0.04 lx, respectively [10]. Diet was formulated to meet minimum industry standards [38]. The light was on from 6 am day 1–3 (23 h light (L):1 h dark (D) (23 L:1 D) _40 lx); then on day 4–7, the photoperiod schedule was changed into 20 L:4 D_20 lx. On day 7, lighting programs were started for 20 lx and VL houses (16 L:8 D, light on 6 am). Environmental enrichments (EEs: board, hut, or ramp) were placed in each section of the houses (3 EEs/92.9 m^2 (1000 sqft), 9 EEs/section).

After the enrichment hut (EH) was selected as an appropriate EE in trial 1, trial 2 was performed to assess the effects of EHs on mental health and liver metabolic functions. Day-old broilers (Cobb 700, mixed sex, 19,200 birds/house) were housed in four broiler houses (Tyson Foods BWRF). Four replicate trials were performed (n = 4 sections/house, 4800 birds/section). Two different light intensity lighting programs (20 lx, and VL) and the same diet were installed as in trial 1. Four treatments were assigned to each house (20 lx_Con, 20 lx_Hut, VL_Con, and VL_Hut). EHs were placed at 4 sections of 20 lx_Hut and VL_Hut house (3 huts/92.9 m^2 (1000 sqft), 9 huts/section). At 42 days of age, birds were selected randomly in each section (n = 16/treatment, male), and transported to the sampling room for brain and liver sampling. The care and experimental use of animals were followed and maintained by the protocol of Tyson Foods BWRF.

2.2. Behavioral Observations Affected by Enrichments

2.2.1. Number of Engaged Birds to Enrichments

In trial 1, a trained observer placed a camera 5 m from an enrichment within a section at the age of 28, 35, and 42 days. The observer then walked the ground adjacent to the enrichment to remove any birds actively using the enrichment. The enrichment was then recorded for 15 min. These data represent the number of birds near, under, and over the enrichment for 15 min (Figure 1D–F).

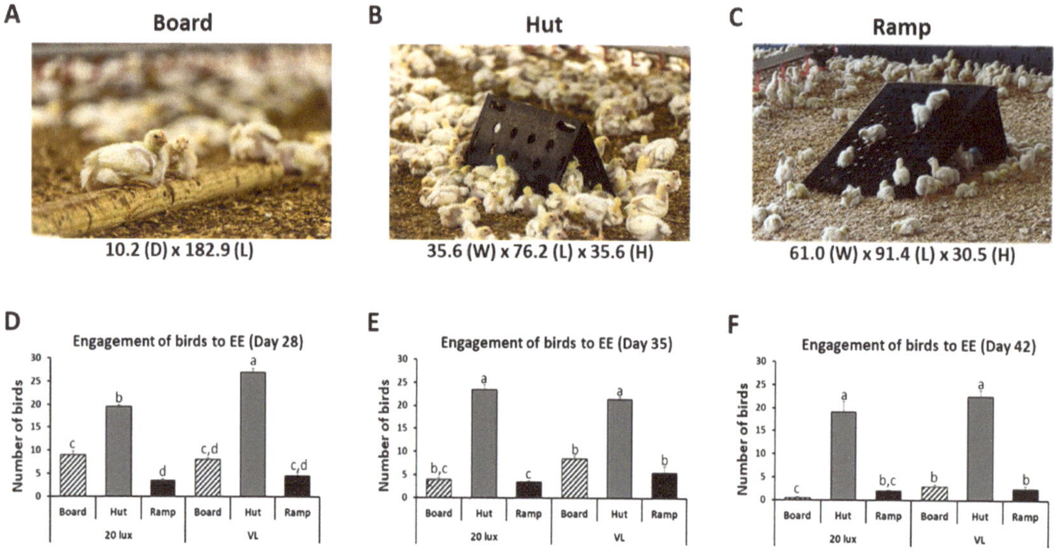

Figure 1. Effects of the three different environment enrichments (board, hut, and ramp) on engagement of birds at different ages. (**A–C**) Three commercially available enrichments placed in trial 1, 20 lx and VL houses. Enrichment dimensions: Board; 10.2 cm (diameter) × 182.9 cm (long), Hut; 35.6 cm (wide) × 76.2 cm (long) × 35.6 cm (height), Ramp; 61.0 cm (wide) × 91.4 cm (long) × 30.5 cm (height). (**D–F**) Weekly number of engaged birds was observed on day 28, 35, and 42 for each enrichment. Data (mean ± SEM) were compared among treatments. Different lower-case letters above the bars denote significant differences ($p < 0.05$) among groups.

2.2.2. Dustbathing and Daily Physical Activity

In trial 1, the number of dustbathing holes was counted within nine identified areas of the section (four sections/house) to see the combined effects of the two different light intensity lighting programs (20 lx and VL) and three EEs (board, hut, and ramp) on dustbathing behavior. In each section, dustbathing holes as evidence of dustbathing behavior were counted at the age of 8, 15, and 22 days [10]. The number of holes per 5 m^2 was determined. Data were compared among treatments. In trial 1, daily physical activity was monitored using a 22 g activity tracker, Animo (www.surepetcare.com (accessed on 10 September 2023), activity and behavior monitor) which monitors animals' activity and behavior including sleep quality, energy burnt, and shaking via tri-axial accelerometer technology [10]. A similar animal activity tracker, Fitbark (www.fitbark.com (accessed on 1 October 2023)), was used for monitoring animals' movement in the behavior study [39]. At 43 days of age, birds were randomly selected, and body weight was measured (n = 8 birds/trt, 4 birds/section, 2 VL houses). An Animo was installed for each bird using a commercially available chicken harness and uninstalled at 48 days of age. Average daily activity (joules/day, 4 days) of each bird was obtained from the installed software. Animo energy calculation is based on an industry standard calculation that considers the bird's weight. The energy burnt is tracked against each movement type for a bird.

2.3. Dissection of Ventral Tegmental Area (VTA) of Midbrain and Liver Sampling

According to previous studies in avian species and a chick brain atlas [8,10,40], the VTA regions from the sampled birds were dissected in cryostat microtome. Dissected section dimensions were 3–3.5 mm (W) × 2–3 mm (H) × 1–1.2 mm (L) for VTA. The thickness (W, H, and L) of the dissected brain tissue block was proportionally increased for 42 days' birds based on brain size and structure. Inside the cryostat, brain areas shown as rectangles were dissected from each flattened brain section using a scalpel handle and blade (#11) and were quickly transferred to the Trizol and then stored at −80 °C until total RNA extraction. The liver tissues from the sampled birds at 42 days of age were dissected, snap-frozen, and stored at −80 °C for total RNA extraction.

2.4. RNA Isolation and Two-Step Real-Time Quantitative RT-PCR

Total RNA was extracted from dissected frozen brain tissue and liver tissues using TRIzol® reagent (Invitrogen Life Technologies, Palo Alto, CA, USA) followed by DNase I treatment and purification of total RNA by the RNeasy mini kit (Qiagen, Valencia, CA, USA). The RNA quality and quantity were determined using agarose gel electrophoresis and NanoDrop 1000 (Thermo Scientific, Wilmington, DE, USA). Two μg of total RNA from sectioned VTA and liver tissue were converted into cDNA with oligo $(dT)_{16}$ primer and SuperScript IV reverse transcriptase (Invitrogen, Grand Island, NY, USA), as previously described [8,10]. The specific oligonucleotide primers were designed using the PRIMERS3 program (http://primer3.ut.ee (accessed on 15 February 2023)). Primer sets for chicken TPH2 (tryptophan hydroxylase 2), TH (tyrosine hydroxylase), GR (glucocorticoid receptor), BDNF, Opn4, CPT1A (carnitine palmitoyl transferase 1), ACCα (acetyl-CoA carboxylase alpha), FBPase, GR, and 11β-HSD1 (11 β-hydroxysteroid dehydrogenease-1) were designed, and conventional RT-PCR performed for optimizing annealing temperature for each primer set (Table 1). The PCR products were analyzed by using agarose gel electrophoresis (3%). Melting curve analysis and PCR efficiency for each selected primer set were validated with the default settings on the ABI 7500 system (Applied Biosystems LLC, Foster, CA, USA). The efficiency of PCR was evaluated by performing a dilution series experiment and the slope of the standard curve was translated into an efficiency value. Efficiency of the PCR within 95–100% was accepted for this study. A portion of the cDNA was subjected to quantitative real-time PCR (qRT-PCR) using an ABI 7500 system with Power SYBR Green PCR Master Mix (Invitrogen, Grand Island, NY, USA). Chicken glyceraldehyde 3-phosphate dehydrogenase (GAPDH), β-actin, and 18S were used as internal controls. Dissociation curves were constructed at the end of amplification for validating the quality of the data. All qRT-PCR experiments were performed in triplicate, and the values of the average cycle threshold (Ct) were determined. Delta-Ct scores for gene transcripts in each sample were normalized using Delta-Ct scores for GAPDH/β-actin/18S and expressed as the relative fold change in gene expression using the equation, $2^{-\Delta\Delta Ct}$. The gene name, NCBI accession numbers, primer sequences, PCR product size, and annealing temperatures used in the present study are shown in Table 1.

Table 1. Primers used for RT-QPCR.

Gene	GenBank #	Primer Sequences (5′-3′)	Size (bp)	Annealing Tm (°C)
TPH2	NM_001001301.1	F: AGGACCTCCGCAGTGATCTA R: CAGCATAAGCAGCTGACAACA	111	58
TH	NM_204805	F: CTTTGATCCTGATGCTGCTG R: CCTCAGCTTGTTTTTGGCAT	103	56
GR	NM_001037826	F: GCCATCGTGAAAAGAGAAGG R: TTTCAACCACATCGTGCAT	95	54

Table 1. *Cont.*

Gene	GenBank #	Primer Sequences (5′-3′)	Size (bp)	Annealing Tm (°C)
BDNF	NM_001031616	F: GACATGGCAGCTTGGCTTAC R: GTTTTCCTCACTGGGCTGGA	167	60
CPT1A	NM_001012898.1	F: GTGGCTGATGATGGTTACGGT R: CCCATGATGTCAACCAATGCT	146	58
ACCα	NM_205505.1	F: GTTGCCATGGATTCGATCGTG R: GGAGTACAGGAAATCGATGCT	128	58
FBPase	AJ276212	F: TTCCATTGGGACCATATTTGG R: ACCCGCTGCCACAAGATTAC	100	58
11β-HSD1	XM_417988	F: CTGGGAACTGTCTGCACAAC R: GATTGCGAGGAACCATTTACAG	96	56
Opn4	AY036061	F: AAGGTTTCGCTGTCATCCAGC R: CTGCTGCTGTTCAAACCAAC	128	58
GAPDH	NM_204305	F: CTTTGGCATTGTGGAGGGTC R: ACGCTGGGATGATGTTCTGG	128	58–60
β-actin	L08165	F: CACAATGTACCCTGGCATTG R: ACATCTGCTGGAAGGTGGAC	158	54–56
18S	AF173612	F: TCCCCTCCCGTTACTTGGAT R: GCGCTCGTCGGCATGTA	60	60

2.5. Statistical Analyses

Statistical analyses were performed using JMP® 14.0 (SAS Institute Inc., Cary, NC, USA). A normal distribution was first tested by the Shapiro–Wilk test and differences among the groups were analyzed using one-way analysis of variance (ANOVA) followed by mean comparison using the Tukey's HSD test at a significance level of $p < 0.05$. Multiple comparisons of group mean by Tukey's HSD test were used to evaluate behavior data including engagement, dustbathing holes, and daily physical activity, and relative changes of gene expression among treatment groups for each gene. Data are presented as the mean ± SEM. A probability level of $p < 0.05$ was considered as statistically significant.

3. Results

3.1. Effects of Environmental Enrichments on Engagement, Dustbathing, and Daily Physical Activity of Broilers

The shape and dimension of the three enrichments (board, hut, and ramp) used in this study are presented in Figure 1A–C. The engagement observed includes climbing or jumping on the objects, sitting next to, and hiding underneath them. On day 28, the first testing day of engagement, the engagement to the hut was the highest in VL houses among treatments and significantly higher compared to 20 lx houses (Figure 1D). On days 35 and 42, similar patterns of engagement were observed (Figure 1E,F). There was no significant difference of engagement between the board and ramp in the 20 lx and VL treated houses. EHs significantly induced the highest engagement of birds compared to boards and ramps on days 28, 35, and 42 of age ($p < 0.05$).

Dustbathing behavior made holes in the floor of the commercial broiler houses [10]. Weekly counting of dustbathing holes in each section of the house was performed and compared among treatments (Figure 2). On days 8, 15 and 22, the numbers of dustbathing holes in 20 lx houses with different enrichments were not different among treatments, excepting the difference between 20 lx_Con and 20 lx_Board on day 15. In the VL houses, VL_Hut sections have the highest number of dustbathing holes. On day 22, the number of dustbathing holes in the VL_Hut section was the highest compared to other treatments including the VL_Con section ($p < 0.05$).

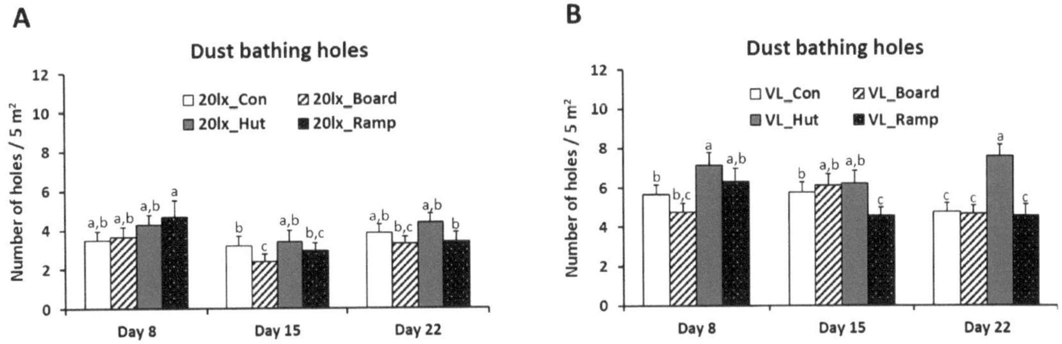

Figure 2. Effects of the three different enrichments (board, hut, and ramp) on the number of dust-bathing holes in different light condition and ages. In each section of trial 1, 20 lx houses (**A**) and VL houses (**B**), dustbathing holes as the evidence of dustbathing behavior were counted at 8, 15, and 22 days of age. Dustbathing holes were observed in nine parts of each section and the number of holes per 5 m^2 was determined. Data (mean ± SEM) were compared among treatments. Different lower-case letters above the bars denote significant differences ($p < 0.05$) among groups.

To select the best enrichment for the broilers' activity in the VL houses, birds of each section in two VL houses were selected and were installed with Animo (activity tracker) on day 43 and uninstalled on day 48 ($n = 4$/section, 2 houses, total birds $n = 32$). Average daily consumed energy (Joule/day) by moving activity was obtained for four days (from day 44 to day 47) (Figure 3). VL_Hut birds burned 33% more energy compared to VL_Con birds ($p > 0.05$) and 66% more energy compared to VL_Board birds for their moving activity ($p < 0.05$).

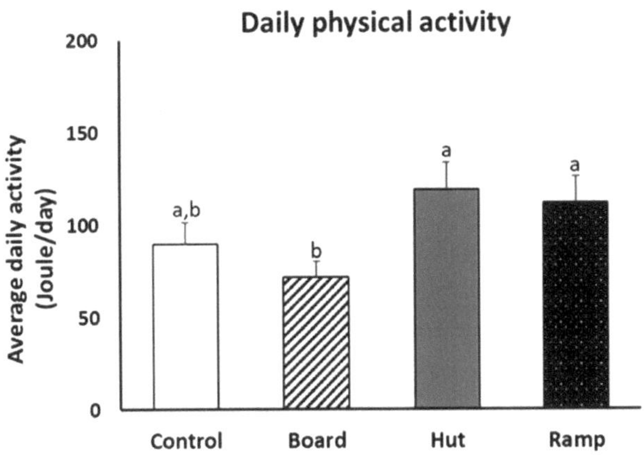

Figure 3. Effects of the different environmental enrichments (board, hut, and ramp) on daily physical activity. In each section of trial 1 VL houses, an activity tracker, Animo, was installed on birds using harness at 43 days of age and uninstalled at 48 days of age ($n = 8$ birds/trt, 4 birds/section, 2 VL houses). Average daily activity (calorie consumption) for each bird from day 44 to day 47 was obtained. Data (mean ± SEM) were compared among treatments (control, board, hut, and ramp). Different lower-case letters above the bars denote significant differences ($p < 0.05$) among groups.

3.2. Effects of Light Programs and EHs on Regulation of Welfare Marker Genes in the Ventral Tegmental Area (VTA) in Commercial Broiler House

To investigate the long-term effects of enrichment huts in the 20 lx and VL houses on the previously identified broiler welfare marker genes in the VTA (Figure 4), expressional changes of TPH2, TH, GR, BDNF, and Opn4 genes were determined in the VTA of broilers' midbrain at 42 days of age (Figures 5 and 6F). Expression of TPH2, an indicator of serotonergic activity in the VTA, in 20 lx birds was downregulated by EHs, and the decreased TPH2 expression in the VL birds compared to 20 lx control birds was further downregulated by EHs ($p < 0.05$). TH expression, an indicator of DAergic activity, was the highest in 20 lx birds (20 lx_Con) among other birds ($p < 0.05$), and there were significant decreases in TH expression in 20 lx_Hut and VL_Con birds ($p < 0.05$). In the VL birds, EHs affected the downregulation of TH expression compared to VL_Con birds ($p < 0.05$). Both VTA-GR expressions, a stress response indicator in VTA, in 20 lx and VL birds were strongly downregulated by EHs treatment, indicating the lower stress status in EHs-treated birds compared to 20 lx and VL control birds ($p < 0.05$). There was no difference in VTA-BDNF expression, a social stress indicator in VTA, in 20 lx birds by EHs treatment ($p > 0.05$), but EHs decreased VTA-BDNF expression significantly in the VL birds. VTA-Opn4 expression was affected by the enriched lighting program as we observed previously [10]. EHs further decreased VTA-Opn4 expression of VL birds ($p < 0.05$).

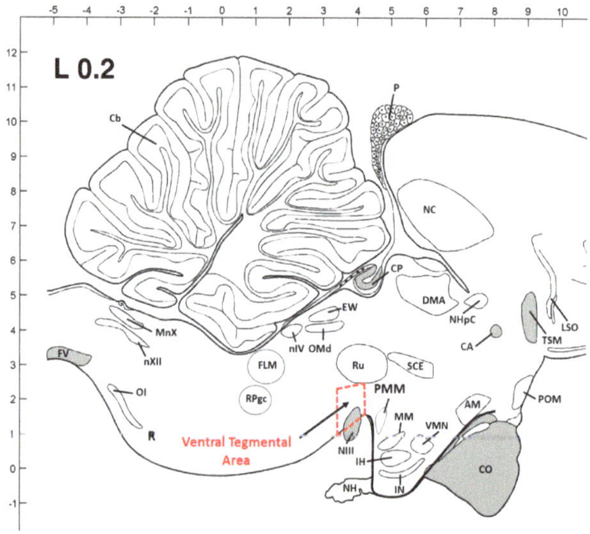

Figure 4. Sagittal view of dissection area of the chicken midbrain (Ventral Tegmental Area). Dimensions of dissected tissue are coronally 3–3.5 mm (W) × 2–3 mm (H) × 1–1.2 mm (L) for VTA. The thickness (W, H, and L) was adjusted proportionally from young birds to older birds based on brain size and structure. Abbreviations used: AM: anterior medial hypothalamic nucleus; CA: anterior commissure; Cb: cerebellum; CO: optic chiasma; CP: posterior commissure; DMA: dorsomedial nucleus; EW: Edinger–Westphal nucleus; FLM: medial longitudinal fasciculus; FV: ventral fasciculus; IH: inferior hypothalamic nucleus; IN: infundibular hypothalamic nucleus; LSO: lateral septal organ; MM: medial mammillary nucleus; MnX: nucleus motorius dorsalis nervi vagi; NC: caudal neostriatum; NH: neurohypophysis; NHpC: nucleus of the hippocampal commissure; NIII: oculomotor nerve; nIV: trochlear nerve nucleus; nXII: hypoglossal nerve nucleus; OI: inferior olivary nucleus; OMd: dorsal oculomotor nucleus; P: pineal gland; POM: medial preoptic nucleus; PVN: paraventricular nucleus; RPgc: nucleus of caudal pontine reticular gigantocellular; Ru: red nucleus; SCE: stratum cellular externum; TSM: septopallio-mesencephalic tract; Top VMN: ventromedial hypothalamic nucleus.

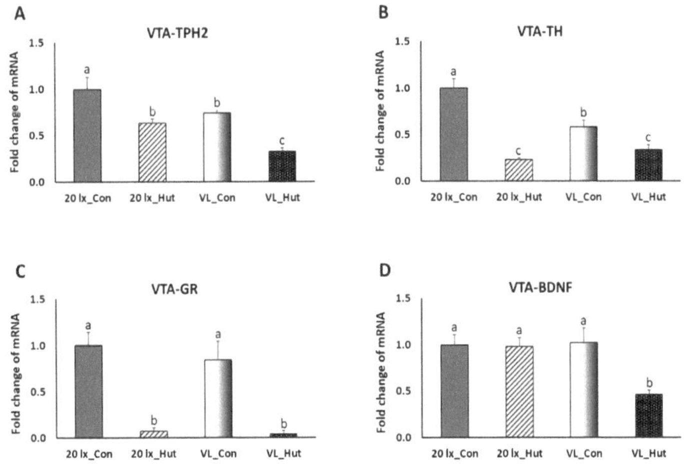

Figure 5. Expression changes of (**A**) TPH2, (**B**) TH (the rate-limiting enzyme of dopamine biosynthesis), (**C**) glucocorticoid receptor (GR), and (**D**) brain-derived neurotropic factor (BDNF) mRNA in the ventral tegmental area (VTA) of birds at 42 days of age. Brains of birds were sampled on day 42 (n = 12/section, 4 sections/house). The VTA of the brainstem from each bird was dissected as described in Materials and Methods. Total RNA was extracted and used for RT-qPCR. Data were set as the relative fold changes of expression levels using the $\Delta\Delta$Ct method with GAPDH and β-actin as internal controls. Data (mean $\pm$ SEM) were expressed from a value set for 1.0 for 20 lx_Con birds for each gene. Different lower-case letters above the bars denote significant differences ($p < 0.05$) among groups.

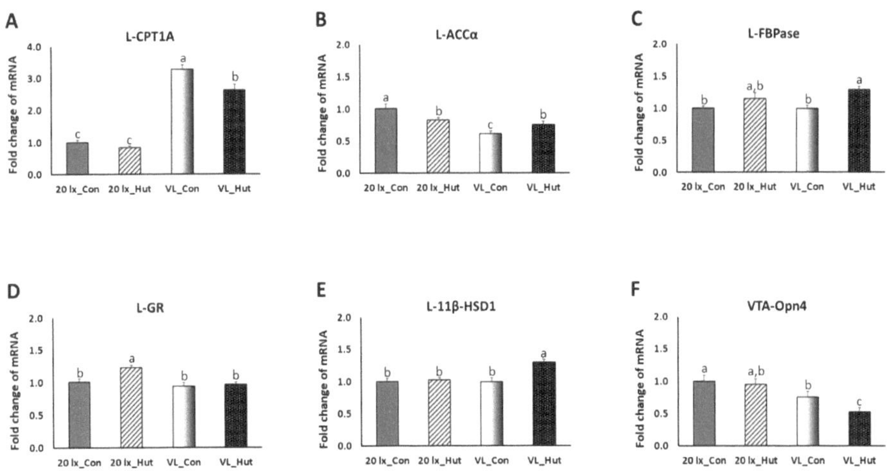

Figure 6. Expression changes of hepatic (**A**) L-CPT1A (liver carnitine palmitoyl transferase 1), (**B**) L-ACCα (liver acetyl-CoA carboxylase alpha), (**C**) L-FBPase (liver Fructose -1,6-bisphosphatase), (**D**) L-GR (liver glucocorticoid receptor), and (**E**) L-11β-HSD1 (liver 11 β-hydroxysteroid dehydrogenease-1) genes and (**F**) VTA Opn4 gene. Total RNA was extracted from liver and macro dissected VTA tissues and used for real time RT-qPCR. Data were set as the relative fold changes of expression levels using the $\Delta\Delta$Ct method with GAPDH, β-actin, and 18S as internal controls. Data (mean $\pm$ SEM) were expressed from a value set for 1.0 for 20 lx_Con birds at 14 days of age. Different lower-case letters above the bars denote significant differences ($p < 0.05$) among groups.

3.3. Combined Effects of Light Programs and EHs on Regulation of Hepatic Metabolic Pathway and Stress Response Genes

To investigate the effects of EHs in the 20 lx and VL houses on the lipid and glucose metabolic and stress response functions in liver (L), expressional changes of L-CPT1A, L-ACCα, L-FBPase, L-GR and L-11β-HSD1 were determined in the liver at 42 days of age (Figure 6). Expression of L-CPT1A, a rate-limiting enzyme of fatty acid β-oxidation in the liver, in 20 lx birds was not affected by EHs treatment, but the expression of L-CPT1A was increased about 3-fold compared to 20 lx in the VL birds.

Interestingly, L-CPT1A expression was decreased in VL_Hut birds compared to VL_Con birds ($p < 0.05$). L-ACCα expression, a rate-limiting enzyme of de novo fatty acid synthesis, was downregulated in VL_Con birds compared to 20 lx_Con birds ($p < 0.05$). EHs decreased the expression of L-ACCα in 20 lx birds, which was slightly increased in VL birds ($p < 0.05$). There was no significant difference in the expression of L-FBPase, a key enzyme of gluconeogenesis, between 20 lx_Con and VL_Con birds. Both L-FBPase expressions in 20 lx and VL birds were upregulated by EHs treatment, and L-FBPase expression was significantly higher in VL_Hut birds compared to VL_Con birds ($p < 0.05$). There was no difference in VTA-GR expression, a social stress indicator in VTA, in between 20 lx_Con and VL_Con birds. EHs increased L-GR expression in 20 lx_Hut birds compared to 20 lx_Con birds ($p < 0.05$). There was no effect of EHs in L-11β-HSD1 expression in 20 lx birds ($p > 0.05$). EHs increased expression of L-11β-HSD1 in the VL_Hut birds compared to VL_Con birds. The avian VTA was suggested as an important area of the midbrain of birds involved in the light perception by Opn4 and might be involved in the welfare of birds [8,10,24]. The expression of the Opn4 gene was determined to investigate the effects of EHs in the 20 lx and VL houses. Opn4 expression was downregulated in VL_Con birds compared to 20 lx_Con birds ($p < 0.05$), and Opn4 expression in VL_Hut birds was significantly lower than that in VL_Con birds ($p < 0.05$).

4. Discussion

The present study was aimed to investigate the optimal environmental enrichments with the enriched lighting program (VL lighting program or gradient lighting program), which may synergically improve the birds' natural behavior, and physiological responses of birds. Environments are perceived by birds as either frightening to survival or to homeostatic interference, leading to behavioral responses and physiological impacts on birds [10]. A lack of environmental complexity in the broiler houses can lead to low levels of activity and the frustration of highly motivated behavior, such as dustbathing [41]. Different enrichment programs in commercial broiler houses may influence diverse behavioral and physiological impacts on birds to improve welfare. Therefore, it is important to determine enrichment(s) that the animals will actively engage with, and several effective enrichment objects for broiler chickens have been suggested including ramps, platforms, and gradient lighting [42]. An enriched lighting program (VL_Con) was effective in improving birds' voluntary natural activity and engagement compared to 20 lx program (20 lx_Con) in this study and the previous study [10], indicating the improvement of comfort natural behavior using the VL lighting program. Broiler performance including daily weight gain and feed conversion ratio was improved, and expressional changes of mental-health-indicating genes suggested better mental health through the enriched lighting program [10]. From the tested EEs in the present study, EHs showed the highest engagement, implying that EHs provide a lower light intensity area that birds can rest inside for safety and mental health compared to control (no enrichment), board, and ramp. Chicken house conditions that do not allow dustbathing behavior of birds were suggested to cause stress, and nonperformance of dustbathing behavior has been suggested to causee the experience of stress [41]. In the present study, results of dustbathing holes from the dustbathing activity of birds were counted in three weekly observations, indicating that EHs in the VL house had a most stimulating effect on dustbathing behavior with the increasing significance of results as the birds got older. These results may suggest that EHs might significantly decrease the stress level of birds in VL birds compared to control (no EE), board, and ramp, but not significantly in 20 lx houses. An activity tracker, Animo, was used to observe daily physical activity of birds at late stage of age (Days 44–47). The daily physical activities showed a significant difference in average daily activity between VL-board birds and VL-hut/ramp birds (Figure 3), suggesting the significant role of huts and ramps for birds' behavior and mental welfare. Environmental light has been recognized as a critical component that can affect animal behavior, and fear is an adaptive emotional and behavior response to potentially harmful stimuli and serves to defend animals. In the present study, it appears that birds engaged huts and ramps for their emotional and behavior responses in the broiler houses that provided different light intensity areas by VL lighting program and enrichment huts/ramps. Interestingly, the energy used by VL-ramp birds was not significantly different from VL-hut birds, indicating that increasing physical activity may not be correlated with dustbathing behavior in broilers. Accordingly, the results of engagement, dustbathing behavior, and daily physical activity studies indicated that EHs would be a preferable EE in the enriched lighting program broiler houses. It may be possible that the behavioral changes of broilers by the VL lighting program and enrichment hut/ramp in the present study is the epigenetic adaptation of broilers' physiology with environmental light intensity. The Opn4 (melanopsin) expressional changes in the VTA of broilers midbrain, in the present study, indicate the possibility of light perception and integration of behavior and physiological responses by Opn4 [24].

The high-quality health of animals is not simply the absence of negative experiences, but rather it is inherently the presence of positive emotions and experiences [16,43,44]. Studies reported that the midbrain VTA contains cell bodies of mesolimbic DAergic neurons as well as the 5-HTergic system in mammals and avian species, and is associated with mental health [8,10,45,46]. To address the mental welfare of broilers under different lighting programs and EHs, we measured TPH2 and TH expressions in the VTA as an indicator of 5-HTergic and DAergic activities. Birds in the VL houses took a rest and slept in the dark-lighting area of the house and actively fed and drank water in the bright feedline area, which may contribute a favorable environment condition. Apparently, providing enrichment huts added more comfortable resting areas for birds to take a break in and around as we observed in the engagement study. Results that EHs affected downregulation of TPH2 and TH expression in both 20 lx house and VL house, suggest that the lower synthesis of these two positive welfare indicators in EHs-treated birds indicates the lower stress-susceptibility in the VL_Hut birds [8,10]. The current study used the GR expression in VTA (VTA-GR) to investigate the effect of two different enrichments (enriched lighting and huts) on stress susceptibility in the broiler brain. Downregulated GR expressions in both 20 lx- and VL-treated birds by EHs indicate that the effect of EHs on stress susceptibility occurred in both houses (20 lx and VL houses), but not specific to the lighting conditions of broiler houses. Enhancing physical activity by EE has been suggested to be a crucial strategy to improve broilers' welfare including emotion and leg health [47–49]. Several brain mechanisms may explain the positive impact of increased activity, including an increase in beneficial neurotropic factors. The stimulation of neurotropic factor BDNF in VTA was suggested to be involved in the long-term social defeat stress, and the deletion of the BDNF gene in the VTA diminished stress-induced behaviors, such as social avoidance in rodents [50,51]. In the present study, downregulation of VTA-BDNF expression occurred only in the VL houses, and these results indicate that the effect of enrichment huts on social stress is specific to the VL house lighting condition. Opn4 has been linked to several behavioral responses to light, including circadian photo-entrainment, light repression of movement in nocturnal animals, and activity in diurnal animals [52]. Avian midbrain VTA was suggested as an important brain region involved in light perception and has been suggested to be associated with avian welfare [8,10]. The result of synergistic downregulation of Opn4 expression by EHs suggests that the VTA-Opn4 might be involved in the direct perception of light information for sensual adaptation and welfare of broilers.

The metabolism of lipids has some essential differences in the biosynthesis and transport of lipids between avians and mammals [53]. In avian species, the liver is the main site of lipid biosynthesis and responsible for 90% of the free fatty acids synthesized de novo [54]. Fatty liver syndrome (FLS) is an augmented hepatic triacylglycerol content and a nutritional disease caused by a metabolic disorder which has effected huge economical losses to the poultry industry, but the pathogenesis remains incompletely elucidated [55,56]. Many factors were suggested as possible causes of FLS in chickens including feed ingredient quality and inappropriate feed formulation, but imbalanced hepatic lipid metabolism may be a critical one [57]. CPT1A is a rate-limiting enzyme for fatty acid β-oxidation that catalyzes the transfer of the long-chain acyl group in acyl-CoA ester to carnitine, enabling fatty acids to enter the mitochondrial matrix for oxidation [58,59]. The deficiency of CPT1A or abnormal regulation can result in diseases like metabolic disorders [60]. Elevated hepatic CPT1A expression in VL birds compared to 20 lx birds indicates the beneficial effect of voluntary activity induced by the enriched lighting program on hepatic fatty acid β-oxidation, and the lower expression of CPT1A in 20 lx houses may be associated with the FLS of broilers. Intriguingly, this upregulated CPT1A expression by the VL lighting program was slightly downregulated by EHs, but still higher than 20 lx-Hut birds. ACCα, a rate-limiting enzyme for de novo fatty acid synthesis, determines the level of hepatic lipid content [61]. In chickens, ACCα expression was significantly upregulated in the FLS chicken, suggesting that altered de novo lipogenesis may be the main pathway of pathogenesis of FLS in chickens [55]. Downregulated ACCα expression in the VL_Con birds compared to 20 lx_Con indicated that fatty acid de novo synthesis decreased using the VL lighting program, but there was no difference in ACCα expression levels between 20 lx_Hut and VL_Hut birds, suggesting that the enriched lighting program is more critical to decreasing de novo fatty acid synthesis in broiler liver than EHs. The deficiency of FBPase, a key enzyme of de novo glucose synthesis (gluconeogenesis), is associated with hepatic steatosis, causing hypoglycemia and fatty liver disease, displaying distinct levels of glucose-derived de novo lipogenesis. [35,61]. Results in this study suggest that the upregulation of FBPase expression in the EHs-treated birds in VL houses (VL_Hut) may contribute to the improvement of hepatic welfare by efficient new glucose production from glycerol in the liver [33,62]. GR and 11β-HSD1 intermediate the regulation of intracellular corticosterone, indicating that they play fundamental roles in the pathogenesis of the metabolic syndrome [63]. Stress changes GR expression in wild birds and broilers [64,65]. The effects of EHs on the GR expression in the broiler liver occurred only in 20 lx birds, suggesting the specificity of the enrichment hut effect on GR expression in the non-light-enriched house. The regeneration of active GCs in cells by 11β-HSD1 is critical to developing the tissue specific phenotype of GC excess [66]. In fact, liver-specific overexpression of 11β-HSD1 increased hepatic lipid synthesis through an upregulation of fatty acid synthase and reduced lipid clearance in mice liver [67]. We observed hepatic 11β-HSD1 expression increased only in the VL_Hut birds that also have the increased expression of de novo fatty acid synthesis gene (ACCα) compared to control birds (VL_Con), suggesting that EHs may be involved in the upregulation of fatty acid synthesis and increase lipid removal in liver.

5. Conclusions

In conclusion, the results suggest that the combined enriched lighting and enrichment hut programs in broiler houses stimulated voluntary activity, and enhanced mental health and hepatic metabolic function of broilers, indicating that the improved hepatic metabolic function may contribute to efficient nutritional support for broilers by EHs. Improved hepatic lipid metabolic functions by the VL lighting program may be associated with the better performance (higher daily weight gain and lower feed conversion ratio), and the combined effects with EHs appear to be synergistic in welfare and mental health.

Author Contributions: Conceptualization, S.W.K. and K.D.C.; Data curation, S.W.K., K.D.C. and M.T.K.J.; Formal analysis, S.W.K. and M.T.K.J.; Funding acquisition, S.W.K. and K.D.C.; Methodology, S.W.K., K.D.C. and M.T.K.J.; Project administration, S.W.K. and K.D.C.; Writing—original draft, S.W.K.; Writing—review and editing, S.W.K., K.D.C., M.T.K.J. and S.K.O. All authors have read and agreed to the published version of the manuscript.

Funding: This research was supported by the RBI (Restaurant Brands International Inc.) Research Grant.

Institutional Review Board Statement: Authors of University of Arkansas were trained for live animal use required by IACUC (Institutional Animal Care and Use Committee). The present animal studies were reviewed and approved by the Tyson Foods BWRF and performed in Tyson Foods BWRF.

Informed Consent Statement: Not applicable.

Data Availability Statement: Data are unavailable due to privacy restrictions.

Acknowledgments: The authors wish to thank Laramie Wieland, Keith Squires, Todd Bucher, and James Clark for their help in Tyson Foods Broiler Welfare Research Farm.

Conflicts of Interest: The authors declare that the research was conducted in the absence of any commercial or financial relationships that could be construed as a potential conflict of interest.

References

1. Dal Bosco, A.; Mattioli, S.; Cartoni Mancinelli, A.; Cotozzolo, E.; Castellini, C. Extensive rearing systems in poultry production: The right chicken for the right farming system. A review of twenty years of scientific research in Perugia university, Italy. *Animals* **2021**, *11*, 1281. [CrossRef]
2. Grzinic, G.; Piotrowicz-Cieslak, A.; Klimkowicz-Pawlas, A.; Gorny, R.L.; Lawniczek-Walczyk, A.; Piechowicz, L.; Olkowska, E.; Potrykus, M.; Tankiewicz, M.; Krupka, M.; et al. Intensive poultry farming: A review of the impact on the environment and human health. *Sci. Total. Environ.* **2023**, *858*, 160014. [CrossRef]
3. Siviy, S.M.; Panksepp, J. In search of the neurobiological substrates for social playfulness in mammalian brains. *Neurosci. Biobehav. Rev.* **2011**, *35*, 1821–1830. [CrossRef]
4. Jacobs, L.; Blatchford, R.A.; de Jong, I.C.; Erasmus, M.A.; Levengood, M.; Newberry, R.C.; Regmi, P.; Riber, A.B.; Weimer, S.L. Enhancing their quality of life: Environmental enrichment for poultry. *Poult. Sci.* **2023**, *102*, 102233. [CrossRef]
5. Blatchford, R.A.; Archer, G.S.; Mench, J.A. Contrast in light intensity, rather than day length, influences the behavior and health of broiler chickens. *Poult. Sci.* **2012**, *91*, 1768–1774. [CrossRef]
6. Olanrewaju, H.A.; Miller, W.W.; Maslin, W.R.; Collier, S.D.; Purswell, J.L.; Branton, S.L. Influence of light sources and photoperiod on growth performance, carcass characteristics, and health indices of broilers grown to heavy weights. *Poult. Sci.* **2018**, *97*, 1109–1116. [CrossRef]
7. Wu, Y.; Huang, J.; Quan, S.; Yang, Y. Light regimen on health and growth of broilers: An update review. *Poult. Sci.* **2022**, *101*, 101545. [CrossRef]
8. Kang, S.W.; Christensen, K.D.; Aldridge, D.; Kuenzel, W.J. Effects of light intensity and dual light intensity choice on plasma corticosterone, central serotonergic and dopaminergic activities in birds, gallus gallus. *Gen. Comp. Endocrinol.* **2020**, *285*, 113289. [CrossRef] [PubMed]
9. Raccoursier, M.; Thaxton, Y.V.; Christensen, K.; Aldridge, D.J.; Scanes, C.G. Light intensity preferences of broiler chickens: Implications for welfare. *Animal* **2019**, *13*, 2857–2863. [CrossRef] [PubMed]
10. Kang, S.W.; Christensen, K.D.; Kidd, M.T., Jr.; Orlowski, S.K.; Clark, J. Effects of a variable light intensity lighting program on the welfare and performance of commercial broiler chickens. *Front. Physiol.* **2023**, *14*, 1059055. [CrossRef] [PubMed]
11. Riber, A.B.; van de Weerd, H.A.; de Jong, I.C.; Steenfeldt, S. Review of environmental enrichment for broiler chickens. *Poult. Sci.* **2018**, *97*, 378–396. [CrossRef]
12. Coria-Avila, G.A.; Pfaus, J.G.; Orihuela, A.; Dominguez-Oliva, A.; Jose-Perez, N.; Hernandez, L.A.; Mota-Rojas, D. The neurobiology of behavior and its applicability for animal welfare: A review. *Animals* **2022**, *12*, 928. [CrossRef]
13. Alcaro, A.; Panksepp, J. The seeking mind: Primal neuro-affective substrates for appetitive incentive states and their pathological dynamics in addictions and depression. *Neurosci. Biobehav. Rev.* **2011**, *35*, 1805–1820. [CrossRef]
14. Derdeyn, P.; Hui, M.; Macchia, D.; Beier, K.T. Uncovering the connectivity logic of the ventral tegmental area. *Front. Neural Circuits* **2021**, *15*, 799688. [CrossRef]
15. Geisler, C.E.; Hayes, M.R. Metabolic hormone action in the vta: Reward-directed behavior and mechanistic insights. *Physiol. Behav.* **2023**, *268*, 114236. [CrossRef]
16. Boissy, A.; Manteuffel, G.; Jensen, M.B.; Moe, R.O.; Spruijt, B.; Keeling, L.J.; Winckler, C.; Forkman, B.; Dimitrov, I.; Langbein, J.; et al. Assessment of positive emotions in animals to improve their welfare. *Physiol. Behav.* **2007**, *92*, 375–397. [CrossRef]
17. Rault, J.L.; Lawrence, A.J.; Ralph, C.R. Brain-derived neurotrophic factor in serum as an animal welfare indicator of environmental enrichment in pigs. *Domest. Anim. Endocrinol.* **2018**, *65*, 67–70. [CrossRef] [PubMed]

18. Phillips, C. Brain-derived neurotrophic factor, depression, and physical activity: Making the neuroplastic connection. *Neural Plast.* **2017**, *2017*, 7260130. [CrossRef] [PubMed]
19. Arosio, B.; Guerini, F.R.; Voshaar, R.C.O.; Aprahamian, I. Blood brain-derived neurotrophic factor (bdnf) and major depression: Do we have a translational perspective? *Front. Behav. Neurosci.* **2021**, *15*, 626906. [CrossRef] [PubMed]
20. Hotting, K.; Roder, B. Beneficial effects of physical exercise on neuroplasticity and cognition. *Neurosci. Biobehav. Rev.* **2013**, *37*, 2243–2257. [CrossRef] [PubMed]
21. Ralph, C.R.; Tilbrook, A.J. Invited review: The usefulness of measuring glucocorticoids for assessing animal welfare. *J. Anim. Sci.* **2016**, *94*, 457–470. [CrossRef]
22. Ferrario, C.R.; Labouebe, G.; Liu, S.; Nieh, E.H.; Routh, V.H.; Xu, S.; O'Connor, E.C. Homeostasis meets motivation in the battle to control food intake. *J. Neurosci.* **2016**, *36*, 11469–11481. [CrossRef]
23. Mizoguchi, A.; Banno, R.; Sun, R.; Yaginuma, H.; Taki, K.; Kobayashi, T.; Sugiyama, M.; Tsunekawa, T.; Onoue, T.; Takagi, H.; et al. Glucocorticoid receptor signaling in ventral tegmental area neurons increases the rewarding value of a high-fat diet in mice. *Sci. Rep.* **2021**, *11*, 12873. [CrossRef] [PubMed]
24. Kang, S.W. Central nervous system associated with light perception and physiological responses of birds. *Front. Physiol.* **2021**, *12*, 723454. [CrossRef] [PubMed]
25. Dumond Bourie, A.; Potier, J.B.; Pinget, M.; Bouzakri, K. Myokines: Crosstalk and consequences on liver physiopathology. *Nutrients* **2023**, *15*, 1729. [CrossRef] [PubMed]
26. Machado, M.V. Aerobic exercise in the management of metabolic dysfunction associated fatty liver disease. *Diabetes Metab. Syndr. Obes.* **2021**, *14*, 3627–3645. [CrossRef] [PubMed]
27. van der Windt, D.J.; Sud, V.; Zhang, H.; Tsung, A.; Huang, H. The effects of physical exercise on fatty liver disease. *Gene Expr.* **2018**, *18*, 89–101. [CrossRef] [PubMed]
28. Heald, P.J.; Badman, H.G. Lipid metabolism and the laying hen. I. Plasma-free fatty acids and the onset of laying in the domestic fowl. *Biochim. Biophys. Acta* **1963**, *70*, 381–388. [CrossRef]
29. Leveille, G.A.; O'Hea, E.K.; Chakbabarty, K. In vivo lipogenesis in the domestic chicken. *Proc. Soc. Exp. Biol. Med.* **1968**, *128*, 398–401. [CrossRef]
30. Hodson, L.; Gunn, P.J. The regulation of hepatic fatty acid synthesis and partitioning: The effect of nutritional state. *Nat. Rev. Endocrinol.* **2019**, *15*, 689–700. [CrossRef]
31. Bechmann, L.P.; Hannivoort, R.A.; Gerken, G.; Hotamisligil, G.S.; Trauner, M.; Canbay, A. The interaction of hepatic lipid and glucose metabolism in liver diseases. *J. Hepatol.* **2012**, *56*, 952–964. [CrossRef]
32. Parhofer, K.G. Interaction between glucose and lipid metabolism: More than diabetic dyslipidemia. *Diabetes Metab. J.* **2015**, *39*, 353–362. [CrossRef] [PubMed]
33. Shah, A.M.; Wondisford, F.E. Tracking the carbons supplying gluconeogenesis. *J. Biol. Chem.* **2020**, *295*, 14419–14429. [CrossRef] [PubMed]
34. Dupont, J.; Tesseraud, S.; Simon, J. Insulin signaling in chicken liver and muscle. *Gen. Comp. Endocrinol.* **2009**, *163*, 52–57. [CrossRef] [PubMed]
35. Gorce, M.; Lebigot, E.; Arion, A.; Brassier, A.; Cano, A.; De Lonlay, P.; Feillet, F.; Gay, C.; Labarthe, F.; Nassogne, M.C.; et al. Fructose-1,6-bisphosphatase deficiency causes fatty liver disease and requires long-term hepatic follow-up. *J. Inherit. Metab. Dis.* **2022**, *45*, 215–222. [CrossRef] [PubMed]
36. Brust, K.B.; Corbell, K.A.; Al-Nakkash, L.; Babu, J.R.; Broderick, T.L. Expression of gluconeogenic enzymes and 11beta-hydroxysteroid dehydrogenase type 1 in liver of diabetic mice after acute exercise. *Diabetes Metab. Syndr. Obes.* **2014**, *7*, 495–504. [PubMed]
37. Chapman, K.; Holmes, M.; Seckl, J. 11beta-hydroxysteroid dehydrogenases: Intracellular gate-keepers of tissue glucocorticoid action. *Physiol. Rev.* **2013**, *93*, 1139–1206. [CrossRef] [PubMed]
38. National Research Council (NRC). *Nutrient Requirements of Poultry*, 9th ed.; National Academy Press: Washington, DC, USA, 1994.
39. Delgado, M.M.; Han, B.S.G.; Bain, M.J. Correction to: Domestic cats (*Felis catus*) prefer freely available food over food that requires effort. *Anim. Cogn.* **2022**, *25*, 493. [CrossRef] [PubMed]
40. Kuenzel, W.J.; Masson, M. *A Stereotaxic Atlas of the Brain of Chick (Gallus domesticus)*; Johns Hopkins University Press: Baltimore, MD, USA, 1988.
41. Vestergaard, K.S.; Skadhauge, E.; Lawson, L.G. The stress of not being able to perform dustbathing in laying hens. *Physiol. Behav.* **1997**, *62*, 413–419. [CrossRef]
42. Grandin, T. A practical approach to providing environmental enrichment to pigs and broiler chickens housed in intensive systems. *Animals* **2023**, *13*, 2372. [CrossRef]
43. Alexander, R.; Aragon, O.R.; Bookwala, J.; Cherbuin, N.; Gatt, J.M.; Kahrilas, I.J.; Kastner, N.; Lawrence, A.; Lowe, L.; Morrison, R.G.; et al. The neuroscience of positive emotions and affect: Implications for cultivating happiness and wellbeing. *Neurosci. Biobehav. Rev.* **2021**, *121*, 220–249. [CrossRef] [PubMed]
44. Marcet Rius, M.; Cozzi, A.; Bienboire-Frosini, C.; Teruel, E.; Chabaud, C.; Monneret, P.; Leclercq, J.; Lafont-Lecuelle, C.; Pageat, P. Providing straw to allow exploratory behaviour in a pig experimental system does not modify putative indicators of positive welfare: Peripheral oxytocin and serotonin. *Animal* **2018**, *12*, 2138–2146. [CrossRef] [PubMed]
45. Carkaci-Salli, N.; Salli, U.; Kuntz-Melcavage, K.L.; Pennock, M.M.; Ozgen, H.; Tekin, I.; Freeman, W.M.; Vrana, K.E. Tph2 in the ventral tegmental area of the male rat brain. *Brain Res. Bull.* **2011**, *84*, 376–380. [CrossRef] [PubMed]

46. Chaudhury, D.; Walsh, J.J.; Friedman, A.K.; Juarez, B.; Ku, S.M.; Koo, J.W.; Ferguson, D.; Tsai, H.C.; Pomeranz, L.; Christoffel, D.J.; et al. Rapid regulation of depression-related behaviours by control of midbrain dopamine neurons. *Nature* **2013**, *493*, 532–536. [CrossRef]
47. Bizeray, D.; Estevez, I.; Leterrier, C.; Faure, J.M. Influence of increased environmental complexity on leg condition, performance, and level of fearfulness in broilers. *Poult. Sci.* **2002**, *81*, 767–773. [CrossRef]
48. Kristensen, H.H.; Aerts, J.M.; LeRoy, T.; Berckmans, D.; Wathes, C.M. Using light to control activity in broiler chickens. *Br. Poult. Sci.* **2004**, *45*, S30–S31. [CrossRef]
49. Reiter, K.; Bessei, W. Effect of locomotor activity on leg disorder in fattening chicken. *Berl. Munch. Tierarztl. Wochenschr.* **2009**, *122*, 264–270.
50. Berton, O.; McClung, C.A.; Dileone, R.J.; Krishnan, V.; Renthal, W.; Russo, S.J.; Graham, D.; Tsankova, N.M.; Bolanos, C.A.; Rios, M.; et al. Essential role of bdnf in the mesolimbic dopamine pathway in social defeat stress. *Science* **2006**, *311*, 864–868. [CrossRef]
51. Krishnan, V.; Han, M.H.; Graham, D.L.; Berton, O.; Renthal, W.; Russo, S.J.; Laplant, Q.; Graham, A.; Lutter, M.; Lagace, D.C.; et al. Molecular adaptations underlying susceptibility and resistance to social defeat in brain reward regions. *Cell* **2007**, *131*, 391–404. [CrossRef]
52. Sikka, G.; Hussmann, G.P.; Pandey, D.; Cao, S.; Hori, D.; Park, J.T.; Steppan, J.; Kim, J.H.; Barodka, V.; Myers, A.C.; et al. Melanopsin mediates light-dependent relaxation in blood vessels. *Proc. Natl. Acad. Sci. USA* **2014**, *111*, 17977–17982. [CrossRef]
53. Hermier, D.; Chapman, M.J.; Leclercq, B. Plasma lipoprotein profile in fasted and refed chickens of two strains selected for high or low adiposity. *J. Nutr.* **1984**, *114*, 1112–1121. [CrossRef]
54. Leveille, G.A.; Romsos, D.R.; Yeh, Y.; O'Hea, E.K. Lipid biosynthesis in the chick. A consideration of site of synthesis, influence of diet and possible regulatory mechanisms. *Poult. Sci.* **1975**, *54*, 1075–1093. [CrossRef]
55. Liu, Z.; Li, Q.; Liu, R.; Zhao, G.; Zhang, Y.; Zheng, M.; Cui, H.; Li, P.; Cui, X.; Liu, J.; et al. Expression and methylation of microsomal triglyceride transfer protein and acetyl-coa carboxylase are associated with fatty liver syndrome in chicken. *Poult. Sci.* **2016**, *95*, 1387–1395. [CrossRef]
56. Zhang, C.; Meng, S.; Li, C.; Yang, Z.; Wang, G.; Wang, X.; Ma, Y. Primary broiler hepatocytes for establishment of a steatosis model. *Vet. Sci.* **2022**, *9*, 316. [CrossRef]
57. Meng, J.; Ma, N.; Liu, H.; Liu, J.; Liu, J.; Wang, J.; He, X.; Zhao, X. Untargeted and targeted metabolomics profiling reveals the underlying pathogenesis and abnormal arachidonic acid metabolism in laying hens with fatty liver hemorrhagic syndrome. *Poult. Sci.* **2021**, *100*, 101320. [CrossRef]
58. Houten, S.M.; Violante, S.; Ventura, F.V.; Wanders, R.J. The biochemistry and physiology of mitochondrial fatty acid beta-oxidation and its genetic disorders. *Annu. Rev. Physiol.* **2016**, *78*, 23–44. [CrossRef]
59. Schlaepfer, I.R.; Joshi, M. Cpt1a-mediated fat oxidation, mechanisms, and therapeutic potential. *Endocrinology* **2020**, *161*, bqz046. [CrossRef]
60. Liang, K. Mitochondrial cpt1a: Insights into structure, function, and basis for drug development. *Front. Pharmacol.* **2023**, *14*, 1160440. [CrossRef]
61. Wang, M.D.; Wu, H.; Fu, G.B.; Zhang, H.L.; Zhou, X.; Tang, L.; Dong, L.W.; Qin, C.J.; Huang, S.; Zhao, L.H.; et al. Acetyl-coenzyme a carboxylase alpha promotion of glucose-mediated fatty acid synthesis enhances survival of hepatocellular carcinoma in mice and patients. *Hepatology* **2016**, *63*, 1272–1286. [CrossRef]
62. Kalemba, K.M.; Wang, Y.; Xu, H.; Chiles, E.; McMillin, S.M.; Kwon, H.; Su, X.; Wondisford, F.E. Glycerol induces g6pc in primary mouse hepatocytes and is the preferred substrate for gluconeogenesis both in vitro and in vivo. *J. Biol. Chem.* **2019**, *294*, 18017–18028. [CrossRef]
63. Whorwood, C.B.; Donovan, S.J.; Flanagan, D.; Phillips, D.I.; Byrne, C.D. Increased glucocorticoid receptor expression in human skeletal muscle cells may contribute to the pathogenesis of the metabolic syndrome. *Diabetes* **2002**, *51*, 1066–1075. [CrossRef]
64. Kang, S.W.; Madkour, M.; Kuenzel, W.J. Tissue-specific expression of DNA methyltransferases involved in early-life nutritional stress of chicken, gallus gallus. *Front. Genet.* **2017**, *8*, 204. [CrossRef]
65. Lattin, C.R.; Romero, L.M. Chronic stress alters concentrations of corticosterone receptors in a tissue-specific manner in wild house sparrows (*Passer domesticus*). *J. Exp. Biol.* **2014**, *217*, 2601–2608. [CrossRef] [PubMed]
66. Morgan, S.A.; McCabe, E.L.; Gathercole, L.L.; Hassan-Smith, Z.K.; Larner, D.P.; Bujalska, I.J.; Stewart, P.M.; Tomlinson, J.W.; Lavery, G.G. 11beta-hsd1 is the major regulator of the tissue-specific effects of circulating glucocorticoid excess. *Proc. Natl. Acad. Sci. USA* **2014**, *111*, E2482–E2491. [CrossRef] [PubMed]
67. Paterson, J.M.; Morton, N.M.; Fievet, C.; Kenyon, C.J.; Holmes, M.C.; Staels, B.; Seckl, J.R.; Mullins, J.J. Metabolic syndrome without obesity: Hepatic overexpression of 11beta-hydroxysteroid dehydrogenase type 1 in transgenic mice. *Proc. Natl. Acad. Sci. USA* **2004**, *101*, 7088–7093. [CrossRef] [PubMed]

animals

MDPI

Article

Effect of In Ovo Supplementation of Slab51 Probiotic Mixture, Associated with Marek's Disease Vaccine, on Growth Performance, Intestinal Morphology and *Eimeria* spp. Infection in Broiler Chickens

Lucia Biagini [1], Livio Galosi [1,*], Adolfo Maria Tambella [1], Alessandra Roncarati [1], Danilo De Bellis [1], Stefano Pesaro [2], Anna-Rita Attili [1], Sara Berardi [1] and Giacomo Rossi [1]

1 School of Biosciences and Veterinary Medicine, University of Camerino, 62024 Matelica, Italy; lucia.biagini@unicam.it (L.B.); adolfomaria.tambella@unicam.it (A.M.T.); alessandra.roncarati@unicam.it (A.R.); danilo.debellis@unicam.it (D.D.B.); annarita.attili@unicam.it (A.-R.A.); sara.berardi@unicam.it (S.B.); giacomo.rossi@unicam.it (G.R.)

2 Department of Agricultural, Food, Environmental and Animal Sciences, University of Udine, 33100 Udine, Italy; stefano.pesaro@uniud.it

* Correspondence: livio.galosi@unicam.it

Simple Summary: In the last decades, several researchers investigated the feasibility of in ovo administration of different compounds for promoting health and productivity in poultry. The use of various substances is described in the literature, including vaccines, drugs, hormones, competitive exclusion cultures, pre-, pro- and postbiotics and nutritional supplements. Injected substances, volumes, dosages or concentrations vary according to bird type, egg size, time and site of injection, incubation system and regimen, with variable and sometimes contrasting results. Several reports indicated that probiotics may be effectively used to fight intestinal bacterial infections and promote growth in chickens. In ovo Marek's disease vaccination is a practice widespread throughout the world, and the association of this vaccine with a probiotic mixture could be performed at the industrial scale, limiting the additional costs. The application of a probiotic mixture, associated with the Marek's disease vaccine and inoculated in ovo, was tested to evaluate benefits to the poultry industry in a pilot scale.

Abstract: The interest for in ovo feeding has grown in the last decades mainly concerning probiotics, live microorganisms that can actively interact with the embryo. The aim of this study was to evaluate the effects of a multi-strain probiotic diluted in Marek's disease vaccine (MDV) on zootechnical performances, intestinal morphology and *Eimeria* spp. infection. One hundred and twenty eggs of Ross 308 broiler chickens were incubated until 18 d, when 105 fertilised and vital eggs were randomly allocated into three groups. A control group (C) was inoculated with MDV; two treated groups (P1 and P2) were inoculated with MDV and different concentrations of probiotics: 1×10^5 CFU/100 µL in P1 and 1×10^6 CFU/100 µL in P2. After hatching, chickens were separated into three replicates (10/replicate). Zootechnical parameters were determined. At the end of the cycle (35 d), chickens were slaughtered, and the intestine was collected for morphological analysis from nine chickens per group (three/replicate). *Eimeria* spp. oocyst shedding was determined weekly, and parasitic lesions were analysed on the histological sample. In ovo treatment with probiotic did not influence hatching rate but significantly improved body weight and positively influenced intestinal morphometric data compared to C. Oocyst shedding in faeces resulted in an increase in C, with significant differences at sampling performed at 14, 21 and 28 d of age. These results suggest that the tested probiotic compound is safe for in ovo supplementation and effectively improves zootechnical performances and coccidian resistance.

Keywords: in ovo inoculation; chicken; probiotics; Marek's disease vaccine; intestinal morphology; coccidiosis

check for updates

Citation: Biagini, L.; Galosi, L.; Tambella, A.M.; Roncarati, A.; De Bellis, D.; Pesaro, S.; Attili, A.-R.; Berardi, S.; Rossi, G. Effect of In Ovo Supplementation of Slab51 Probiotic Mixture, Associated with Marek's Disease Vaccine, on Growth Performance, Intestinal Morphology and *Eimeria* spp. Infection in Broiler Chickens. *Animals* **2024**, *14*, 3435. https://doi.org/10.3390/ani14233435

Academic Editor: Juan David Latorre

Received: 26 October 2024
Revised: 20 November 2024
Accepted: 26 November 2024
Published: 27 November 2024

1. Introduction

During incubation, the chicken embryo is completely dependent on the contents of the egg. A rapid change in its metabolism takes place in the last third of incubation, starting with the ingestion of amniotic fluid at around embryonic day 17 and the internalisation of the yolk in the abdominal cavity [1,2]. In the modern poultry industry, access to feed after hatching is usually delayed by 48–72 h, and during this period, the yolk represents the only source of nutrients to support gastrointestinal tract development, considering that at the end of incubation glycogen stores are almost completely depleted [3,4]. At the same time, the developing chicken embryo is able to produce an immune response to a pathogen prior to hatch; a feature that is widely exploited for modern large-scale poultry production with the routine administration of in ovo vaccination for multiple pathogens, including Marek's Disease (MD) and Infectious Bursal Disease (IBD) [5]. Another fundamental change that occurs between hatching and housing is the microbial colonisation of the gastrointestinal tract [6]. In fact, unlike mammals or natural-hatched birds, which receive a natural microbial inoculation during parturition or contact within the nest, in poultry production, the interaction between chickens and the hen is completely excluded, and the main source of bacteria is the environment [7]. Vertical transmission of intestinal pathogens and commensal bacteria from hens to eggs and offspring has been demonstrated, identifying a number of core genera that may be vertically transmitted from the hen's intestine to the chick, with a contribution of both the hen's intestine and oviduct [8–10]. However, after hatching, chickens' gut microbiota will be strongly influenced by the environmental microbes, and pathogens can easily be introduced in this process [11,12].

A balanced microbiota is essential to achieve one of the poultry industry's objectives, which is to minimise the prevalence of infections while optimising productive performance [13]. Poultry meat production accounts for approximately one third of the world's total meat production [14] and there is now strong evidence that the misuse of antibiotics, especially as growth promoters, has contributed to the emergence of multi-drug resistant bacteria in animals and humans, leading to a restriction in the use of antibiotics and the search for effective alternatives [15,16].

For all these reasons, in ovo feeding (IOF), the provision of nutrients to chicken embryo during the incubation, has gained increasing interest. This administration is usually performed in the amniotic fluid at a late incubation stage [17]. This represents an optimal delivery site because it is ingested by the chicken during hatching and comes into direct contact with enterocytes [18]. Various elements such as vitamins, carbohydrates, hormones, probiotic and prebiotics have been supplemented with IOF, not only to provide nutrients but also to promote correct chicken development [19,20]. Probiotics, defined as live microorganisms capable of positively modifying the health of the host [21], can actively interact with the gastrointestinal system and its microbiota, bringing multiple benefits [22–25]. Thus, the administration of probiotics in ovo during the late incubation stage allows an early interaction with the chicken, leading to a direct impact on the composition of microbiota [26,27], intestinal development and growth performances [28–30], immune system functionality [31–33] and resistance to infectious agents [28,31,34–37]. Among these, coccidiosis, caused by *Eimeria* spp., is considered one of the most important challenges in the poultry industry, as it causes intestinal damage with impaired digestive processes, loss of nutrient absorption capacity, dehydration and increased susceptibility to other infections [38]. The use of probiotics, both inoculated in ovo or administered in water, is a natural alternative to reduce the infection rate and *Eimeria* spp. oocysts shedding [39,40]. Most of the effects observed after probiotics administration are dependent on concentration, bacterial strain, age and site of administration [19,41]; therefore, each new probiotic compound needs to be tested to verify its safety and efficacy, especially concerning in ovo administration. The aim of this study is to evaluate the effects of two different concentrations of a multi-strain probiotic administered in ovo in the amniotic fluid on hatching rate, zootechnical performances, intestinal morphology and *Eimeria* spp. infection.

2. Materials and Methods

2.1. Probiotic and Vaccine Preparation

A commercial combined live, cell-associated Marek's Disease vaccine (MDV, Nobilis® Rismavac + CA126, MSD Animal Health S.r.l., Milano, Italy) was used for in ovo immunisation of chickens against Marek's Disease. One ampoule of the vaccine (1000 doses) was diluted in 100 mL of sterile diluent.

The commercially available multi-strain probiotic Slab51 (Ormendes SA, Jouxtens-Mezery, Switzerland) was used. This product contains 200 billion units of lactic acid bacteria per 1.5 g of product, composed of the following strains: *Streptococcus thermophilus* DSM 32245, *Bifidobacterium lactis* DSM 32246, *Bifidobacterium lactis* DSM 32247, *Lactobacillus acidophilus* DSM 32241, *Lactobacillus helveticus* DSM32242, *Lactobacillus paracasei* DSM 32243, *Lactobacillus plantarum* DSM 32244 and *Lactobacillus brevis* DSM 27961. The product was properly diluted using the MDV diluent. The last dilution was performed directly into the reconstituted MDV. The treatments included the following: the control group (C), inoculated with 100 μL of MDV; the treated group (P1), inoculated with 100 μL of MDV and 1×10^5 CFU/100 μL of probiotics and the second treated group (P2), inoculated with 100 μL of MDV and 1×10^6 CFU/100 μL of probiotics. Concentration was assessed according to previous trials [37].

2.2. Birds and In Ovo Treatments

One hundred and twenty eggs of Ross 308 broiler chickens were obtained from a commercial hatchery (Avizoo-Euroagricola s.s., Longiano, FC, Italy) and incubated under standard conditions (37.5 °C and 54% RH) in a MG100/150 incubator (FIEM S.r.l., Como, Italy). Dead embryos or unfertilised eggs were excluded from incubation by candling. At day 18 of embryonic life, 105 fertilised and vital eggs were randomly allocated into three experimental groups (35 eggs/group) and subjected to the in ovo procedure. After checking the exact position of the air chamber, all eggs were sanitised at the blunt end with 70% ethanol. A pilot hole was made with an 18-gauge needle to pierce only the shell without entering the air chamber. The needle was disinfected with 70% ethanol between injections. Administration into the amnion was performed with a 25-gauge needle (2.5 cm long) fitted to a 1 mL syringe, using a new sterile needle and syringe for each egg. After injection, the hole was sealed with glue and all eggs were transferred to separate hatching baskets. At hatching, the number of live hatched or unhatched chickens was counted to calculate the percentage of hatchability for each group.

2.3. Animal Housing and Diets

After hatching, the 3 groups were separated into 3 replicates of 10 animals each. Chickens were housed in 3 adjacent sheds (12 m^2 each), separated in 3 pens by nets, and a replicate of each group was placed randomly in every shed to avoid environmental interference in this study. All the animals received the same commercial feed (Cruciani, Montappone, MC, Italy) ad libitum throughout the trial (Table 1) and tap water. All the animals were weighted weekly with an electronic balance (mod. ACS-A9, My Scale, Foggia, Italy), while feed consumption was registered daily. From these data, body weight (BW), average weekly body weight gain (BWG), average weekly feed intake (FI) and the feed conversion ratio (FCR) were determined. More specifically, for each replicate, the average weekly weight gain was calculated as the cumulative weight gain of the animals within a replicate/number of animals per replicate; weekly feed intake was obtained as the total weekly consumption/number of animals per replicate and the feed conversion ratio was obtained as the total weekly feed consumption/total weight gain.

Table 1. Proximate composition of the feeds administered to the three groups during the trial.

	Feed I *Starter*	**Feed II** *Grower*
Age of broiler chickens (days)	1–15	16–35
Proximate composition (%) wet weight basis:		
Protein	22	21
Lipids	5	5.8
Ash	7.4	7.5
Fibre	4.4	4.5
Calcium	1	1.04
Phosphorum	1	0.85
Sodium	0.2	0.19
Lysine	1.2	1.2
Methionine	0.6	0.58
Phytasis (FTY)	1500	750
Endo-1.4-Beta-Xylanasis (FXU)	200	---
Vitamin A E672 (UI/kg)	12,000	4000
Vitamin D3 E671 (UI/kg)	2000	1250
Vitamin E (apha toc. 91%) (mg/kg)	40	20
Cupper E4 (mg/kg)	16	10
Selenium E8 (mg/kg)	0.16	0.2
Luthein E161b (g/kg)	---	41
Zeaxanthin E161 (g/kg)	---	8.4

2.4. Sample Collection and Processing

At the end of the zootechnical cycle (35 days of age), chickens were slaughtered by electrical stunning and bleeding in an authorised slaughterhouse, and 9 chicken per group (3/replicate) were randomly selected for sampling. For each bird, ∼3 cm long portions of the different intestinal segments (duodenum, jejunum, ileum and caecum) were collected and 10% buffered-formalin fixed for 24 h. The samples were dehydrated through a graded series of alcohol, cleared with a ethanol/xylene solution, before embedding in paraffin wax. Serial sections (3 µm thick) were cut, stained with Haematoxylin-Eosin and mounted for the observation under the optical microscope (Leica DM2500, Wetzlar, Germany). Serum samples were obtained at slaughtering and frozen at −80 °C for serological analysis. Marek's disease antibodies were assessed using a commercial ELISA test kit (ABBEXA, Cambridge, UK).

2.5. Histology and Morphometric Measurement

Haematoxylin–Eosin-stained sections of 5 well-oriented villi of the duodenum, jejunum, ileum and cecum from each sampled animal were measured considering villus height (VH), villus width (VW), crypt depth (CD) and lamina propria width (LPW), using an image processing and analysis system (Leica Imaging Systems Ltd., Cambridge, UK). VH was measured from the top of the villus to the top of the lamina propria [42]. CD was measured from the base upward to the region of transition between the crypt and villus [43]. VW was measured at the widest area of each villus, whereas the villus width–crypt depth ratio was determined as the ratio of VH to CD. Villus surface area (VSA) was calculated using the formula $(2\pi)(VW/2)(VL)$ [44].

2.6. *Eimeria* Oocyst Shedding and Histological Lesions Score

Considering that the birds were raised in a commercial poultry farm, natural coccidian infection could occur. To assess *Eimeria* spp. oocyst shedding, weekly, starting from 7 days of age until the end of the trial, 10 fresh faecal samples were collected per pen, pooled and kept in separate airtight plastic bags. After homogenisation, samples were stored at 4 °C until assessed for oocyst count, which was determined using a McMaster counting chamber and stated as oocysts per gram of excreta, as previously described [45]. Histological lesions caused by *Eimeria* spp. were scored with a 0–3 scoring system in the duodenum, jejunum, ileum and caecum, considering both distribution and severity of infection, adapting a model previously described [46]. Briefly, a sum of the distribution (A) and severity (B) of protozoal infection was performed, where A represents the distribution of developmental stages of *Eimeria* spp. along the examined intestinal segment (0 = no parasites; 1 = parasites in one 10× field; 2 = parasites in two 10× fields and 3 = parasites in three 10× fields), and B represents the severity of the infection within the examined fields (0 = parasites in 0% of villi; 1 = parasites in <25% of villi; 2 = parasites in 25 to 50% of villi and 3 = parasites in >50% of villi). Then, the sum of the two parameters was divided by 2, to obtain a final total score 0–3.

In the same intestinal segments, enterocyte proliferation and presence of mixed inflammatory infiltrate in the lamina propria of examined villi was performed. A score from 0 to 3 points was assigned for the following parameters: 0 = no inflammation/proliferation; 1 = mild inflammation/proliferation; 2 = moderate inflammation/proliferation and 3 = severe inflammation/proliferation. The score was applied in three microscopic fields and the average value was calculated.

2.7. Statistical Analysis

Cardinal data were summarised using the arithmetic mean and the standard error of the mean; they were assessed for the assumption of normality of the data distribution using the Shapiro–Wilk test. All normally distributed cardinal data were analysed with one-way ANOVA (Analysis of Variance) and the Holm–Šidák post hoc test; otherwise, statistical analyses were performed with a non-parametric approach. Ordinal data were summarised using the median and range and compared between groups using the Kruskal–Wallis test and Dunn's multiple comparison test. Differences with p values < 0.05 were considered statistically significant. All data were analysed using GraphPad Prism 10 statistical software for MacOS, version 10.1.1-270 (GraphPad Software Inc., San Diego, CA, USA).

3. Results

3.1. Marek's Disease Virus Antibodies

All animals in the three groups were positive for the qualitative Marek's disease antibody test.

3.2. Hatchability and Post Hatch Performance

No relevant differences were recorded in hatching rates between the three groups (94.2% for groups P2 and C; 97.1% for group P1). Table 2 reports the body weight of the chickens evaluated weekly. Starting from 7 days of age and up to the end of the cycle at 35 days, in ovo treatment with probiotic significantly improved body weight compared to C. No differences are observed regarding hatching weight.

Table 2. Effect of in ovo probiotic supplementation on body weight (g).

Days of Age	C	P1	P2	Statistical Data
0	46.32 ± 0.6196	45.84 ± 0.4692	46.10 ± 0.6080	F = 0.1756, p = 0.8391, p^* = 1.0
7	104.5 ± 2.539 ^{a}	156.6 ± 4.798 ^{b}	147.0 ± 2905 ^{b}	F = 60.74, p < 0.001, p^* = 0.0006
14	297.2 ± 6.854 ^{a}	353.0 ± 12.70 ^{b}	326.2 ± 6.826	F = 9.159, p = 0.0002, p^* = 0.0012
21	614.7 ± 13.50 ^{a}	750.9 ± 12.82 ^{b}	755.7 ± 16.82 ^{b}	F = 22.41, p < 0.000, p^* = 0.0006
28	1021 ± 26.79 ^{a}	1223 ± 32.23 ^{b}	1269 ± 26.19 ^{b}	F = 20.97, p < 0.000, p^* = 0.0006
35	1741 ± 31.1 ^{a}	1839 ± 32.88 ^{b}	1955 ± 32.84 ^{c}	F = 10.98, p < 0.000, p^* = 0.006

Data are reported as mean ± standard error of the mean. Different pairs of superscript letters on the same row indicate significant differences found at the post hoc multiple comparisons test. C: control group, MDV/100 µL; P1: MDV + 1 × 10^5 CFU/100 µL of probiotic mixture; P2: MDV + 1 × 10^6 CFU/100 µL of probiotic mixture. F: result of one-way ANOVA test between groups; p: p-value; p^*: adjusted p-value with Bonferroni–Šidák correction.

Growth performances (BWG, FI and FCR) are listed in Table 3, with significant differences evident mainly until 21 days of age, for all the parameters.

Table 3. Effect of in ovo probiotic supplementation on body weight gain, feed intake and feed conversion ratios.

	C	P1	P2	Statistical Data
Body Weight Gain (g)				
Day of hatch to 7 days	58.22 ± 2.552 [a]	110.8 ± 3.162 [b]	100.9 ± 4.260 [b]	F = 67.49, p < 0.0001, p* = 0.0005
7 to 14 days	192.7 ± 2.503	196.4 ± 0.5508 [a]	179.3 ± 5.345 [b]	F = 6.941, p = 0.0275, p* = 0.1301
14 to 21 days	317.4 ± 10.38 [a]	397.9 ± 10.90 [b]	429.4 ± 15.48 [b]	F = 21.46, p = 0.0018, p* = 0.0090
21 to 28 days	404.2 ± 42.55	472.2 ± 22.96	513.0 ± 55.39	F = 1.678, p = 0.2637, p* = 0.7836
28 to 35 days	723.3 ± 100.6	616.1 ± 16.04	686.5 ± 21.52	F = 0.8228, p = 0.4833, p* = 0.9632
Feed intake (g)				
Day of hatch to 7 days	123.8 ± 1.618 [b]	104.7 ± 6.819 [a]	131.3 ± 2.179 [b]	F = 10.47, p = 0.0111, p* = 0.0543
7 to 14 days	283.1 ± 6.823 [a]	238.7 ± 1.953 [b]	304.9 ± 0.7796 [c]	F = 66.84, p < 0.0001, p* = 0.0005
14 to 21 days	370.5 ± 11.66 [a]	488.3 ± 2.998 [b]	508.1 ± 4.070 [b]	F = 102.9, p < 0.0001, p* = 0.0005
21 to 28 days	816.3 ± 79.77	836.5 ± 1.770	830.5 ± 12.73	F = 0.0498, p = 0.9518, p* = 1.0
28 to 35 days	1111 ± 17.92	1121 ± 17.14	1062 ± 8.906	F = 4.340, p = 0.0683, p* = 0.2979
Feed conversion ratio				
Day of hatch to 7 days	2.136 ± 0.1189 [a]	0.9495 ± 0.08403 [b]	1.308 ± 0.07395 [c]	F = 41.70, p = 0.0003, p* = 0.0018
7 to 14 days	1.469 ± 0.02469 [a]	1.216 ± 0.01235 [ac]	1.704 ± 0.05283 [bc]	F = 50.32, p = 0.0002, p* = 0.0012
14 to 21 days	1.168 ± 0.02550	1.229 ± 0.02612	1.186 ± 0.03364	F = 1.197, p = 0.3651, p* = 0.9345
21 to 28 days	2.079 ± 0.3462	1.779 ± 0.07964	1.657 ± 0.1788	F = 0.8944, p = 0.4571, p* = 0.9744
28 to 35 days	1.608 ± 0.2569	1.822 ± 0.06035	1.550 ± 0.05723	F = 0.8442, p = 0.4753, p* = 0.9791
Day of hatch to 35 days	1.603 ± 0.1012	1.556 ± 0.0214	1.490 ± 0.05435	F = 0.7098, p = 0.5288, p* = 0.9891

Data are reported as the mean ± standard error of the mean. Different pairs of superscript letters on the same row indicate significant differences found at the post hoc multiple comparisons test. C: control group, MDV/100 µL; P1: MDV + 1 × 10^5 CFU/100 µL of probiotic mixture; P2: MDV + 1 × 10^6 CFU/100 µL of probiotic mixture. F: result of one-way ANOVA test between groups; p: p-value; p*: adjusted p-value with Bonferroni–Šidák correction.

3.3. Intestinal Morphometry

The results of morphometric analysis of the different intestinal tracts are shown in Table 4. Briefly, in the duodenum all the parameters showed no significant differences ($p > 0.05$). In the jejunum, the P2 group presented an increase in VW, CD and VSA compared both to C and P1 and a reduced VH/CD ratio compared with P1. In the ileum, probiotic treatment increased VH and reduced VW especially in P1, compared to both of the other groups ($p < 0.01$). Lastly, cecal CD and LPW of the C group are increased compared to both p groups ($p < 0.01$).

Table 4. Morphometric measurement of different intestinal tracts (µm) at 35 days of age.

	C	P1	P2	Statistical Data
DUODENUM				
villus height (VH)	1349 ± 60.29	1448 ± 71.90	1469 ± 33.53	F = 1.238, p = 0.3078
villus width (VW)	117.3 ± 6.896	108.4 ± 5.112	100.3 ± 4.447	F = 2.339, p = 0.1180
crypt depth (CD)	196.7 ± 13.13	194.2 ± 11.00	223.7 ± 13.50	F = 1.695, p = 0.2048
lamina propria width (LPW)	209.7 ± 13.60	230.2 ± 17.58	256.8 ± 16.18	F = 2.210, p = 0.1316
villus surface area (VSA)	498.4 ± 39.41	493.3 ± 33.62	463.9 ± 25.54	F = 0.3119, p = 0.7350
VH/CD ratio	7.278 ± 0.7143	7.633 ± 0.4124	6.743 ± 0.2921	F = 0.7862, p = 0.4670
JEJUNUM				
villus height (VH)	946.3 ± 46.84	1035 ± 71.37	1139 ± 61.47	F = 2.521, p = 0.1014
villus width (VW)	115.3 ± 5.855 [a]	121.1 ± 4.698 [b]	126.6 ± 8.130 [b]	F = 3.991, p = 0.0319
crypt depth (CD)	200.6 ± 9.506	186.7 ± 14.48 [a]	250.3 ± 18.79 [b]	F = 5.136, p = 0.0139
lamina propria width (LPW)	248.8 ± 13.38 [a]	250.9 ± 20.71	322.1 ± 26.04 [b]	F = 4.063, p = 0.0302
villus surface area (VSA)	339.6 ± 18.59 [a]	397.5 ± 26.82 [b]	449.9 ± 34.70 [b]	F = 6.005, p = 0.0077
VH/CD ratio	4.870 ± 0.1904 [b]	5.783 ± 0.2950 [a]	4.762 ± 0.3480 [b]	F = 3.865, p = 0.0351
ILEUM				
villus height (VH)	613.5 ± 27.26 [b]	754.2 ± 33.76 [a]	648.4 ± 25.60 [b]	F = 6.347, p = 0.0061
villus width (VW)	130.7 ± 7.242 [b]	101.2 ± 2.998 [a]	132.5 ± 5.500 [b]	F = 10.10, p = 0.0007
crypt depth (CD)	165.1 ± 8.229	160.3 ± 8.542	154.5 ± 6.815	F = 0.4551, p = 0.6398
lamina propria width (LPW)	208.9 ± 10.03	190.2 ± 13.36	181.3 ± 6.968	F = 1.823, p = 0.1832
villus surface area (VSA)	248.8 ± 11.01	240.6 ± 16.11	207.6 ± 16.68	F = 1.094, p = 0.3511
VH/CD ratio	3.869 ± 0.33	4.827 ± 0.2521	4.345 ± 0.2991	F = 2.628, p = 0.0929
CECUM				
villus height (VH)	338.2 ± 21.38	289.7 ± 14.00	344.6 ± 24.44	F = 2.159, p = 0.1373
villus width (VW)	149.5 ± 6.602	137.5 ± 9.463	169.3 ± 33.93	F = 0.6010, p = 0.5563
crypt depth (CD)	160.5 ± 15.07 [a]	113.8 ± 2.500 [b]	129.2 ± 6.305	F = 6.225, p = 0.0066
lamina propria width (LPW)	218.1 ± 13.26 [a]	158.0 ± 6.118 [b]	160.6 ± 10.53 [b]	F = 10.67, p = 0.0005
villus surface area (VSA)	160.4 ± 15.27	126.2 ± 13.26	183.8 ± 40.34	F = 1.237, p = 0.3081
VH/CD ratio	2.245 ± 0.2089	2.566 ± 0.08593	2.707 ± 0.1237	F = 2.535, p = 0.1003

Data are reported as the mean ± standard error of the mean. The statistical significance was set at $p < 0.05$. Different pairs of superscript letters on the same row indicate significant differences found at the post hoc multiple comparisons test. C: control group, MDV/100 µL; P1: MDV + 1 × 10^5 CFU/100 µL of probiotic mixture; P2: MDV + 1 × 10^6 CFU/100 µL of probiotic mixture; p: p-value.

3.4. Oocyst Shedding and Histological Lesions Score

Oocyst count in faeces showed a significant increase in C at sampling performed at 14, 21 and 28 days of age. No significant differences were observed at the end of the cycle (Figure 1).

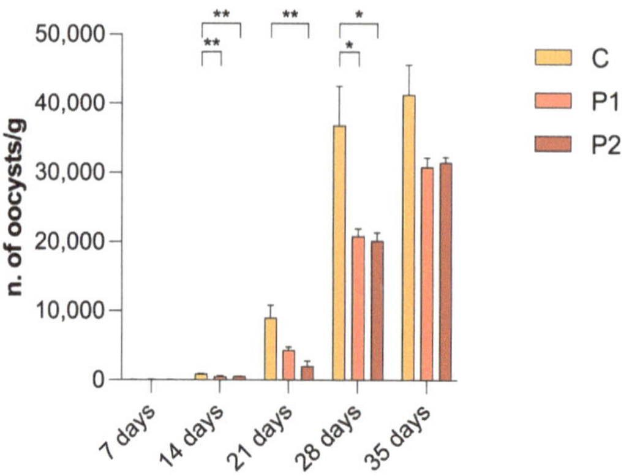

Figure 1. Mean number (and standard error) of coccidian oocysts per gram of faeces in the three groups; C: control group, MDV/100 μL; P1: MDV + 1 × 10^5 CFU/100 μL of probiotic mixture; P2: MDV + 1 × 10^6 CFU/100 μL of probiotic mixture. *p*-values, *: $p < 0.05$; **: $p < 0.01$.

At histological examination, the presence of *Eimeria* spp. was only revealed in the duodenum and jejunum of P1 and P2, with a median score of 1 (Figures 2 and 3). Results of enterocyte proliferation and lamina propria inflammation are described in Figure 3.

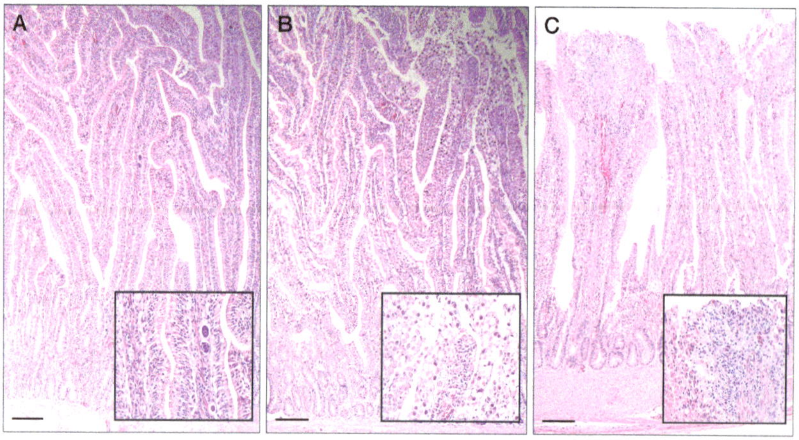

Figure 2. Coccidian infection in duodenum of broiler chickens. (**A**) In P1, mild infiltration of zygotes/macrogametes in the lamina propria was observed (insert), associated with the absence of lamina propria inflammation or enterocyte proliferation. (**B**) In P2, severe localised area of cellular swelling and hyperplasia (insert) of duodenal villous epithelium and lamina propria by zygotes/macrogametes was noted, associated with mild lamina propria inflammation. (**C**) In C, severe tissue damage was recorded, characterised by some areas of denudation of the villus apex due to apoptosis/rupture of the enterocytes (insert), associated with an obvious inflammation of the lamina propria, compatible with a strong proliferation of coccidia and the passage of zygotes/macrogametes in the faecal content immediately before slaughtering. Hematoxylin & Eosin, scale bar = 200 μm.

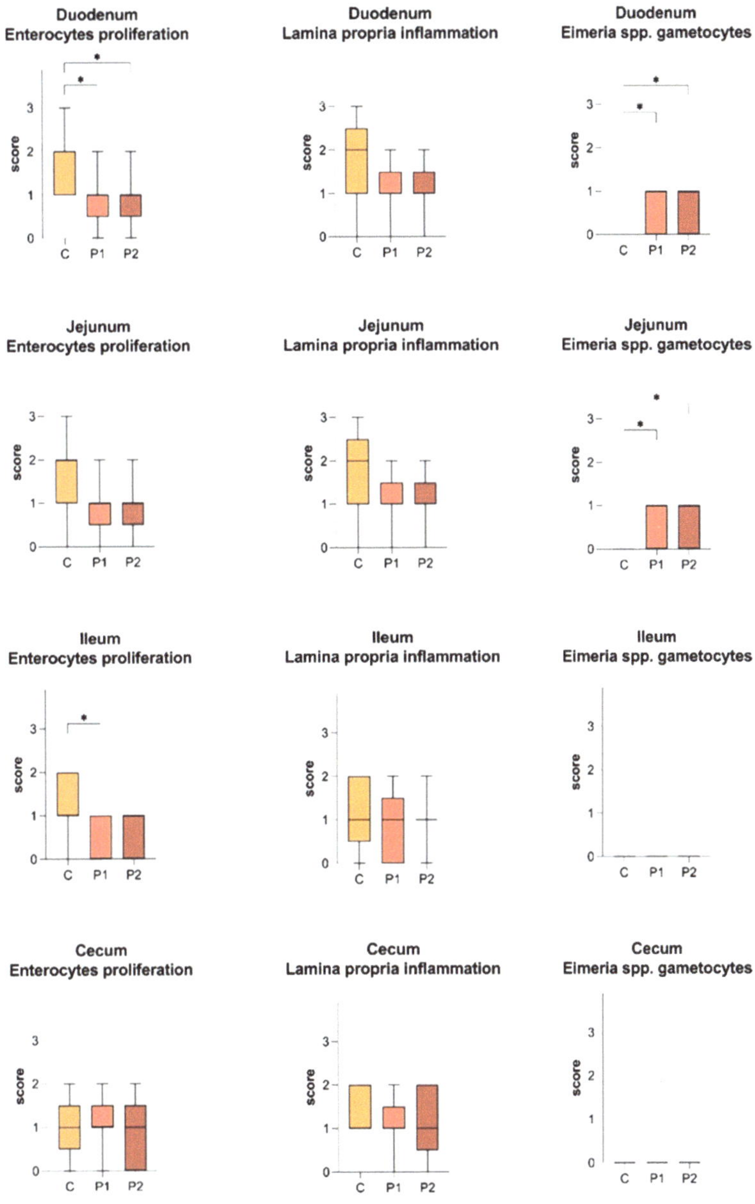

Figure 3. Boxplot showing the results of the scoring system detected in different segments of the intestinal tract (duodenum, jejunum, ileum and cecum) in chickens slaughtered at 35 days of age. C: control group, MDV/100 µL; P1: MDV + 1 × 10^5 CFU/100 µL of probiotic mixture; P2: MDV + 1 × 10^6 CFU/100 µL of probiotic mixture. The ends of the whiskers show minimum and maximum score values; boxes show the median, the first and the third quartile. Asterisks indicate significant differences between groups. p-values, *: $p < 0.05$.

4. Discussion

This study aimed to investigate the effect of the in ovo administration of two different concentrations of the probiotic mixture Slab51. This multi-strain probiotic has already been successfully tested as a feed supplement in avian species [47], but it is likely that not all probiotic bacteria commonly used as feed supplements are suitable also for in ovo injection. Therefore, a study was needed to evaluate the optimal concentration that could be safe for in ovo use and to assess the main effects of this administration over a 35-day life cycle.

In this trial, both the in ovo procedure and the probiotic supplementation in the amnios did not affect hatchability compared to the control group. As shown by previous works, no significant differences between controls and probiotic groups were found in the vast majority of studies reporting hatching rate [7,27,28,37,48–52], with a few reporting a negative hatching percentage [26,49]. These different results can be attributed to several influencing factors related to both the in ovo technique (volume of the injected solution, inoculation site, embryonic day of incubation and dilution vehicle) and to the probiotic characteristics (e.g., strain and concentration) [53]. For example, *Lactobacillus* spp.-based probiotics have already been proven to be safer for in ovo administration than those based on *Bacillus* spp. [49] due to a possible competition for nutrients of certain bacteria with developing embryonic cells. Most studies, including the present one, evaluated inoculation on incubation day 17 or 18, by which time most embryonic development has occurred, which allows both vaccination and in ovo feeding to be performed with the same injection, thus avoiding the increased risk of contamination due to the double injections necessary in the case of early administration (e.g., embryonic day 12). Concerning this trial, the combination of the probiotic with an MDV could represent another possible risk factor. Marek's Disease is a lymphoproliferative disease of domestic chickens caused by an oncogenic α-herpesvirus with lymphotropic properties of gamma-herpesviruses, associated with lymphomas, neurological manifestations and immunosuppression [54], representing a major concern to the poultry industry. In ovo vaccination against this virus represents the only effective control method, due to the abundance and stability of the virus in the environment [55,56]. Previous studies showed that there is no negative effect on hatching rate after the combination of the vaccine with other compounds, such as probiotics. On the contrary, the early administration of the probiotic can promote the development of immunity [27,28]. Our results showed a favourable response to vaccination in all three groups suggesting both that there was no negative effect on MD vaccine efficacy following probiotic administration and that there was no direct stimulation of in ovo-administered probiotics in the production of Marek disease antibodies.

In terms of hatching weight, no significant differences between the different groups were observed in our work. Previous studies indicated that in ovo administration of *Bacillus subtilis* combined with *Bacillus amyloliquefaciens* [50], lactic acid bacteria [51], *Lactobacillus plantarum* + raffinose [26] and *Lactobacillus lactis* + *Bacillus subtilis* [57] resulted in an increased hatching weight, while others observed a reduction after the injection with *Lactobacillus animalis* + *Bifidobacterium animalis* or *Lactobacillus plantarum* + *Lactobacillus salivarius* [49,52]. However, the majority of the data reported are consistent with no significant results [7,27,28,48,49,58].

Among the after-hatching parameters, BW increased significantly in the two probiotic-injected groups throughout this study, with a gain of more than 100 g at 35 days of age compared with C. This stimulatory effect on body weight also emerges from a comprehensive review regarding the use of probiotics in ovo [53], and it is reported as an overall improvement in zootechnical performance (body weight, feed consumption and FCR) [26,28,57,58]. In fact, body weight data are often correlated with both feed consumption and intestinal morphology. For this specific probiotic mixture, dietary supplementation of Slab51 in Guinea fowls improved both BW and FCR [47]. Similarly, in this study, FCR is reduced in probiotic-treated chicken, with values of FI and FCR higher in the C group until 14 days of age, without statistically significant differences in the second part of the trial (14–35 days). All these data are consistent with the BW, indicating a better absorption capacity in the probiotic-treated groups. At the end of this study, P1 and P2 reached the higher mean weight with a lower total FCR, in line with other studies which applied in ovo probiotic supplementation [36,50,58,59].

Among the various factors influenced by the administration of probiotics that may explain the positive effects on animal performance are those related to intestinal morphology. The assessment of intestinal morphology is considered a reliable indicator of gut functionality, particularly in terms of the absorptive surface and subsequent transformation of nutrients [60], with studies describing both positive [47,61] or slight/absent [62,63] effects of oral probiotic supplementation on these parameters. With regard to in ovo administration, an improvement in gut morphology was observed in this study, which is consistent with previous data [27,28,35,48]. The variable extent of this improvement is attributable to sampling age, in relation to which a proportional increase occurs which makes it impossible to compare net data in relation to villus length or other parameters (e.g CD and VW). In our study, histomorphometric analysis confirmed that the use of probiotics induced an increase in villus height, with a gain of over 100 μm compared to group C, from the duodenum to ileum. Although not statistically significant, this difference observed at 35 days of age is a remarkable result given that it is achieved by a single administration during incubation and without any boost during the breeding period. In the jejunum, which is the main site of nutrient absorption, there was also an increase in VSA, further highlighting the improvement in the digestive capacity of the treated subjects. In addition, the HE analysis showed that the jejunum in the P1 and P2 groups exhibited intact histological structure, orderly arrangement, and well-grown intestinal villi with no obvious tissue damage and pathological changes, while the ileum in the C group appeared to be in the process of some structural alterations/lower VSA. Previously, Kim et al. (2012) [64] also reported that the administration of multi-microbe probiotic products increased the ratio of villus height to crypt depth in the jejunum. Consistently, this study provided evidence supporting the notion that supplementation with a multi-species probiotic mixture improves the intestinal structure by increasing the villus length. Other significant differences in the parameters evaluated for intestinal morphology were evidenced in this study, but without homogeneous results. It can be assumed that all these morphological changes are not only a direct effect of the interaction of probiotic bacteria with the intestinal mucosa, but also a result of the positive manipulation of intestinal microbiota by probiotic bacteria involved in the metabolism of nutrients as well as in the intestinal morphological development [65,66]. The gut microbiota constitutes the biological barrier that prevents pathogens from colonising the intestine and contributes to the maturation of the gastrointestinal tract and immune system [67]. The physical barrier of the intestinal epithelium, and especially the intestinal tight junction, confers the direct property of selective permeability to the gut [68]. In newly hatched chickens, the gastrointestinal tract is structurally and functionally immature, and maturation is induced by many factors, one of which is the presence and composition of gut microbiota. In both mammals and birds, studies have shown that the gut microbiota has a significant effect on gastrointestinal tract development: gut villus architecture, crypt depth, stem cell proliferation, blood vessel density, mucus layer properties and maturation of mucosal-associated lymphoid tissues are reduced in germ-free animals compared to those that are conventionally reared. Among the various stimulating factors produced by microbiota, in birds, enterocyte proliferation has been particularly linked to bacterial fermentation and the production of short-chain fatty acids (SCFAs), which are essential for enterocyte development and proliferation. [69–72]. Early intestinal colonisation with a positive microflora can, therefore, promote the proper maturation of the various intestinal components, including the related mucosal immune system, especially when this stimulation occurs before hatching in the final stage of incubation when organ development is almost complete.

Among the various factors that can negatively influences animal growth performance, there is *Eimeria* spp. infection. Many different probiotic compounds administered through the diet have been shown to provide a protective effect in chickens that were naturally infected or challenged with different *Eimeria* species [73–76]. Whereas, to date, the beneficial effects of probiotic bacteria administered in ovo to reduce the impact of coccidia on the intestinal mucosa have only been demonstrated in a limited number of studies in chickens and associated to exclusive probiotics effect [36,77] or to their combination with a coccidiosis vaccine [78]. In the present study, it was hypothesised that the beneficial effects of in ovo administration of a multi-strain probiotic mixture would lead to a stabilisation of the gut microflora by competitive exclusion of their pathogenic counterparts and, in addition, that the treatment would not lead to an increase in the severity of the disease induced by natural infection with *Eimeria* species, favouring, on the contrary, a reduction in clinical signs due to its beneficial effect on the development of the gut microbiota and immunity. One of the main mechanisms of action of probiotics is the challenge for receptor sites, which prevents the perforation and secretion of *Eimeria* sporozoites into the intestinal mucosa, resulting in reduced intestinal damage, proliferation and oocyst shedding [79]. Competition for receptors is even more limited in the case of IOF as the administration of the probiotic occurs before possible contact with the pathogen, which will, therefore, have limited receptor availability. In our study, from 14 days of age until the end of the trial, oocyst shedding was significantly reduced following in ovo probiotic supplementation, with no relevant differences among P1 and P2. The severity of the infection was assessed by histological evaluation of various parameters [80], and, in all the groups, the infection was limited to the upper gastrointestinal tract, without histological lesions in the ileum or caecum. The involvement of specific intestinal tracts depends on the specificity for the site of infection of the different *Eimeria* species that could be naturally present in the facility. Interestingly, in group C, which had the higher number of oocyst *per gram* of faeces, no macrogametes were histologically detected. This could be attributed to a different stage of the parasite life cycle, with oocysts being released from the ruptured epithelial cells and shed into the environment with the faeces at the time of organ sampling. In addition, although the protective mechanisms of probiotics are not fully understood, it has been shown that probiotics can significantly enhance mucosal-associated immune responses, increase the production of anti-Eimeria antibodies and reduce oocyst shedding [81]. Some studies have shown that the presence of IgA antibodies in the intestinal mucus, directed against some sporozoite-associated antigens, is able to modulate the degree of infection by coccidia and delay the release of oocysts [82]. In this case, it can be postulated that early intestinal colonisation with multi-species probiotic bacteria increased non-specific and specific mucosal defences, as well as the production of specific antibodies against protozoan antigens, leading to a delay in the sexual phase of the *Eimeria* species cycle and the release of oocysts, thus reducing the impact of coccidia on intestinal morphology. The results of the scores suggest a condition of severe inflammation of the lamina propria associated with enterocyte proliferation in the C group, especially in the duodenum, serving as further evidence of improved gut health and control of coccidian infection in probiotics-treated chickens. However, these histological changes could also be related to other causes of inflammation than just coccidian infection. In the caecum of group C, both the attributed scores and the histomorphometric measurements showed an increased enterocytic proliferation, CD, LPW and VW, suggesting a condition of widespread inflammation, with an increasing cellular turnover which is essential to replace the cellular damage.

The concentration of probiotics was chosen based on previous works that tested different compounds and concentrations for in ovo administration. In this study, no clear differences were observed on chicken's performance or coccidian infection between P1 and P2; however, a slight reduction in hatchability percentage was observed in P2. Consequently, the concentration of 1×10^5 CFU/100 µL could be more suitable for in ovo administration, even if further studies on an industrial scale are needed.

5. Conclusions

In conclusion, our data demonstrate that in ovo supplementation with Slab51 can be considered safe, and its association with MDV does not affect hatchability. Despite the lack of information on the microbiota composition of the animals included in this study, the efficacy of supplementation is demonstrated by the effects observed on productive performances, histological parameters and coccidian resistance, which were improved by the in ovo treatment.

Author Contributions: Conceptualisation, L.B., L.G. and G.R.; methodology, A.M.T., A.R., D.D.B., S.P. and A.-R.A.; formal analysis, L.B., L.G., D.D.B. and S.B.; investigation, L.B., L.G. and A.M.T.; data curation, L.B.; writing—original draft preparation, L.B., L.G. and A.M.T.; writing—review and editing, A.R., D.D.B., S.P., A.-R.A., S.B. and G.R.; supervision, G.R. All authors have read and agreed to the published version of the manuscript.

Funding: This research received no external funding.

Institutional Review Board Statement: This study was approved by the Animal Welfare Body of the University of Camerino (protocol n. 2/2024).

Informed Consent Statement: Not applicable.

Data Availability Statement: All data are included in this manuscript.

Conflicts of Interest: The authors declare no conflicts of interest.

References

1. Oliveira, J.E.D.; Uni, Z.; Ferket, P.R. Important Metabolic Pathways in Poultry Embryos Prior to Hatch. *Worlds Poult. Sci. J.* **2008**, *64*, 488–499. [CrossRef]
2. Omede, A.A.; Bhuiyan, M.M.; Lslam, A.F.; Iji, P.A. Physico-Chemical Properties of Late-Incubation Egg Amniotic Fluid and a Potential In Ovo Feed Supplement. *Asian-Australas. J. Anim. Sci.* **2017**, *30*, 1124–1134. [CrossRef] [PubMed]
3. Geyra, A.; Uni, Z.; Sklan, D. The Effect of Fasting at Different Ages on Growth and Tissue Dynamics in the Small Intestine of the Young Chick. *Br. J. Nutr.* **2001**, *86*, 53–61. [CrossRef] [PubMed]
4. Givisiez, P.E.N.; Moreira Filho, A.L.B.; Santos, M.R.B.; Oliveira, H.B.; Ferket, P.R.; Oliveira, C.J.B.; Malheiros, R.D. Chicken Embryo Development: Metabolic and Morphological Basis for In Ovo Feeding Technology. *Poult. Sci.* **2020**, *99*, 6774–6782. [CrossRef]
5. Steel, J.; Burmakina, S.V.; Thomas, C.; Spackman, E.; García-Sastre, A.; Swayne, D.E.; Palese, P. A Combination In-Ovo Vaccine for Avian Influenza Virus and Newcastle Disease Virus. *Vaccine* **2008**, *26*, 522–531. [CrossRef]
6. Seal, B.S.; King, D.J.; Sellers, H.S. The Avian Response to Newcastle Disease Virus. *Dev. Comp. Immunol.* **2000**, *24*, 257–268. [CrossRef]
7. De Oliveira, J.E.; van der Hoeven-Hangoor, E.; van de Linde, I.B.; Montijn, R.C.; van der Vossen, J.M.B.M. In Ovo Inoculation of Chicken Embryos with Probiotic Bacteria and Its Effect on Posthatch *Salmonella* Susceptibility. *Poult. Sci.* **2014**, *93*, 818–829. [CrossRef]
8. Lee, S.; La, T.-M.; Lee, H.-J.; Choi, I.-S.; Song, C.-S.; Park, S.-Y.; Lee, J.-B.; Lee, S.-W. Characterization of Microbial Communities in the Chicken Oviduct and the Origin of Chicken Embryo Gut Microbiota. *Sci. Rep.* **2019**, *9*, 6838. [CrossRef]
9. Ding, J.; Dai, R.; Yang, L.; He, C.; Xu, K.; Liu, S.; Zhao, W.; Xiao, L.; Luo, L.; Zhang, Y. Inheritance and Establishment of Gut Microbiota in Chickens. *Front. Microbiol.* **2017**, *8*, 1967. [CrossRef]
10. Shterzer, N.; Rothschild, N.; Sbehat, Y.; Stern, E.; Nazarov, A.; Mills, E. Large Overlap Between the Intestinal and Reproductive Tract Microbiomes of Chickens. *Front. Microbiol.* **2020**, *11*, 1508. [CrossRef]
11. Akinyemi, F.T.; Ding, J.; Zhou, H.; Xu, K.; He, C.; Han, C.; Zheng, Y.; Luo, H.; Yang, K.; Gu, C.; et al. Dynamic Distribution of Gut Microbiota During Embryonic Development in Chicken. *Poult. Sci.* **2020**, *99*, 5079–5090. [CrossRef] [PubMed]
12. Cisek, A.A.; Binek, M. Chicken Intestinal Microbiota Function with a Special Emphasis on the Role of Probiotic Bacteria. *Pol. J. Vet. Sci.* **2014**, *17*, 385–394. [CrossRef] [PubMed]
13. De Mesquita Souza Saraiva, M.; Lim, K.; do Monte, D.F.M.; Givisiez, P.E.N.; Alves, L.B.R.; de Freitas Neto, O.C.; Kariuki, S.; Júnior, A.B.; de Oliveira, C.J.B.; Gebreyes, W.A. Antimicrobial Resistance in the Globalized Food Chain: A One Health Perspective Applied to the Poultry Industry. *Braz. J. Microbiol.* **2022**, *53*, 465–486. [CrossRef] [PubMed]
14. FAO. *Meat Market Review: Emerging Trends and Outlook, 2023*; Meat Market Review; FAO: Rome, Italy, 2023.
15. Abd El-Hack, M.E.; El-Saadony, M.T.; Salem, H.M.; El-Tahan, A.M.; Soliman, M.M.; Youssef, G.B.A.; Taha, A.E.; Soliman, S.M.; Ahmed, A.E.; El-kott, A.F.; et al. Alternatives to Antibiotics for Organic Poultry Production: Types, Modes of Action and Impacts on Bird's Health and Production. *Poult. Sci.* **2022**, *101*, 101696. [CrossRef]
16. Muaz, K.; Riaz, M.; Akhtar, S.; Park, S.; Ismail, A. Antibiotic Residues in Chicken Meat: Global Prevalence, Threats, and Decontamination Strategies: A Review. *J. Food Prot.* **2018**, *81*, 619–627. [CrossRef]
17. Uni, Z.; Ferket, P.R. Enhancement of Development of Oviparous Species by In Ovo Feeding. European Patent EP1307230A2, 7 May 2003.
18. Uni, Z.; Ferket, P.R.; Tako, E.; Kedar, O. In Ovo Feeding Improves Energy Status of Late-Term Chicken Embryos. *Poult. Sci.* **2005**, *84*, 764–770. [CrossRef]

19. Das, R.; Mishra, P.; Jha, R. In Ovo Feeding as a Tool for Improving Performance and Gut Health of Poultry: A Review. *Front. Vet. Sci.* **2021**, *8*, 754246. [CrossRef]
20. Hou, T.; Tako, E. The In Ovo Feeding Administration (Gallus Gallus)—An Emerging In Vivo Approach to Assess Bioactive Compounds with Potential Nutritional Benefits. *Nutrients* **2018**, *10*, 418. [CrossRef]
21. Fuller, R. Probiotics in Man and Animals. *J. Appl. Bacteriol.* **1989**, *66*, 365–378.
22. Clavijo, V.; Flórez, M.J.V. The Gastrointestinal Microbiome and Its Association with the Control of Pathogens in Broiler Chicken Production: A Review. *Poult. Sci.* **2018**, *97*, 1006–1021. [CrossRef]
23. Grozina, A.A.; Ilina, L.A.; Laptev, G.Y.; Yildirim, E.A.; Ponomareva, E.S.; Filippova, V.A.; Tyurina, D.G.; Fisinin, V.I.; Kochish, I.I.; Griffin, D.K.; et al. Probiotics as an Alternative to Antibiotics in Modulating the Intestinal Microbiota and Performance of Broiler Chickens. *J. Appl. Microbiol.* **2023**, *134*, lxad213. [CrossRef] [PubMed]
24. He, T.; Long, S.; Mahfuz, S.; Wu, D.; Wang, X.; Wei, X.; Piao, X. Effects of Probiotics as Antibiotics Substitutes on Growth Performance, Serum Biochemical Parameters, Intestinal Morphology, and Barrier Function of Broilers. *Animals* **2019**, *9*, 985. [CrossRef] [PubMed]
25. Shaufi, M.A.M.; Sieo, C.C.; Chong, C.W.; Geok Hun, T.; Omar, A.R.; Han Ming, G.; Wan Ho, Y. Effects of Phage Cocktail, Probiotics, and Their Combination on Growth Performance and Gut Microbiota of Broiler Chickens. *Animals* **2023**, *13*, 1328. [CrossRef] [PubMed]
26. Dunislawska, A.; Slawinska, A.; Stadnicka, K.; Bednarczyk, M.; Gulewicz, P.; Jozefiak, D.; Siwek, M. Synbiotics for Broiler Chickens—In Vitro Design and Evaluation of the Influence on Host and Selected Microbiota Populations Following In Ovo Delivery. *PLoS ONE* **2017**, *12*, e0168587. [CrossRef]
27. Teague, K.D.; Graham, L.E.; Dunn, J.R.; Cheng, H.H.; Anthony, N.; Latorre, J.D.; Menconi, A.; Wolfenden, R.E.; Wolfenden, A.D.; Mahaffey, B.D.; et al. In Ovo Evaluation of FloraMax®-B11 on Marek's Disease HVT Vaccine Protective Efficacy, Hatchability, Microbiota Composition, Morphometric Analysis, and Salmonella Enteritidis Infection in Broiler Chickens. *Poult. Sci.* **2017**, *96*, 2074–2082. [CrossRef]
28. Castañeda, C.D.; Dittoe, D.K.; Wamsley, K.G.; McDaniel, C.D.; Blanch, A.; Sandvang, D.; Kiess, A.S. In Ovo Inoculation of an Enterococcus Faecium–Based Product to Enhance Broiler Hatchability, Live Performance, and Intestinal Morphology. *Poult. Sci.* **2020**, *99*, 6163–6172. [CrossRef]
29. Miśta, D.; Króliczewska, B.; Pecka-Kiełb, E.; Kapuśniak, V.; Zawadzki, W.; Graczyk, S.; Kowalczyk, A.; Łukaszewicz, E.; Bednarczyk, M. Effect of In Ovo Injected Prebiotics and Synbiotics on the Caecal Fermentation and Intestinal Morphology of Broiler Chickens. *Anim. Prod. Sci.* **2016**, *57*, 1884–1892. [CrossRef]
30. Oladokun, S.; Adewole, D. The Effect of Bacillus Subtilis and Its Delivery Route on Hatch and Growth Performance, Blood Biochemistry, Immune Status, Gut Morphology, and Microbiota of Broiler Chickens. *Poult. Sci.* **2023**, *102*, 102473. [CrossRef]
31. Alizadeh, M.; Shojadoost, B.; Astill, J.; Taha-Abdelaziz, K.; Karimi, S.H.; Bavananthasivam, J.; Kulkarni, R.R.; Sharif, S. Effects of In Ovo Inoculation of Multi-Strain Lactobacilli on Cytokine Gene Expression and Antibody-Mediated Immune Responses in Chickens. *Front. Vet. Sci.* **2020**, *7*, 105. [CrossRef]
32. Madej, J.P.; Bednarczyk, M. Effect of In Ovo-Delivered Prebiotics and Synbiotics on the Morphology and Specific Immune Cell Composition in the Gut-Associated Lymphoid Tissue. *Poult. Sci.* **2016**, *95*, 19–29. [CrossRef]
33. Sławinska, A.; Siwek, M.Z.; Bednarczyk, M.F. Effects of Synbiotics Injected In Ovo on Regulation of Immune-Related Gene Expression in Adult Chickens. *Am. J. Vet. Res.* **2014**, *75*, 997–1003. [CrossRef] [PubMed]
34. Broom, L.J.; Kogut, M.H. The Role of the Gut Microbiome in Shaping the Immune System of Chickens. *Vet. Immunol. Immunopathol.* **2018**, *204*, 44–51. [CrossRef] [PubMed]
35. Oladokun, S.; Koehler, A.; MacIsaac, J.; Ibeagha-Awemu, E.M.; Adewole, D.I. Bacillus Subtilis Delivery Route: Effect on Growth Performance, Intestinal Morphology, Cecal Short-Chain Fatty Acid Concentration, and Cecal Microbiota in Broiler Chickens. *Poult. Sci.* **2021**, *100*, 100809. [CrossRef] [PubMed]
36. Pender, C.M.; Kim, S.; Potter, T.D.; Ritzi, M.M.; Young, M.; Dalloul, R.A. Effects of In Ovo Supplementation of Probiotics on Performance and Immunocompetence of Broiler Chicks to an Eimeria Challenge. *Benef. Microbes* **2016**, *7*, 699–705. [CrossRef]
37. Pender, C.M.; Kim, S.; Potter, T.D.; Ritzi, M.M.; Young, M.; Dalloul, R.A. In Ovo Supplementation of Probiotics and Its Effects on Performance and Immune-Related Gene Expression in Broiler Chicks. *Poult. Sci.* **2017**, *96*, 1052–1062. [CrossRef]
38. McDougald, L.R.; Cervantes, H.M.; Jenkins, M.C.; Hess, M.; Beckstead, R. Protozoal Infections. In *Diseases of Poultry*; Wiley: Hoboken, NJ, USA, 2020; pp. 1192–1254.
39. Biagini, L.; Galosi, L.; Roncarati, A.; Attili, A.-R.; Mangiaterra, S.; Rossi, G. The Role of Nutraceuticals and Phytonutrients in Chickens' Gastrointestinal Diseases. *Animals* **2022**, *12*, 892. [CrossRef]
40. Lee, Y.; Lu, M.; Lillehoj, H.S. Coccidiosis: Recent Progress in Host Immunity and Alternatives to Antibiotic Strategies. *Vaccines* **2022**, *10*, 215. [CrossRef]
41. Saeed, M.; Babazadeh, D.; Naveed, M.; Alagawany, M.; Abd El-Hack, M.E.; Arain, M.A.; Tiwari, R.; Sachan, S.; Karthik, K.; Dhama, K.; et al. In Ovo Delivery of Various Biological Supplements, Vaccines and Drugs in Poultry: Current Knowledge. *J. Sci. Food Agric.* **2019**, *99*, 3727–3739. [CrossRef]
42. Yitbarek, A.; Rodriguez-Lecompte, J.C.; Echeverry, H.M.; Munyaka, P.; Barjesteh, N.; Sharif, S.; Camelo-Jaimes, G. Performance, Histomorphology, and Toll-like Receptor, Chemokine, and Cytokine Profile Locally and Systemically in Broiler Chickens Fed Diets Supplemented with Yeast-Derived Macromolecules. *Poult. Sci.* **2013**, *92*, 2299–2310. [CrossRef]

43. Biloni, A.; Quintana, C.F.; Menconi, A.; Kallapura, G.; Latorre, J.; Pixley, C.; Layton, S.; Dalmagro, M.; Hernandez-Velasco, X.; Wolfenden, A. Evaluation of Effects of EarlyBird Associated with FloraMax-B11 on *Salmonella* Enteritidis, Intestinal Morphology, and Performance of Broiler Chickens. *Poult. Sci.* **2013**, *92*, 2337–2346. [CrossRef]
44. Murakami, A.E.; Sakamoto, M.I.; Natali, M.R.M.; Souza, L.M.G.; Franco, J.R.G. Supplementation of Glutamine and Vitamin E on the Morphometry of the Intestinal Mucosa in Broiler Chickens. *Poult. Sci.* **2007**, *86*, 488–495. [CrossRef] [PubMed]
45. Abdelrahman, W.; Mohnl, M.; Teichmann, K.; Doupovec, B.; Schatzmayr, G.; Lumpkins, B.; Mathis, G. Comparative Evaluation of Probiotic and Salinomycin Effects on Performance and Coccidiosis Control in Broiler Chickens. *Poult. Sci.* **2014**, *93*, 3002–3008. [CrossRef] [PubMed]
46. Goodwin, M.A.; Brown, J.; Bounous, D.I. Use of Microscopic Lesion Scores, Gross Lesion Scores and Oocyst Count Scores to Detect *Eimeria maxima* in Chickens. *Avian Pathol.* **1998**, *27*, 405–408. [CrossRef] [PubMed]
47. Galosi, L.; Desantis, S.; Roncarati, A.; Robino, P.; Bellato, A.; Nebbia, P.; Ferrocino, I.; Santamaria, N.; Biagini, L.; Filoni, L.; et al. Positive Influence of a Probiotic Mixture on the Intestinal Morphology and Microbiota of Farmed Guinea Fowls (*Numida meleagris*). *Front. Vet. Sci.* **2021**, *8*, 743889. [CrossRef]
48. Beck, C.N.; McDaniel, C.D.; Wamsley, K.G.S.; Kiess, A.S. The Potential for Inoculating *Lactobacillus animalis* and *Enterococcus faecium* Alone or in Combination Using Commercial In Ovo Technology Without Negatively Impacting Hatch and Post-Hatch Performance. *Poult. Sci.* **2019**, *98*, 7050–7062. [CrossRef]
49. Triplett, M.D.; Zhai, W.; Peebles, E.D.; McDaniel, C.D.; Kiess, A.S. Investigating Commercial In Ovo Technology as a Strategy for Introducing Probiotic Bacteria to Broiler Embryos. *Poult. Sci.* **2018**, *97*, 658–666. [CrossRef]
50. Arreguin-Nava, M.A.; Graham, B.D.; Adhikari, B.; Agnello, M.; Selby, C.M.; Hernandez-Velasco, X.; Vuong, C.N.; Solis-Cruz, B.; Hernandez-Patlan, D.; Latorre, J.D.; et al. Evaluation of *In Ovo Bacillus* Spp. Based Probiotic Administration on Horizontal Transmission of Virulent *Escherichia Coli* in Neonatal Broiler Chickens. *Poult. Sci.* **2019**, *98*, 6483–6491, reprinted in *Poult. Sci.* **2020**, *99*, 2301. https://doi.org/10.1016/j.psj.2020.02.001. [CrossRef]
51. Arreguin-Nava, M.A.; Graham, B.D.; Adhikari, B.; Agnello, M.; Selby, C.M.; Hernandez-Velasco, X.; Vuong, C.N.; Solis-Cruz, B.; Hernandez-Patlan, D.; Latorre, J.D.; et al. In Ovo Administration of Defined Lactic Acid Bacteria Previously Isolated from Adult Hens Induced Variations in the Cecae Microbiota Structure and *Enterobacteriaceae* Colonization on a Virulent *Escherichia coli* Horizontal Infection Model in Broiler Chickens. *Front. Vet. Sci.* **2020**, *7*, 489. [CrossRef]
52. Khaligh, F.; Hassanabadi, A.; Nassiri-Moghaddam, H.; Golian, A.; Kalidari, G.A. Effect of Probiotic Administration Route and Dietary Nutrient Density on Growth Performance, Gut Health, and Some Hematological Variables in Healthy or Eimeria Infected Broiler Chickens. *Iran. J. Appl. Anim. Sci.* **2019**, *9*, 473–485.
53. Leão, A.P.A.; Alvarenga, R.R.; Zangeronimo, M.G. In Ovo Inoculation of Probiotics for Broiler Chickens: Systematic Review and Meta-Analysis. *Anim. Feed Sci. Technol.* **2021**, *280*, 115080. [CrossRef]
54. Calnek, B.W. Pathogenesis of Marek's Disease Virus Infection. In *Marek's Disease*; Hirai, K., Ed.; Current Topics in Microbiology and Immunology; Springer: Berlin/Heidelberg, Germany, 2001; Volume 255, pp. 25–55. ISBN 978-3-540-67798-7.
55. Dunn, J.R.; Witter, R.L.; Silva, R.F.; Lee, L.F.; Finlay, J.; Marker, B.A.; Kaneene, J.B.; Fulton, R.M.; Fitzgerald, S.D. The Effect of the Time Interval Between Exposures on the Susceptibility of Chickens to Superinfection with Marek's Disease Virus. *Avian Dis.* **2010**, *54*, 1038–1049. [CrossRef] [PubMed]
56. Sharma, J.M.; Burmester, B.R. Resistance of Marek's Disease at Hatching in Chickens Vaccinated as Embryos with the Turkey Herpesvirus. *Avian Dis.* **1982**, *26*, 134–149. [CrossRef] [PubMed]
57. Rizk, Y.S.; Beshara, M.M.; Al-Mwafy, A.A. Effect of In-Ovo Injection with Probiotic on Hatching Traits and Subsequent Growth And Physiological Response of Hatched Sinai Chicks. *Egypt. Avian Dis.* **2018**, *38*, 439–450.
58. Abdel-Moneim, A.-M.E.; Elbaz, A.M.; Khidr, R.E.-S.; Badri, F.B. Effect of In Ovo Inoculation of *Bifidobacterium* Spp. on Growth Performance, Thyroid Activity, Ileum Histomorphometry, and Microbial Enumeration of Broilers. *Probiotics Antimicrob. Proteins* **2020**, *12*, 873–882. [CrossRef]
59. Pruszynska-Oszmalek, E.; Kolodziejski, P.A.; Stadnicka, K.; Sassek, M.; Chalupka, D.; Kuston, B.; Nogowski, L.; Mackowiak, P.; Maiorano, G.; Jankowski, J. In Ovo Injection of Prebiotics and Synbiotics Affects the Digestive Potency of the Pancreas in Growing Chickens. *Poult. Sci.* **2015**, *94*, 1909–1916. [CrossRef]
60. Samanya, M.; Yamauchi, K. Histological Alterations of Intestinal Villi in Chickens Fed Dried *Bacillus subtilis* Var. natto. *Comp. Biochem. Physiol. A Mol. Integr. Physiol.* **2002**, *133*, 95–104. [CrossRef]
61. Awad, W.; Ghareeb, K.; Böhm, J. Intestinal Structure and Function of Broiler Chickens on Diets Supplemented with a Synbiotic Containing Enterococcus Faecium and Oligosaccharides. *Int. J. Mol. Sci.* **2008**, *9*, 2205–2216. [CrossRef]
62. Bai, S.P.; Wu, A.M.; Ding, X.M.; Lei, Y.; Bai, J.; Zhang, K.Y.; Chio, J.S. Effects of Probiotic-Supplemented Diets on Growth Performance and Intestinal Immune Characteristics of Broiler Chickens. *Poult. Sci.* **2013**, *92*, 663–670. [CrossRef]
63. Rahimi, S.; Kathariou, S.; Grimes, J.L.; Siletzky, R.M. Effect of Direct-Fed Microbials on Performance and Clostridium Perfringens Colonization of Turkey Poults. *Poult. Sci.* **2011**, *90*, 2656–2662. [CrossRef]
64. Kim, J.S.; Ingale, S.L.; Kim, Y.W.; Kim, K.H.; Sen, S.; Ryu, M.H.; Lohakare, J.D.; Kwon, I.K.; Chae, B.J. Effect of Supplementation of Multi-microbe Probiotic Product on Growth Performance, Apparent Digestibility, Cecal Microbiota and Small Intestinal Morphology of Broilers. *J. Anim. Physiol. Anim. Nutr.* **2012**, *96*, 618–626. [CrossRef]
65. Liao, X.; Shao, Y.; Sun, G.; Yang, Y.; Zhang, L.; Guo, Y.; Luo, X.; Lu, L. The Relationship Among Gut Microbiota, Short-Chain Fatty Acids, and Intestinal Morphology of Growing and Healthy Broilers. *Poult. Sci.* **2020**, *99*, 5883–5895. [CrossRef] [PubMed]

66. Pan, D.; Yu, Z. Intestinal Microbiome of Poultry and Its Interaction with Host and Diet. *Gut Microbes* **2014**, *5*, 108–119. [CrossRef]
67. Kelly, J.R.; Kennedy, P.J.; Cryan, J.F.; Dinan, T.G.; Clarke, G.; Hyland, N.P. Breaking down the Barriers: The Gut Microbiome, Intestinal Permeability and Stress-Related Psychiatric Disorders. *Front. Cell. Neurosci.* **2015**, *9*, 166028. [CrossRef] [PubMed]
68. Johansson, M.E.; Hansson, G.C. Immunological Aspects of Intestinal Mucus and Mucins. *Nat. Rev. Immunol.* **2016**, *16*, 639–649. [CrossRef] [PubMed]
69. Sommer, F.; Bäckhed, F. The Gut Microbiota—Masters of Host Development and Physiology. *Nat. Rev. Microbiol.* **2013**, *11*, 227–238. [CrossRef]
70. Khan, S.; Moore, R.J.; Stanley, D.; Chousalkar, K.K. The Gut Microbiota of Laying Hens and Its Manipulation with Prebiotics and Probiotics To Enhance Gut Health and Food Safety. *Appl. Environ. Microbiol.* **2020**, *86*, e00600. [CrossRef]
71. Rychlik, I. Composition and Function of Chicken Gut Microbiota. *Animals* **2020**, *10*, 103. [CrossRef]
72. Shehata, A.M.; Paswan, V.K.; Attia, Y.A.; Abdel-Moneim, A.-M.E.; Abougabal, M.S.; Sharaf, M.; Elmazoudy, R.; Alghafari, W.T.; Osman, M.A.; Farag, M.R.; et al. Managing Gut Microbiota through In Ovo Nutrition Influences Early-Life Programming in Broiler Chickens. *Animals* **2021**, *11*, 3491. [CrossRef]
73. Dalloul, R.; Lillehoj, H.; Shellem, T.; Doerr, J. Enhanced Mucosal Immunity Against Eimeria Acervulina in Broilers Fed a Lactobacillus-Based Probiotic. *Poult. Sci.* **2003**, *82*, 62–66. [CrossRef]
74. Erdoğmuş, S.Z.; Gülmez, N.; Findik, A.; Hüseyin, Ş.; Gülmez, M. Efficacy of Probiotics on Health Status and Growth Performance of Eimeria Tenella Infected Broiler Chickens. *Kafkas Üniversitesi Vet. Fakültesi Derg.* **2019**, *25*, 311–320.
75. Giannenas, I.; Papadopoulos, E.; Tsalie, E.; Triantafillou, E.; Henikl, S.; Teichmann, K.; Tontis, D. Assessment of Dietary Supplementation with Probiotics on Performance, Intestinal Morphology and Microflora of Chickens Infected with *Eimeria tenella*. *Vet. Parasitol.* **2012**, *188*, 31–40. [CrossRef] [PubMed]
76. Mohsin, M.; Zhang, Z.; Yin, G. Effect of Probiotics on the Performance and Intestinal Health of Broiler Chickens Infected with Eimeria Tenella. *Vaccines* **2022**, *10*, 97. [CrossRef] [PubMed]
77. Angwech, H.; Tavaniello, S.; Ongwech, A.; Kaaya, A.N.; Maiorano, G. Efficacy of In Ovo Delivered Prebiotics on Growth Performance, Meat Quality and Gut Health of Kuroiler Chickens in the Face of a Natural Coccidiosis Challenge. *Animals* **2019**, *9*, 876. [CrossRef] [PubMed]
78. Ritzi, M.M.; Abdelrahman, W.; van-Heerden, K.; Mohnl, M.; Barrett, N.W.; Dalloul, R.A. Combination of Probiotics and Coccidiosis Vaccine Enhances Protection Against an Eimeria Challenge. *Vet. Res.* **2016**, *47*, 111. [CrossRef]
79. Madlala, T.; Okpeku, M.; Adeleke, M.A. Understanding the Interactions between Eimeria Infection and Gut Microbiota, towards the Control of Chicken Coccidiosis: A Review. *Parasite* **2021**, *28*, 48. [CrossRef]
80. Belote, B.L.; Soares, I.; Tujimoto-Silva, A.; Sanches, A.W.D.; Kraieski, A.L.; Santin, E. Applying I See inside Histological Methodology to Evaluate Gut Health in Broilers Challenged with *Eimeria*. *Vet. Parasitol.* **2019**, *276*, 100004. [CrossRef]
81. Travers, M.-A.; Florent, I.; Kohl, L.; Grellier, P. Probiotics for the Control of Parasites: An Overview. *J. Parasitol. Res.* **2011**, *2011*, 610769. [CrossRef]
82. Steiner, K.L.; Kabir, M.; Priest, J.W.; Hossain, B.; Gilchrist, C.A.; Cook, H.; Ma, J.Z.; Korpe, P.S.; Ahmed, T.; Faruque, A.S.G. Fecal Immunoglobulin A Against a Sporozoite Antigen at 12 Months Is Associated with Delayed Time to Subsequent Cryptosporidiosis in Urban Bangladesh: A Prospective Cohort Study. *Clin. Infect. Dis.* **2020**, *70*, 323–326. [CrossRef]

animals

MDPI

Article

Effects of Sea-Buckthorn Flavonoids on Growth Performance, Serum Inflammation, Intestinal Barrier and Microbiota in LPS-Challenged Broilers

Kexin Zhi [1], Fanwen Gong [1], Lele Chen [1], Zezheng Li [1], Xiang Li [1], Huadi Mei [1], Chenxing Fu [1], Yurong Zhao [1], Zhuying Liu [2,*] and Jianhua He [1,*]

1 College of Animal Science and Technology, Hunan Agricultural University, Changsha 410128, China; 13276901791@163.com (K.Z.); 13667920590@163.com (F.G.); chen1830909@163.com (L.C.); l15093587113@163.com (Z.L.); lxffy0123@163.com (X.L.); m15973699341@163.com (H.M.); chenxing1110@outlook.com (C.F.); zyr@hunau.edu.cn (Y.Z.)

2 College of Animal Science and Technology, Hunan Biological and Electromechanical Polytechnic, Changsha 410128, China

* Correspondence: liuzhuying@hnbemc.edu.cn (Z.L.); jianhuahy@hunau.net (J.H.)

Simple Summary: High-density intensive farming easily induces immune stress in broilers, leading to significant impairment of growth performance and intestinal health. Sea-buckthorn flavone (SF), extracted from the fruits, leaves, and branches of sea buckthorn, exhibits anti-oxidative, anti-inflammatory, anti-obesity, liver protective, and lipid metabolism-regulating properties. This study examined the effects of SF on growth performance, serum inflammation, the intestinal barrier, and the microbiota in lipopolysaccharide (LPS)-challenged broilers. Our findings demonstrate that SF alleviates LPS-induced inflammatory responses in broilers while improving intestinal health and enhancing growth performance. These findings serve as a valuable reference for poultry production.

Abstract: The experiment investigated the effects of sea-buckthorn flavonoids (SF) on lipopolysaccharide (LPS)-challenged broilers. A total of 288 one-day-old male broilers were randomly assigned to 4 groups, with 6 replicates of 12 broilers each. The experiment lasted for 20 days. The diet included two levels of SF (0 or 1000 mg/kg) and broilers intraperitoneally injected with 500 μg/kg LPS on 16, 18, and 20 days, or an equal amount of saline. LPS challenge decreased final body weight, average daily gain, and average daily feed intake, increased feed-to-gain ratio, and elevated serum IL-1β, IL-2, TNF-α, D-LA, and endotoxin levels. Moreover, it resulted in a reduction in the IL-10 level. LPS impaired the intestinal morphology of the duodenum, jejunum, and ileum, down-regulated the mRNA relative expression of Occludin, ZO-1, and MUC-2 in the jejunum mucosa, up-regulated the mRNA relative expression of TLR4, MyD88, NF-κB, and IL-1β, and increased the relative abundance of *Erysipelatoclostridium* in broilers ($p < 0.05$). However, SF supplementation mitigated the decrease in growth performance, reduced serum IL-1β, IL-2, and D-LA levels, increased IL-10 levels, alleviated intestinal morphological damage, up-regulated mRNA expression of Occludin and ZO-1, down-regulated the mRNA expression of TLR4, NF-κB, and IL-1β in jejunum mucosal ($p < 0.05$), and SF supplementation presented a tendency to decrease the relative abundance of proteobacteria ($0.05 < p < 0.1$). Collectively, incorporating SF can enhance the growth performance, alleviate serum inflammation, and improve the intestinal health of broilers, effectively mitigating the damage triggered by LPS-challenges.

Keywords: sea-buckthorn flavone; lipopolysaccharide; inflammation; intestinal health; broilers

check for updates

Citation: Zhi, K.; Gong, F.; Chen, L.; Li, Z.; Li, X.; Mei, H.; Fu, C.; Zhao, Y.; Liu, Z.; He, J. Effects of Sea-Buckthorn Flavonoids on Growth Performance, Serum Inflammation, Intestinal Barrier and Microbiota in LPS-Challenged Broilers. *Animals* **2024**, *14*, 2073. https://doi.org/10.3390/ani14142073

Academic Editor: Janghan Choi

Received: 14 June 2024
Revised: 10 July 2024
Accepted: 13 July 2024
Published: 15 July 2024

1. Introduction

Under modern intensive breeding conditions, high breeding density, pathogenic microorganisms, temperature, and humidity factors can induce immune stress in broilers, which results in damage to growth performance and the intestinal barrier in poultry [1,2]. LPS is a common endotoxin and is considered to be one of the most potent immune activators [3]. Previous studies demonstrated that intrabitoneal injection of LPS in broilers could seriously compromise the integrity of the intestinal mucosal barrier, elevate inflammatory factors, alter intestinal flora, and consequently impact growth performance [4,5]. LPS has been shown to activate the NF-κB signaling pathway by binding to the TLR4 receptor on the cell surface to generate a number of pro-inflammatory cytokines [6]. Notably, natural plant extracts contain a variety of nutrients, which can enhance the growth performance and feed utilization of livestock and poultry [7]. Therefore, it is crucial to develop a natural and botanic extract as a feed additive to ameliorate the immune stress of poultry and maintain intestinal health.

Sea-buckthorn (*Hippophae rhamnoides* L.) is a versatile medicinal and edible plant known for its strong adaptability and high yield [8]. It has been reported to possess typical physiological functions, including anti-inflammatory properties [9], antioxidant activity [10], immunomodulatory effects [11], and lipid-lowering capabilities [12]. Studies have indicated the beneficial impact of sea-buckthorn leaves or extracts on growth performance, feed efficiency, and the development of immune organs in Turkey poults, with long-term feeding being deemed safe and reliable [13]. Sea-buckthorn flavonoids (SF) are the major active components extracted from the fruits, roots, and leaves of sea-buckthorn. They may serve as natural feed additives, enhancing growth performance, rumen fermentation, and serum antioxidant capacity in lambs [14]. Furthermore, evidence suggests that key components of sea-buckthorn flavonoids such as quercetin, isorhamnetin, and kaempferol could alleviate inflammation challenged by LPS in mice [15]. Studies have shown that these compounds fight inflammation through different mechanisms [16–18]. It is currently attracting attention from researchers due to its extensive physiological functions. However, there is limited data available regarding the regulatory effects of SF on inflammatory responses, the intestinal barrier, and the microbiota in broilers under LPS-challenged immune stress. The present investigation aims to explore the effects of SF on the performance, serum inflammation, intestinal barrier, and microbiota of LPS challenge broilers via the LPS challenge immune stress model.

2. Materials and Methods

2.1. Experimental Design

A total of 288 healthy one-day-old yellow-feathered male broilers (32.94 g ± 0.10 g, Tang Renshen Group Co., Ltd., Zhuzhou, China) were randomly assigned to 4 groups, including the CON group (basal diet + saline), the LPS group (basal diet+ LPS), the SF group (basal diet + 1000 mg/kg SF + saline), and the SF + LPS group (basal diet + 1000 mg/kg SF + LPS), with 6 replicates of 12 broilers each. SF (purity ≥ 30%) was provided by Baoji Fangsheng Biological Development Co., Ltd. (Xi'an, China) and fed from 1 day of age; the dose of SF supplemented was referred to in previous studies [19]. LPS (*Escherichia coli* O55: L2880) was obtained from Sigma Aldrich (St. Louis, MO, USA). LPS was dissolved in saline, and 500 µg LPS/mL was prepared before use. LPS-challenged broilers were intraperitoneally injected with 500 µg LPS /kg on 16, 18, and 20 d; non-LPS-challenged broilers were injected with an equal amount of saline, and the frequency and dosage of LPS administration were selected based on previous studies [20,21]. The feeding trial lasted for 20 days. The basal diet (Table 1) was formulated based on the feeding standards of chickens (NY/T33-2004). All broilers were housed on the net floor (cage size: 2.5 m × 1.5 m), which had free access to feed and water. The artificial continuous lighting system was used, with temperatures ranging from 33 °C to 34 °C in the first week and 27 °C to 31 °C in the second week. Heating was applied depending on the specific temperature conditions. Natural ventilation in chicken coops. Maintain a relative humidity between 55 and 65%.

Table 1. Composition and nutrient levels of basic diet (air dried basis, %).

Ingredients, %	Content	Nutrient Levels, % [2]	Content
Corn	56.78	Metabolizable energy, MJ/kg	2879
Soybean meal	29.00	Crude protein	20.06
Corn gluten meal	5.00	Calcium	0.921
Wheat bran	3.60	Phosphorus	0.55
Soybean oil	1.20	Lysine	1.20
$CaHPO_4$	1.30	Methionine	0.54
Limestone	1.20	Methionine + Cysteine	0.88
Sodium chloride	0.30	Threonine	0.77
L-Lysine	0.54	Tryptophan	0.21
DL-Methionine	0.18		
Choline (60%)	0.10		
Premix [1]	0.80		
Total	100		

[1] Premix provided the following per kg of the diet: vitamin A 8000 IU, vitamin D3 4000 IU, vitamin E 15 IU, vitamin K 15 mg, vitamin B1 2.0 mg, vitamin B2 4.0 mg, vitamin B6 3.0 mg, vitamin B12 0.015 mg, folic acid 0.9 mg, nicotinic acid 35 mg, pantothenic acid 15 mg, biotin 0.2 mg, Mn 90 mg, Fe 90 mg, Zn 60 mg, Cu 10 mg, Se 0.20 mg, and I 0.40 mg. [2] The content of crude protein, calcium, and total phosphorus in the basic dietary nutrition level are all measured values, while the rest are calculated values.

2.2. Sample Collection

At the age of 21 days, one healthy bird with similar weight was selected from each replicate (fasting 8–12 h in advance, and live weight was recorded) and euthanized. Blood was collected using common tubes (Liuyang Sanli Medical Science and Technology Development Co., Ltd., Liuyang, China), centrifuged at 3500 r /min at 4 °C for 10 min, taken as the supernatant, and stored in −20 °C freezers. Open the abdominal cavity; the whole intestinal segment was removed and separated (duodenum, jejunum, and ileum). The middle segment was taken about 2 cm and fixed in 4% paraformaldehyde solution (Wuhan Kanos Technology Co., Ltd., Wuhan, China), scraped portions of the mucosa, and stored in −80 °C freezers. Appropriate amounts of cecal chyme were collected and stored in the refrigerator at −80 °C.

2.3. Index Determination Method

2.3.1. Measurement of Growth Performance

Initial weight (IW), final weight (FW), and feed consumption of each cage were recorded to calculate the average daily gain (ADG), average daily feed intake (ADFI), and feed-to-gain ratio (FCR) in the non-LPS challenged period and LPS challenged period.

2.3.2. Determination of Serum Indicators

The concentrations of interleukin-1β (IL-1β), interleukin-2 (IL-2), interleukin-10 (IL-10), tumor necrosis factor-α (TNF-α), corticosterone (CORT), D-lactate (D-LA), and endotoxin were determined by ELISA kits. The kits were purchased from Jiangsu Enzyme-Free Industrial Co., Ltd. (Yancheng, China), as described by Cheng et al. [22].

2.3.3. Gut Histomorphological Analysis

The duodenum, jejunum, and ileum samples were removed from a 4% paraformaldehyde fixing solution, dehydrated and embedded with paraffin, and cut into 5 μm slices, as described by Chang et al. [23]. Villi height (VH) and crypt depth (CD) were measured by optical microscopy in combination with software (Image Analysis 1.6.1) after staining with hematoxylin eosin (HE), magnifying 40×, and calculating the ratio of villus height to crypt depth (V/C) from the results. Six good villi and crypts were selected from each section for measurement, and the average value was taken.

2.3.4. Gene Expression Using Quantitative Real-Time PCR

Total RNA was extracted from jejunal mucosal tissue samples using AG reagents (Hangzhou Aikerui Biotechnology Co., Ltd., Hangzhou, China). RNA integrity was evaluated by agarose gel. and RNA concentration and purity were measured by a Nanodrop ND-2000 spectrophotometer (Thermo Scientific, Ottawa, ON, Canada). Reverse transcribed into cDNA using AG reagents (Hangzhou Aikerui Biotechnology Co., Ltd., Hangzhou, China). All primers (Table 2) were purchased from Changsha Biotechnology Co., Ltd. Real-time quantitative PCR detection by the SYBR Green PCR kit (Hangzhou Aikerui Biotechnology Co., Ltd., Hangzhou, China) in CFX Connect real-time PCR detection system, measured β-actin, Zonula occludens-1 (ZO-1), Occludin, Claudin-1, MUC-2, Toll-like receptor 4 (TLR4), myeloid differentiation primary response 88 (MyD88), and nuclear factor kappa B (NF-κB) relative mRNA expression levels. Results were calculated using the $2^{-\Delta\Delta CT}$ method, as described by Xing et al. [24].

Table 2. Real time PCR primer sequences.

Genes	Primer [1]	Sequence	Product Size (bp)	GenBank Accession Number
Occludin	F:5′-3′ R:5′-3′	GATGGACAGCATCAACGACC CTTGCTTTGGTAGTCTGGGC	142	NM_205128.1
Claudin-1	F:5′-3′ R:5′-3′	ACACCCGTTAACACCAGATTT GCATTTTTGGGGTAGCCTCG	152	NM_001013611.2
ZO-1	F:5′-3′ R:5′-3′	GCCTACTGCTGCTCCTTACAACTC GCTGGATCTATATGCGGCGGTAAG	129	XM_040680630.1
MUC-2	F:5′-3′ R:5′-3′	GTGAATGGCACTACGAGCCT CTGGGGTAGCAACCTTCCAG	106	XM_040701656.2
TLR4	F:5′-3′ R:5′-3′	TGGATCTTTCAAGGTGCCACA AGTGTCCGATGGGTAGGTCA	198	NM_001030693.2
NF-κB	F:5′-3′ R:5′-3′	TCAACGCAGGACCTAAAGACAT GCAGATAGCCAAGTTCAGGATG	162	NM_205134.1
MyD88	F:5′-3′ R:5′-3′	GGATGATCCGTATGGGCATGG ATGGACCACACACACGTTCC	171	NM_001030962.5
IL-1β	F:5′-3′ R:5′-3′	ACTGGGCA TCAAGGGCTA GGTAGAAGA TGAAGCGGGTC	154	NM_214005.1
β-actin	F:5′-3′ R:5′-3′	TGCGTGACATCAAGGAGAAG TGCCAGGGTACATTGTGGTA	199	L08165

[1] F = forward primer; R = reverse primer.

2.3.5. 16S rRNA Sequencing

Genomic DNA was extracted using a commercial kit (Qiagen, Hilden, Germany). The 16S rRNA was amplified by 338F (ACTCCTACGGGAGGCAGCAG) and 806R (GGACTACHVGGGTWTCTAAT) primers, then sequenced using Illumina. Bioinformatics analysis was performed using the online platform (https://cloud.majorbio.com, accessed on 11 September 2023) provided by Shanghai Majorbio Bio-Pharm Technology Co. Ltd. (Shanghai, China).

2.4. Statistical Analysis

All data were carried out using the Shapiro-Wilk test, analyzed by two-way ANOVA, and then Duncan multiple comparisons were performed with IBM SPSS statistical software (26.0 version, Chicago, IL, USA). The main effects include the LPS challenge and SF treatment, as well as the interactions between LPS and SF. Data were expressed as mean ± standard error of the mean (SEM). Values of $p < 0.05$ were regarded as statistically remarkable.

3. Results

3.1. Growth Performance

The growth performance results were presented in Table 3. There was no remarkable difference in growth performance between broilers fed with SF and without SF ($p > 0.05$) before the LPS challenge (1–15 days old). However, LPS lowered ADG, ADFI, and FW ($p < 0.05$) and increased FCR ($p < 0.05$) during the LPS challenge (16–20 days old) when compared with that of the control. SF supplementation significantly enhanced ADG and FW ($p < 0.05$), and the ADFI presented a tendency to increase ($0.05 < p < 0.1$), but the FCR had no significant effect ($p > 0.05$). However, there was no remarkable interaction between SF and LPS on the growth performance of broilers ($p > 0.05$). The details were presented in Table 4.

Table 3. Effects of sea-buckthorn flavonoids supplementation on growth performance of broilers [1].

Items [2]	Group [3]		SEM [4]	*p*-Value
	CON	SF		
	Before LPS challenge (1 to 15 day of age)			
IW (g)	32.93	32.96	0.069	0.968
ADG (g)	15.70	15.73	0.177	0.561
ADFI (g)	25.07	25.17	0.179	0.826
FCR	1.60	1.61	0.014	0.175
FW (g)	268.43	266.35	2.556	0.602

[1] Data represent mean values of 12 replicates per group (n = 12). [2] IW: initial weight; ADG: average daily gain; ADFI: average daily feed intake; FCR: feed-to-gain ratio [3] CON: basal diet; SF: basal diet +1000 mg/kg SF. [4] SEM: standard error of the mean.

Table 4. Effects of sea-buckthorn flavonoids supplementation on the growth performance of lipopolysaccharide-challenged broilers [1].

Items [2]	Group [3]				SEM [4]	*p*-Value		
	CON	LPS	SF	SF + LPS		SF	LPS	SF × LPS
	During LPS challenge (16 to 20 days of age)							
ADG(g)	26.58	23.43	28.03	25.58	0.759	0.028	0.001	0.652
ADFI(g)	49.90	47.54	51.20	50.31	0.882	0.079	0.032	0.414
FCR	1.88	2.04	1.83	1.97	0.062	0.346	0.022	0.878
FW(g)	400.26	386.61	404.10	396.68	3.177	0.041	0.003	0.338

[1] Data represent mean values of six replicates per group (n = 6). [2] IW: initial weight; ADG: average daily gain; ADFI: average daily feed intake; FCR: feed-to-gain ratio [3] CON: basal diet + saline; LPS: basal diet +500 μg/kg LPS; SF: 1000 mg/kg SF + saline; SF + LPS: 1000 mg/kg SF + 500 μg/kg LPS. [4] SEM: standard error of the mean.

3.2. Serum Inflammation

As presented in Figure 1, the LPS challenge remarkably elevated the serum concentration of IL-1β, IL-2, TNF-α, and CORT ($p < 0.05$) and decreased the concentration of IL-10 ($p < 0.05$) when compared with that of the control. SF supplementation markedly attenuated IL-1β, IL-2, and CORT concentrations and increased IL-10 concentrations ($p < 0.05$). In addition, the multiple comparisons show that the LPS group had a significantly lower serum concentration of IL-10 and a higher concentration of TNF-α compared with the SF + LPS group ($p < 0.05$).

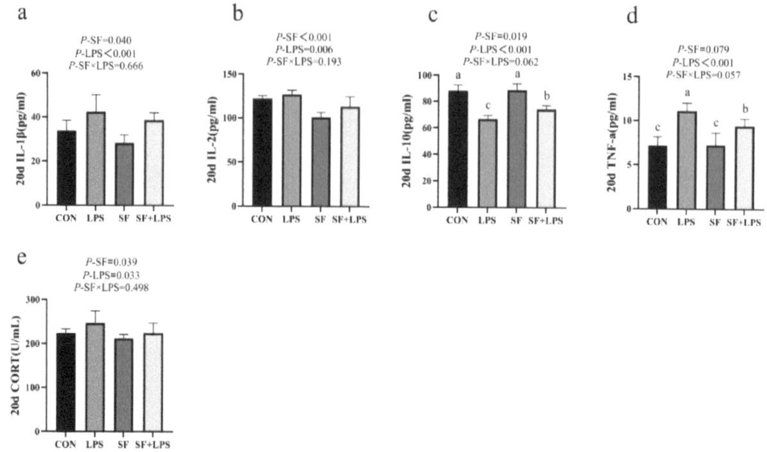

Figure 1. Serum inflammation and immune indexes in 20 d broilers (**a–e**). Data represent mean values of six replicates per group (n = 6). [a–c] Means in the same row with different superscripts differ (p < 0.05). CON: basal diet + saline; LPS: basal diet +500 µg/kg LPS; SF: 1000 mg/kg SF + saline; SF + LPS: 1000 mg/kg SF + 500 µg/kg LPS.

3.3. Intestinal Morphology

Severe damage to the intestinal morphology of broilers was observed by the LPS challenge (Figure 2a). As shown in Figure 2, LPS challenge significantly lowered duodenum VH and jejunum V/C (p < 0.05) compared with that of control broilers, while an opposite trend was observed for jejunum and ileum CD (p < 0.05) and tended to lower duodenum and ileum V/C (0.05 < p < 0.1). SF addition significantly reduced jejunal CD (p < 0.05) and elevated duodenal VH, V/C, and jejunal V/C (0.05 < p < 0.1). Multiple comparisons have shown that the duodenal VH of the SF + LPS group was significantly higher than that of the LPS group (p < 0.05). SF supplementation could alleviate these negative effects of broilers in duodenal VH stimulated by LPS.

3.4. Intestinal Permeability and Expression of Jejunal Mucosa-Associated Gene mRNA

As shown in Figure 3, LPS stress significantly increased the endotoxin, D-LA levels in serum, and TLR4, MyD88, NF-κB, and IL-1β relative gene expression in the jejunal mucosa (p < 0.05) when compared with that of the control, and signally decreased the MUC-2, Occludin, and ZO-1 relative gene expression in the jejunal mucosa in broilers (p < 0.05). On the contrary, SF supplementation reduced D-LA level and TLR4, NF-κB, and IL-1β relative gene expression in broilers (p < 0.05) and increased the relative expression of Occludin and ZO-1. However, no major changes in Claudin-1, MUC-2, or MyD88 mRNA expression in the jejunal mucosa were observed after SF supplementation (p > 0.05). There was a remarkable interaction between SF and LPS on D-LA level and IL-1β mRNA expression (p < 0.05). Multiple comparisons have shown that the relative expression levels of D-LA and IL-1β mRNA in the SF + LPS group were significantly lower than those in the LPS group (p < 0.05). SF supplementation could reverse these adverse effects in the serum D-LA level and jejunal mucosa IL-1β relative mRNA expression of broilers stimulated by LPS.

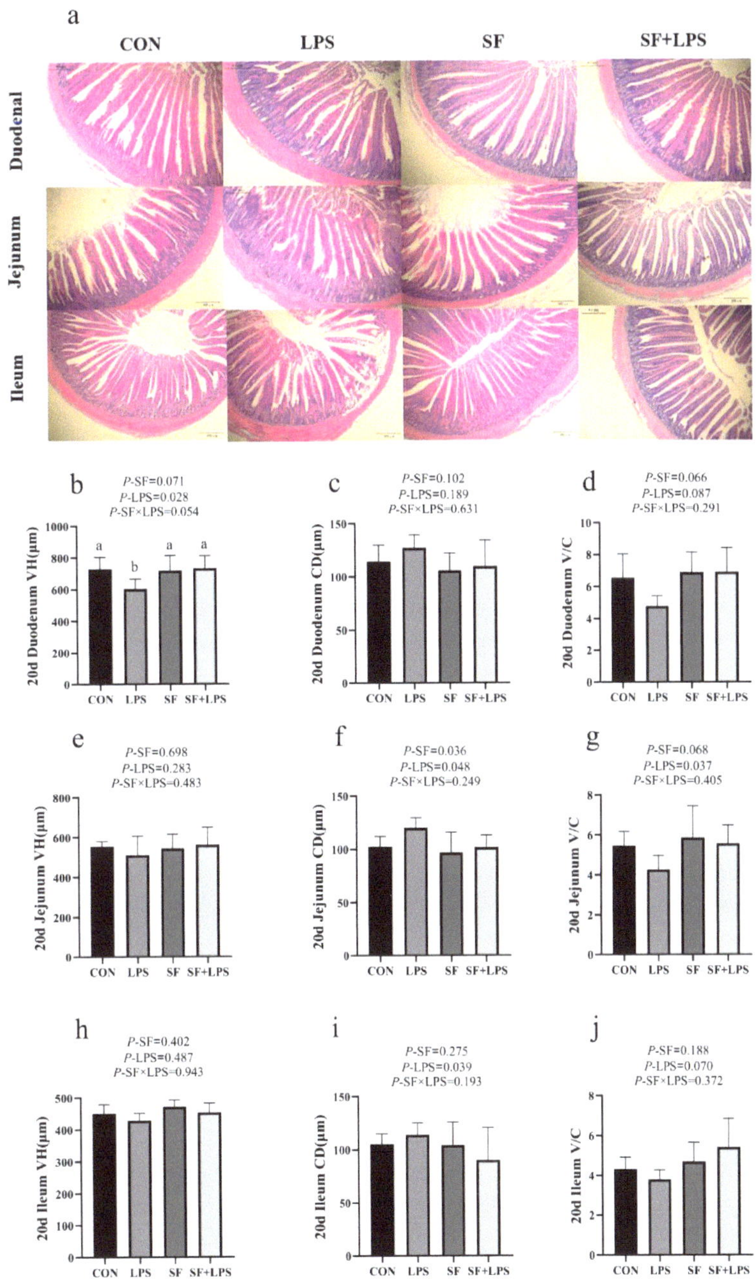

Figure 2. Histological morphology of duodenum, jejunum, and ileum in 20 d broilers (**a**–**j**). Data represent mean values of six replicates per group ($n = 6$). [ab] Means in the same row with different superscripts differ ($p < 0.05$). CON: basal diet + saline; LPS: basal diet +500 μg/kg LPS; SF: 1000 mg/kg SF + saline; SF + LPS: 1000 mg/kg SF + 500 μg/kg LPS.

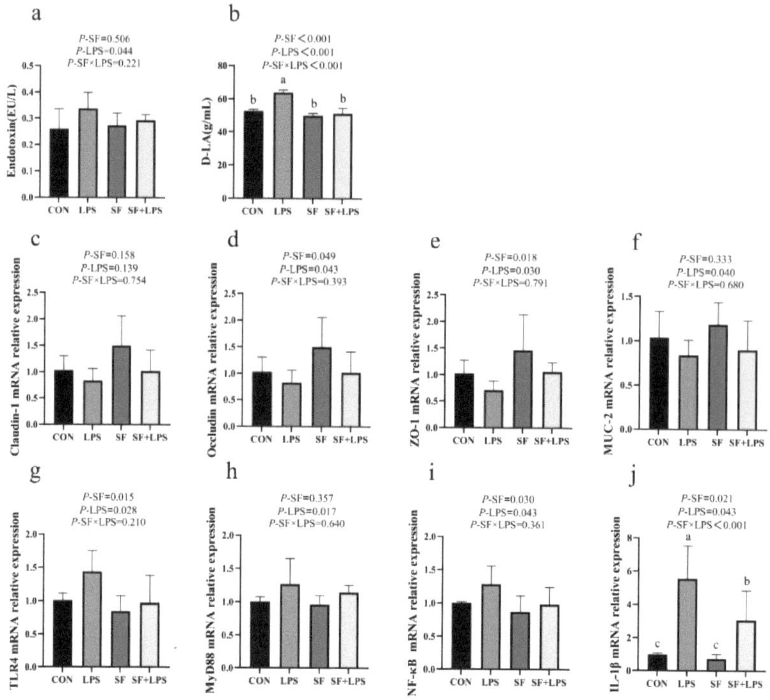

Figure 3. Intestinal permeability and relative expression of jejunal mucosa-associated gene mRNA in 20 d broilers (**a–j**). Data represent mean values of six replicates per group (n = 6). $^{a–c}$ Means in the same row with different superscripts differ ($p < 0.05$). CON: basal diet + saline; LPS: basal diet +500 μg/kg LPS; SF: 1000 mg/kg SF + saline; SF + LPS: 1000 mg/kg SF + 500 μg/kg LPS.

3.5. Gut Microbiota Analysis

At the 97% similarity level, 969,264 high-throughput sequence reads were obtained from all cecal content samples (n = 6), with an average of 40,386 sequences per sample. There were 640 core OTUs shared by the cecal flora in each group, and there were 943 OTUs in the cecal flora of the CON group, 954 OTUs in the cecal flora of the LPS group, 901 OTUs in the cecal flora of the SF group, and 959 OTUs in the cecal flora of the SF + LPS group (Figure 4a). No significant difference in community richness of cecal microbiota was observed in broilers, including Ace, Chao, Shannon, Coverage, and Simpson ($p > 0.05$, Figure 4c–f). The PCA map based on the genus level shows no difference between the four groups of cecal flora (Figure 4b). At the phylum level, the cecal flora consisted mainly of Firmicutes, Proteobacteria, Bacteroidota, and Actinobacteriota in broilers (Figure 4g). Firmicutes, Proteobacteria, and Bacteroidota were the dominant phyla, and their relative abundances were analyzed. SF supplementation presented a tendency to decrease the relative abundance of proteobacteria ($0.05 < p < 0.1$, Figure 4i). At the genus level, the cecal flora consisted mainly of *Faecalibacterium*, *Escherichia-Shigella*, Norank_f_norank_o_Clostridia_UCG-014, *Ruminococcus_torques*_group, Unclassified_f_Lachnospiraceae, *Bacteroides*, *Erysipelatoclostridium*, Unclassified_f_Ruminococcaceae, and *Lactobacillus* in broilers (Figure 4h). Compared with nonchallenged broilers, LPS presented a tendency to lower the relative abundance of *Faecalibacterium* ($0.05 < p < 0.1$, Figure 4j) and elevate the relative abundance of *Erysipelatoclostridium* ($p < 0.05$, Figure 4l). In addition, SF supplementation presented a tendency to increase the relative abundance of Unclassified_f_Lachnospiraceae ($0.05 < p < 0.1$, Figure 4k). Spearman correlation analysis confirmed that changes in intestinal flora were associated with LPS treatment and SF supplementation (Figure 5). There was an obviously negative correlation between IL-10 and Proteobacteria ($p < 0.05$), and a positive correlation between IL-1β, CORT, D-LA, and Proteobacteria ($p < 0.05$). In addition, TNF-α and *Erysipelatoclostridium* were positively correlated ($p < 0.05$).

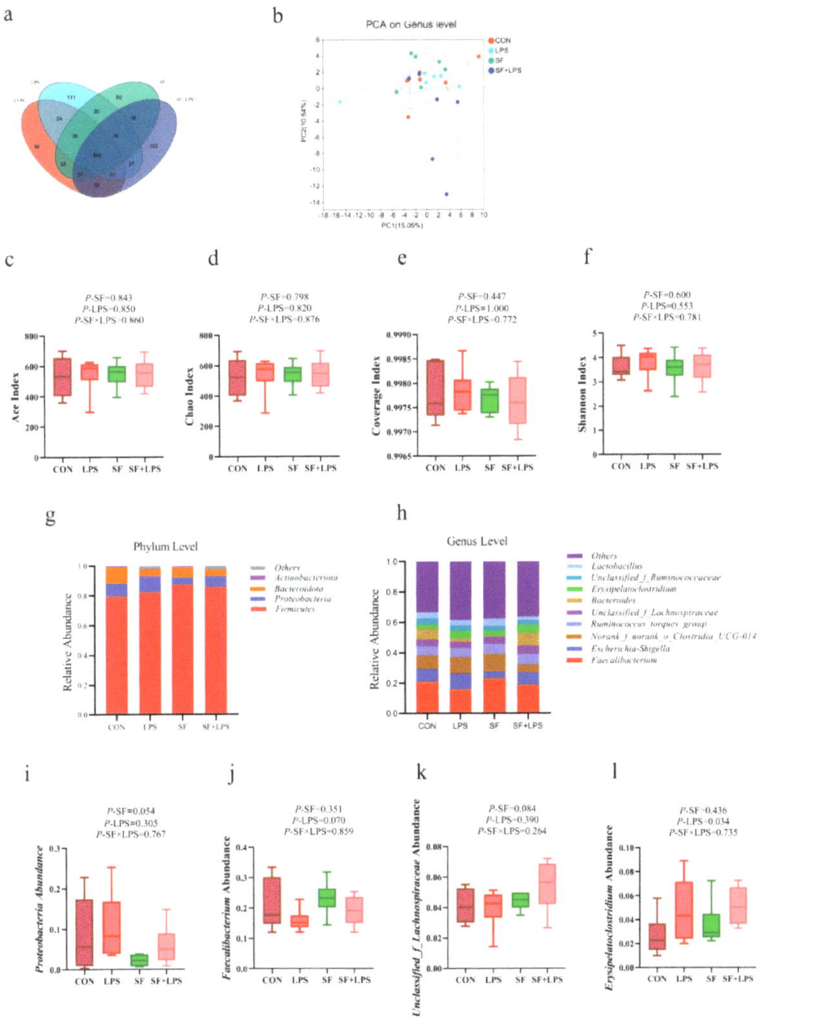

Figure 4. Venn diagram (**a**). Alpha diversity of cecal flora (**b–e**). Principal component analysis (PCA) of cecal flora (**f**). The relative abundance of cecal microbiota composition at the phylum and genus levels (**g–l**). CON: basal diet + saline; LPS: basal diet +500 μg/kg LPS; SF: 1000 mg/kg SF + saline; SF + LPS: 1000 mg/kg SF + 500 μg/kg LPS.

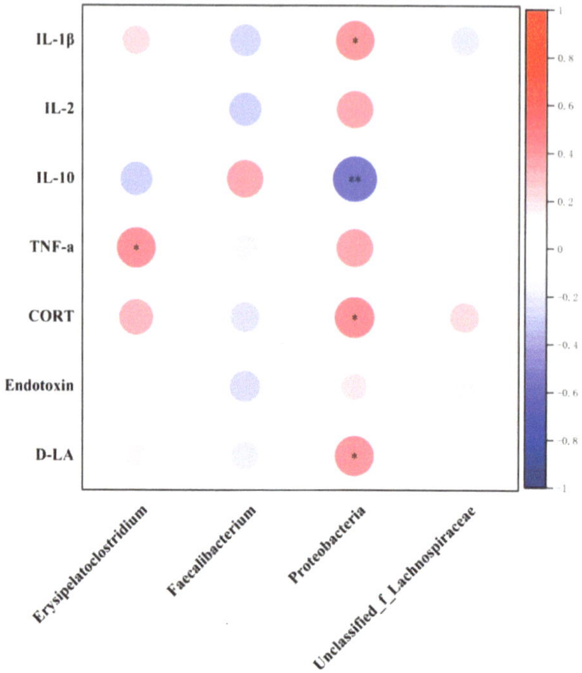

Figure 5. Correlation analysis of cecal flora and inflammation indexes. Data represent mean values of six replicates per group ($n = 6$). * Indicates $p < 0.05$. ** Indicates $p < 0.01$.

4. Discussion

CORT serves as a crucial indicator of hypothalamus–pituitary–adrenal axis activity. Elevated CORT levels directly inhibit growth under immune stress conditions in broilers [25]. In our study, the CORT level of broilers was significantly increased after injection of LPS, validating successful modeling for subsequent tests. In a similar experiment, SF supplementation markedly enhanced the ADG and FW of broilers at days 21, 42, and played a role in improving nutrient digestion and absorption [26]. Quercetin, the primary active component of SF, upregulated the expression of nutrient transporter genes in the intestine, such as glucose transporter 2 and peptide transporter 1, promoted intestinal digestion and absorption, improved intestinal morphology, protected the intestinal barrier, and improved growth performance [27]. Dietary incorporation of 1000 mg/kg SF significantly influenced ADG and FW in broilers, which may be closely related to its active components. The utilization rate of feed and the digestion and absorption of nutrients are critical factors influencing growth performance, especially in the intestine, which plays an important role as the main site for the digestion and absorption of nutrients [28]. The specific reasons need to be further studied.

Intestinal morphology is a key indicator for evaluating digestion [29]. It is primarily characterized by VH, CD, and V/C indexes; longer VH and shallower CD can provide a larger area of intestinal nutrient absorption, on the contrary, which may cause intestinal inflammation and impair intestinal function. Therefore, greater V/C signifies higher absorption capacity; a normal intestinal structure is essential for growth and development [30]. In this study, it was observed that LPS stimulation significantly impaired the intestinal morphology of broilers, which aligns with previous research findings [31,32]. Notably, dietary SF supplementation alleviated the adverse effects of LPS on intestinal morphology by reducing jejunal CD and increasing duodenal VH as well as V/C and jejunal V/C ratios. Thus, SF supplementation-enhanced intestinal VH and V/C may improve intestinal morphology by promoting the proliferation and differentiation of intestinal epithelial cells while increasing the surface area for nutrient absorption in the intestine [33].

D-LA and endotoxin are commonly utilized for evaluating intestinal permeability and intestinal mucosa damage. D LA and endotoxin are the byproducts of lactic acid fermentation and cleavage of bacteria in the intestine, exclusive to this tissue [34,35]. Under normal conditions, levels of D-LA and endotoxin in the blood of livestock and poultry are minimal, increasing only when there is intestinal mucosal damage [36,37]. Increased intestinal permeability due to damaged integrity leads to decreased nutrient digestion and absorption in livestock and poultry, impacting poultry growth performance [38]. Our study also revealed that intraperitoneal injection of LPS raised the levels of D-LA and endotoxin in broilers, indicating LPS-induced damage to the broiler's intestinal mucosa barrier function. However, SF supplementation significantly reduced the D-LA level in broilers. These results suggest that SF mitigated LPS-induced impairment of broiler intestinal permeability while positively influencing their mucosal barrier.

The tight junction is the crucial connection between intestinal epithelial cells. Tight junction proteins are essential components of the intestinal barrier protein, playing a crucial role in its composition and function [39]. Numerous studies have shown that during an inflammatory response, various inflammatory factors enter the intestine, leading to intestinal lesions, damage to the intestinal tissue, disruption of the barrier function, and reducing the Occludin, ZO-1, and Claudin-1 mRNA expression levels [40,41]. In this experiment, LPS stimulation significantly decreased the relative gene expression of MUC-2, Occludin, and ZO-1 in the jejunum mucosa of broilers. The expression of intestinal mucosal barrier genes in broilers was lowered, and intestinal permeability was impaired by intraperitoneal injection of LPS in broilers, which was also supported by elevated serum D-LA and endotoxin. It has been reported that dietary supplementation with flavonoids up-regulates the expression of tight junction proteins (especially ZO-1, Claudin-1, E-cadherin, and Occludin) in broilers at days 21 and provides protective benefit against LPS-challenged dysfunction in their intestinal mucosal barriers [42]. The study yielded similar findings: dietary SF could effectively up-regulate the relative gene expression of Occludin and ZO-1 in the jejunum mucosal of broilers. In summary, we recommend that SF supplementation at 1000 mg/kg enhances the expression of tight junctions in the small intestine challenged by LPS and maintains the integrity of the intestinal mucosal barrier.

When poultry is exposed to LPS, LPS binds to TLR4 and initiates a cascade reaction that stimulates the TLR4/NF-κB signaling pathway, leading to the secretion of a number of pro-inflammatory cytokines (IL-1β, IL-2, IL-6, TNF-α, et al.) by monocytes and macrophages, resulting in tissue damage [43]. NF-κB plays a crucial role in activating inflammation through the synthesis and release of pro-inflammatory cytokines [44,45]. Consistent with previous findings, this study observed that LPS significantly elevated serum inflammatory, and up-regulated the expression of genes related to the TLR4/NF-κB signaling pathway in the jejunum mucosa, confirming the LPS-mediated inflammatory response. It has been reported that flavonoids can reduce the occurrence of disease by negatively regulating the inflammatory factor IL-1β [46]. Quercetin, kaempferol, and isorhamnetin are the main active components of SF, which can enhance animal immunity and reduce inflammation [47]. SF shows great potential for treating inflammation, as does quercetin [48] and kaempferol [49]. and isorhamnetin [50] can inhibit the activation of the TLR4/NF-κB signaling pathway, thereby suppressing the production of pro-inflammatory cytokines. The results demonstrated that the addition of SF significantly reduced the levels of IL-1β, IL-2, and CORT in serum, as well as the relative expression levels of TLR4, NF-κB, and IL-1β mRNA in the jejunum mucosa. Additionally, it markedly elevated the concentration of IL-10 in serum. The results indicated that the anti-inflammatory effect of SF was closely related to the TLR4/NF-κB signaling pathway.

Intestinal flora is a complex ecosystem that plays a vital role in the digestion and absorption of nutrients, metabolism, immune regulation, and other processes. Numerous studies have demonstrated the vital importance of intestinal flora in maintaining overall body health [51]. Intestinal microbes can regulate the whole immune system, strengthen the intestinal barrier function, and thus improve the production performance of poultry [52]. It has been reported that the abundant proteobacteria could easily lead to bacterial imbalance and inflammatory responses [53]. The relative abundance of proteobacteria in mice significantly increased after long-term mildronate treatment, leading to a bacterial imbalance [54]. Xie et al. [55] found that quercetin can improve the intestinal environment and down-regulate the relative abundance of proteobacteria and other microorganisms. Supplementation with SF reduced the relative abundance of proteobacteria, which is basically consistent with previous studies. Furthermore, Spearman correlation analysis also revealed that IL-10 was significantly negatively correlated with Proteobacteria, while IL-1β, CORT, and D-LA were significantly positively correlated with Proteobacteria, indicating that Proteobacteria increased inflammation and reduced immune response. *Faecalibacterium* is a crucial intestinal bacterium with various probiotic effects, such as maintaining intestinal homeostasis and anti-inflammatory properties [56]. LPS challenge enriched potentially harmful microbes and led to an increase in *Erysipelatoclostridium*, which is indicative of intestinal disorders causing inflammation [57]. According to the study, Fuzhuan brick tea crude polysaccharides (FBTPs) restored the microbial imbalance induced by cyclophosphamide (Cy) by elevating several beneficial bacteria, such as lactic acid bacteria, Unclassified_f_Lachnospiraceae, while reducing Bacteroides and *Helicobacter pylori*, thus playing an immune protective role in Cy-induced mice [58]. In this experiment, LPS lowered the relative abundance of *Faecalibacterium* and elevated the relative abundance of *Erysipelatoclostridium*. In addition, SF supplementation presented a tendency to increase the relative abundance of Unclassified_f_Lachnospiraceae. In addition, Spearman correlation analysis showed that TNF-α and *Erysipelatoclostridium* were positively correlated, which supported this point. Therefore, we found that SF can up-regulate the relative abundance of beneficial bacteria and down-regulate the relative abundance of harmful bacteria, maintain intestinal health, and improve the production performance of broilers.

5. Conclusions

In conclusion, dietary supplementation with 1000 mg/kg of SF could enhance the growth performance of broilers by mitigating the serum inflammatory response, improving intestinal health via enhancing intestinal morphology and regulating the intestinal barrier, and effectively alleviating the damage caused by LPS.

Author Contributions: Conceptualization, J.H., Z.L. (Zhuying Liu), K.Z. and F.G.; Methodology, K.Z.; Validation, L.C., Z.L. (Zezheng Li) and X.L.; Writing–Original Draft Preparation, K.Z.; Writing–Review and Editing, K.Z. and H.M.; Project Administration, C.F., Y.Z. and J.H. All authors have read and agreed to the published version of the manuscript.

Funding: This research was funded by National Key R&D Program of China (2023YFD1301200).

Institutional Review Board Statement: The study was conducted in accordance with the Declaration of Helsinki, and approved by the Institutional Animal Care and Use Committee of Hunan Agricultural University (protocol number CACAHU, 2022-00145).

Informed Consent Statement: Not applicable.

Data Availability Statement: All datasets collected and analyzed during the current study are available from the corresponding author by request, the availability of the data is restricted to investigators based at academic institutions.

Acknowledgments: The author would like to acknowledge Tang Renshen Group Co., Ltd., Zhuzhou, China for providing the test site to ensure the smooth conduct of our test. We would also like to acknowledge the hard work of all the students who assisted in the processing of the samples and data.

Conflicts of Interest: The authors declare no conflicts of interest.

References

1. Gao, L.A.; Er, M.W.; Li, L.H.; Wen, P.; Jia, Y.C.; Huo, L.M. Microclimate environment model construction and control strategy of enclosed laying brooder house. *Poult Sci.* **2022**, *101*, 101843. [CrossRef]
2. Liu, Y.H.; Zhang, Y.; Bai, D.Y.; Li, Y.Q.; He, X.L.; Ito, K.; Liu, K.; Tan, H.; Zhen, W.; Zhang, B.; et al. Dietary Supplementation with Chlorogenic Acid Enhances Antioxidant Capacity, Which Promotes Growth, Jejunum Barrier Function, and Cecum Microbiota in Broilers under High Stocking Density Stress. *Animals* **2023**, *13*, 303. [CrossRef]
3. Yao, M.; Gao, W.H.; Tao, H.X.; Yang, J.; Huang, T.H. The regulation effects of danofloxacin on pig immune stress induced by LPS. *Res. Vet. Sci.* **2017**, *110*, 65–71. [CrossRef]
4. Wang, M.Y.; Zhang, Y.; Tong, Y.X.; Guo, P.T.; Zhang, J.; Wang, C.K.; Gao, Y.Y. Effects of lutein on jejunal mucosal barrier function and inflammatory responses in lipopolysaccharide-challenged yellow-feather broilers. *Poult Sci.* **2022**, *101*, 102191. [CrossRef]
5. Dias, K.M.M.; Oliveira, C.H.; Calderano, A.A.; Rostagno, H.S.; Gomes, K.M.; O'Connor, K.E.; Davis, R.; Walsh, M.; Britton, J.; Altieri, E.A.; et al. Dietary Hydroxytyrosol Supplementation on Growth Performance, Gut Morphometry, and Oxidative and Inflammatory Status in LPS-Challenged Broilers. *Animals* **2024**, *14*, 871. [CrossRef] [PubMed]
6. Ding, M.; Yu, Y.; Zhu, Z.; Tian, H.; Guo, Y.; Zan, R.; Tian, Y.; Jiang, R.; Li, K.; Sun, G.; et al. Regulation of the MyD88 gene in chicken spleen inflammation induced by stress. *J. Anim. Sci.* **2023**, *101*, skad060. [CrossRef]
7. Yang, Z.; Yang, J.J.; Zhu, P.J.; Han, H.M.; Wan, X.L.; Yang, H.M.; Wang, Z.Y. Effects of betaine on growth performance, intestinal health, and immune response of goslings challenged with lipopolysaccharide. *Poult Sci.* **2022**, *101*, 102153. [CrossRef]
8. Ciesarová, Z.; Murkovic, M.; Cejpek, K.; Kreps, F.; Tobolková, B.; Koplík, R.; Belajová, E.; Kukurová, K.; Daško, Ľ.; Panovská, Z.; et al. Why is sea buckthorn (*Hippophae rhamnoides* L.) so exceptional? A review. *Food Res Int.* **2020**, *133*, 109170. [CrossRef]
9. Sireswar, S.; Biswas, S.; Dey, G. Adhesion and anti-inflammatory potential of Lactobacillus rhamnosus GG in a sea buckthorn based beverage matrix. *Food Funct.* **2020**, *11*, 2555–2572. [CrossRef]
10. Sławińska, N.; Żuchowski, J.; Stochmal, A.; Olas, B. Extract from Sea Buckthorn Seeds-A Phytochemical, Antioxidant, and Hemostasis Study; Effect of Thermal Processing on Its Chemical Content and Biological Activity In Vitro. *Nutrients* **2023**, *15*, 686. [CrossRef]
11. Jayashankar, B.; Singh, D.; Tanwar, H.; Mishra, K.P.; Murthy, S.; Chanda, S.; Mishra, J.; Tulswani, R.; Misra, K.; Singh, S.B.; et al. Augmentation of humoral and cellular immunity in response to Tetanus and Diphtheria toxoids by supercritical carbon dioxide extracts of *Hippophae rhamnoides* L. leaves. *Int. Immunopharmacol.* **2017**, *44*, 123–136. [CrossRef]
12. Xiao, P.T.; Liu, S.Y.; Kuang, Y.J.; Jiang, Z.M.; Lin, Y.; Xie, Z.S.; Liu, E.H. Network pharmacology analysis and experimental validation to explore the mechanism of sea buckthorn flavonoids on hyperlipidemia. *J Ethnopharmacol.* **2021**, *264*, 113380. [CrossRef]
13. Sharma, A.; Shukla, P.K.; Bhattacharyya, A.; Kumar, U.; Roy, D.; Yadav, B.; Prakash, A. Effect of dietary supplementation of sea buckthorn and giloe leaf meal on the body weight gain, feed conversion ratio, biochemical attributes, and meat composition of turkey poults. *Vet. World* **2018**, *11*, 93–98. [CrossRef]
14. Hao, X.Y.; Zhang, X.Z.; Yang, D.Y.; Xie, Y.Z.; Mu, C.T.; Zhang, J.X. Effects of sea-buckthorn flavonoids on growth performance, nutrient digestibility, microbial protein synthesis, and plasma antioxidant capacity of finishing lambs. *Anim. Feed. Sci. Technol.* **2023**, *305*, 115783. [CrossRef]
15. Ren, Q.C.; Li, X.H.; Li, Q.Y.; Yang, H.L.; Wang, H.L.; Zhang, H.; Zhao, L.; Jiang-Yong, S.L.; Meng, X.L.; Zhang, Y.; et al. Total flavonoids from sea buckthorn ameliorates lipopolysaccharide/cigarette smoke-induced airway inflammation. *Phytother Res.* **2019**, *33*, 2102–2117. [CrossRef]
16. Xu, Y.; Li, J.; Lin, Z.; Liang, W.; Qin, L.; Ding, J.; Chen, S.; Zhou, L. Isorhamnetin Alleviates Airway Inflammation by Regulating the Nrf2/Keap1 Pathway in a Mouse Model of COPD. *Front Pharmacol.* **2022**, *13*, 860362. [CrossRef]
17. Sul, O.J.; Ra, S.W. Quercetin Prevents LPS-Induced Oxidative Stress and Inflammation by Modulating NOX2/ROS/NF-kB in Lung Epithelial Cells. *Molecules* **2021**, *26*, 6949. [CrossRef]
18. Bian, Y.; Lei, J.; Zhong, J.; Wang, B.; Wan, Y.; Li, J.; Liao, C.; He, Y.; Liu, Z.; Ito, K.; et al. Kaempferol reduces obesity, prevents intestinal inflammation, and modulates gut microbiota in high-fat diet mice. *J. Nutr. Biochem.* **2022**, *99*, 108840. [CrossRef]
19. Ma, J.S.; Chang, W.H.; Liu, G.H.; Zhang, S.; Zheng, A.J.; Li, Y.; Xie, Q.; Liu, Z.Y.; Cai, H.Y. Effects of flavones of sea buckthorn fruits on growth performance, carcass quality, fat deposition and lipometabolism for broilers. *Poult Sci.* **2015**, *94*, 2641–2649. [CrossRef]
20. An, J.S.; Shi, J.J.; Liu, K.B.; Li, A.K.; He, B.B.; Wang, Y.; He, J.H. Effects of Solid-State Fermented Wheat Bran on Growth Performance, Immune Function, Intestinal Morphology and Microflora in Lipopolysaccharide-Challenged Broiler Chickens. *Animals* **2022**, *12*, 1100. [CrossRef]
21. Zhang, P.F.; Shi, B.L.; Su, J.L.; Yue, Y.X.; Cao, Z.X.; Chu, W.B.; Li, K.; Yan, S.M. Relieving effect of Artemisia argyi aqueous extract on immune stress in broilers. *J. Anim. Physiol. Anim. Nutr.* **2017**, *101*, 251–258. [CrossRef]
22. Cheng, Y.; Liu, S.; Wang, F.; Wang, T.; Yin, L.; Chen, J.; Fu, C. Effects of Dietary Terminalia chebula Extract on Growth Performance, Immune Function, Antioxidant Capacity, and Intestinal Health of Broilers. *Animals* **2024**, *14*, 746. [CrossRef]
23. Chang, W.Y.; Yu, Y.H. Effect of Bacillus species-fermented products and essential oils on growth performance, gut morphology, cecal short-chain fatty acid levels, and microbiota community in broilers. *Poult. Sci.* **2022**, *101*, 101970. [CrossRef]

24. Xing, Y.Y.; Zheng, Y.K.; Yang, S.; Zhang, L.H.; Guo, S.W.; Shi, L.L.; Xu, Y.Q.; Jin, X.; Yan, S.M.; Shi, B.L. Artemisia ordosica Polysaccharide Alleviated Lipopolysaccharide-induced Oxidative Stress of Broilers via Nrf2/Keap1 and TLR4/NF-κB Pathway. *Ecotoxicol. Environ. Saf.* **2021**, *223*, 112566. [CrossRef]
25. Li, X.; Bian, J.; Xing, T.; Zhao, L.; Li, J.L.; Zhang, L.; Gao, F. Effects of guanidinoacetic acid supplementation on growth performance, hypothalamus-pituitary-adrenal axis, and immunity of broilers challenged with chronic heat stress. *Poult Sci.* **2023**, *102*, 103114. [CrossRef]
26. Kalia, S.; Bharti, V.K.; Giri, A.; Kumar, B.; Arora, A.; Balaje, S.S. *Hippophae rhamnoides* as novel phytogenic feed additive for broiler chickens at high altitude cold desert. *Sci Rep.* **2018**, *8*, 5954. [CrossRef]
27. Abdel-Latif, M.A.; Elbestawy, A.R.; El-Far, A.H.; Noreldin, A.E.; Emam, M.; Baty, R.S.; Albadrani, G.M.; Abdel-Daim, M.M.; Abd El-Hamid, H.S. Quercetin Dietary Supplementation Advances Growth Performance, Gut Microbiota, and Intestinal mRNA Expression Genes in Broiler Chickens. *Animals* **2021**, *11*, 2302. [CrossRef]
28. Liu, X.; Huang, X.Y.; Fu, Y.; Wang, Y.Z.; Lu, Z.Q. The Positive Effects of Exogenous Pancreatin on Growth Performance, Nutrient Digestion and Absorption, and Intestinal Microbiota in Piglets. *Front. Physiol.* **2022**, *13*, 906522. [CrossRef]
29. Song, X.; Anas, M.A.; Kurniawati, A.; Hanim, C.; Muhlisin, A.M.A.; Madani, A.M.A.; Wang, Q.; Chen, H. Effects of reduced-protein diets with protease supplementation on growth, carcass yield, intestinal morphology, organ development, nutrient digestibility, and blood biochemical of broiler chickens. *Transl. Anim. Sci.* **2023**, *7*, txad098. [CrossRef]
30. Zou, T.D.; Deng, C.X.; Wang, Z.R.; Ye, Y.L.; You, J.M. Dietary alanyl-glutamine improves growth performance of weaned piglets through maintaining intestinal morphology and digestion-absorption function. *Animal* **2019**, *13*, 1826–1833. [CrossRef]
31. Xing, Y.; Zheng, Y.; Yang, S.; Zhang, L.; Guo, S.; Shi, L.; Xu, Y.; Jin, X.; Yan, S.; Shi, B.L. Artemisia ordosica polysaccharide ameliorated LPS-induced growth inhibition and intestinal injury in broilers through enhancing immune-regulation and antioxidant capacity. *J. Nutr. Biochem.* **2023**, *115*, 109284. [CrossRef] [PubMed]
32. Sun, H.; Zheng, X.; Yang, B.; Yan, M.; Wang, H.; Yang, S.; Shi, D.; Guo, S.; Liu, C. Effect of Wu Zhi San supplementation in LPS-induced intestinal inflammation and barrier damage in broilers. *Front. Vet. Sci.* **2023**, *10*, 1234769. [CrossRef] [PubMed]
33. Zhu, X.; Shang, X.; Lin, G.; Li, H.; Feng, X.; Zhang, H. Effects of Zinc Glycinate on Growth Performance, Serum Biochemical Indexes, and Intestinal Morphology of Yellow Feather Broilers. *Biol. Trace Elem. Res.* **2022**, *200*, 4089–4097. [CrossRef] [PubMed]
34. Liu, D.Y.; Lou, W.J.; Zhang, D.Y.; Sun, S.Y. ROS Plays a Role in the Neonatal Rat Intestinal Barrier Damages Induced by Hyperoxia. *Biomed. Res. Int.* **2020**, *2020*, 8819195. [CrossRef] [PubMed]
35. Zhang, J.C.; Ma, L.; Wan, X.Y.; Shan, J.J.; Qu, Y.G.; Hashimoto, K.J. (R)-Ketamine attenuates LPS-induced endotoxin-derived delirium through inhibition of neuroinflammation. *Psychopharmacology* **2021**, *238*, 2743–2753. [CrossRef] [PubMed]
36. Zhi, Y.; Li, T.; Li, Y.; Zhang, T.; Du, M.; Zhang, Q.; Wang, X.; Hu, G. Protective role of Cecropin AD against LPS-induced intestinal mucosal injury in chickens. *Front Immunol.* **2023**, *14*, 1290182. [CrossRef] [PubMed]
37. Zhang, L.; Zhang, L.; Zhan, X.; Zeng, X.; Zhou, L.; Cao, G.; Chen, A.; Yang, C. Effects of dietary supplementation of probiotic, Clostridium butyricum, on growth performance, immune response, intestinal barrier function, and digestive enzyme activity in broiler chickens challenged with Escherichia coli K88. *J. Anim. Sci. Biotechnol.* **2016**, *7*, 3. [CrossRef]
38. Wu, Z.; Yang, K.; Zhang, A.; Chang, W.; Zheng, A.; Chen, Z.; Cai, H.; Liu, G.H. Effects of Lactobacillus acidophilus on the growth performance, immune response, and intestinal barrier function of broiler chickens challenged with Escherichia coli O157. *Poult. Sci.* **2021**, *100*, 101323. [CrossRef]
39. Guo, Y.; Yu, Y.; Li, H.; Ding, X.; Li, X.; Jing, X.; Chen, J.; Liu, G.; Lin, Y.; Jiang, C.; et al. Inulin supplementation ameliorates hyperuricemia and modulates gut microbiota in Uox-knockout mice. *Eur. J. Nutr.* **2021**, *60*, 2217–2230. [CrossRef]
40. Zhang, L.H.; Wang, J.; Piao, X.S. Potential Effects of 25-Hydroxycholecalciferol on the Growth Performance, Blood Antioxidant Capacity, Intestinal Barrier Function and Microbiota in Broilers under Lipopolysaccharide Challenge. *Antioxidants* **2022**, *11*, 2094. [CrossRef]
41. Zhang, B.L.; Zhong, Q.Z.; Liu, N.; Song, P.; Zhu, P.; Zhang, C.C.; Sun, Z.W. Dietary Glutamine Supplementation Alleviated Inflammation Responses and Improved Intestinal Mucosa Barrier of LPS-Challenged Broilers. *Animals* **2022**, *12*, 1729. [CrossRef] [PubMed]
42. Gao, M.; Wang, J.; Lv, Z. Supplementing Genistein for Breeder Hens Alters the Growth Performance and Intestinal Health of Offspring. *Life* **2023**, *13*, 1468. [CrossRef] [PubMed]
43. Miao, F.; Shan, C.; Geng, S.; Ning, D. Oleocanthal alleviated lipopolysaccharide-induced acute lung injury in chickens by inhibiting TLR4/NF-κB pathway activation. *Poult. Sci.* **2023**, *102*, 102458. [CrossRef] [PubMed]
44. Dong, N.; Li, X.; Xue, C.; Zhang, L.; Wang, C.; Xu, X.; Shan, A. Astragalus polysaccharides alleviates LPS-induced inflammation via the NF-κB/MAPK signaling pathway. *J. Cell Physiol.* **2020**, *235*, 5525–5540. [CrossRef] [PubMed]
45. Cui, Y.; Qu, Y.Y.; Yin, K.; Zhang, X.T.; Lin, H.J. Selenomethionine ameliorates LPS-induced intestinal immune dysfunction in chicken jejunum. *Metallomics* **2021**, *13*, mfab003. [CrossRef] [PubMed]
46. Kothari, P.; Dhaniya, G.; Sardar, A.; Sinha, S.; Girme, A.; Rai, D.; Chutani, K.; Hingorani, L.; Trivedi, R. A glucuronated flavone TMMG spatially targets chondrocytes to alleviate cartilage degeneration through negative regulation of IL-1β. *Biomed. Pharmacother.* **2023**, *163*, 114809. [CrossRef]
47. Xia, C.X.; Gao, A.X.; Zhu, Y.; Dong, T.T.; Tsim, K.W. Flavonoids from Seabuckthorn (*Hippophae rhamnoides* L.) restore CUMS-induced depressive disorder and regulate the gut microbiota in mice. *Food Funct.* **2023**, *14*, 7426–7438. [CrossRef]

48. Jiang, M.; Lv, Z.; Huang, Y.; Cheng, Z.; Meng, Z.; Yang, T.; Yan, Q.; Lin, M.; Zhan, K.; Zhao, G. Quercetin Alleviates Lipopolysaccharide-Induced Inflammatory Response in Bovine Mammary Epithelial Cells by Suppressing TLR4/NF-κB Signaling Pathway. *Front. Vet. Sci.* **2022**, *9*, 915726. [CrossRef] [PubMed]
49. Qu, Y.; Li, X.; Xu, F.; Zhao, S.; Wu, X.; Wang, Y.; Xie, J. Kaempferol Alleviates Murine Experimental Colitis by Restoring Gut Microbiota and Inhibiting the LPS-TLR4-NF-κB Axis. *Front. Immunol.* **2021**, *12*, 679897. [CrossRef]
50. He, C.; Yu, W.; Yang, M.; Li, Z.; Yu, J.; Zhong, D.; Deng, S.; Song, Z.; Cheng, S. Qi Fu Yin ameliorates neuroinflammation through inhibiting RAGE and TLR4/NF-κB pathway in AD model rats. *Aging* **2023**, *15*, 13239–13264. [CrossRef]
51. Song, B.; Li, P.; Yan, S.; Liu, Y.; Gao, M.; Lv, H.; Lv, Z.; Guo, Y.M. Effects of Dietary Astragalus Polysaccharide Supplementation on the Th17/Treg Balance and the Gut Microbiota of Broiler Chickens Challenged with Necrotic Enteritis. *Front. Immunol.* **2022**, *13*, 781934. [CrossRef] [PubMed]
52. Adewole, D.; Akinyemi, F. Gut Microbiota Dynamics, Growth Performance, and Gut Morphology in Broiler Chickens Fed Diets Varying in Energy Density with or without Bacitracin Methylene Disalicylate (BMD). *Microorganisms* **2021**, *9*, 787. [CrossRef] [PubMed]
53. Zhao, C.; Hu, X.; Bao, L.; Wu, K.; Zhao, Y.; Xiang, K.; Li, S.; Wang, Y.; Qiu, M.; Feng, L.; et al. Gut dysbiosis induces the development of mastitis through a reduction in host anti-inflammatory enzyme activity by endotoxemia. *Microbiome* **2022**, *10*, 205. [CrossRef] [PubMed]
54. Gureev, A.P.; Shaforostova, E.A.; Vitkalova, I.Y.; Sadovnikova, I.S.; Kalinina, Y.I.; Cherednichenko, V.R.; Reznikova, K.A.; Valuyskikh, V.V.; Popov, V.N. Long-term mildronate treatment increased Proteobacteria level in gut microbiome, and caused behavioral deviations and transcriptome change in liver, heart and brain of healthy mice. *Toxicol. Appl. Pharmacol.* **2020**, *398*, 115031. [CrossRef] [PubMed]
55. Xie, L.; Chi, X.; Wang, H.; Dai, A.; Dong, J.; Liu, S.; Zhang, D. Mechanism of action of buckwheat quercetin in regulating lipid metabolism and intestinal flora via Toll-like receptor 4 or nuclear factor κB pathway in rats on a high-fat diet. *Nutrition* **2023**, *115*, 112148. [CrossRef] [PubMed]
56. Zhou, L.X.; Zhang, M.M.; Wang, Y.M.; Dorfman, R.G.; Liu, H.; Yu, T.; Chen, X.; Tang, D.; Xu, L.; Yin, Y.; et al. Faecalibacterium prausnitzii Produces Butyrate to Maintain Th17/Treg Balance and to Ameliorate Colorectal Colitis by Inhibiting Histone Deacetylase 1. *Inflamm. Bowel Dis.* **2018**, *24*, 1926–1940. [CrossRef] [PubMed]
57. Cai, S.; Yang, Y.; Kong, Y.; Guo, Q.; Xu, Y.; Xing, P.; Sun, Y.; Qian, J.; Xu, R.; Xie, L.; et al. Gut Bacteria Erysipelatoclostridium and Its Related Metabolite Ptilosteroid a Could Predict Radiation-Induced Intestinal Injury. *Front. Public Health* **2022**, *10*, 862598. [CrossRef]
58. Sun, Y.; Wang, F.; Liu, Y.; Liu, S.; An, Y.; Xue, H.; Wang, J.; Xia, F.; Chen, X.; Cao, Y.G. Microbiome-metabolome responses of Fuzhuan brick tea crude polysaccharides with immune-protective benefit in cyclophosphamide-induced immunosuppressive mice. *Food Res. Int.* **2022**, *157*, 111370. [CrossRef]

animals MDPI

Article

Effects of β-Mannanase Supplementation and Soyhull Inclusion on Production Performance, Economics, Egg Quality, Blood Biochemicals, Nutrient Digestibility, and Intestinal Morphology in Golden Brown Hens (RIR × Fayoumi) during Late Peak Production

Muhammad Shuaib [1,2,*], Abdul Hafeez [2], Muhammad Tahir [3], Abubakar Sufyan [4], Obaid Ullah [5], Muhammad Adnan Shams [2], Shahrood Ahmed Siddiqui [6] and Ayman A. Swelum [7,*]

1 College of Animal Science and Technology, Yangzhou University, Yangzhou 225009, China
2 Department of Poultry Science, Faculty of Animal Husbandry & Veterinary Sciences, The University of Agriculture, Peshawar 25000, Pakistan; hafeezpak@hotmail.com (A.H.); drmuhammadadnanshams@gmail.com (M.A.S.)
3 Department of Animal Nutrition, Faculty of Animal Husbandry & Veterinary Sciences, The University of Agriculture, Peshawar 25000, Pakistan; tahir065@yahoo.co.uk
4 Department of Livestock and Poultry Production, Bahauddin Zakariya University, Multan 60800, Pakistan; linksufyan@bzu.edu.pk
5 Department of Veterinary Clinical Sciences, Faculty of Veterinary and Animal Sciences, University of Poonch, Rawalkot 12350, Pakistan; obaid.vet@upr.edu.pk
6 Vaccine Production Unit Sindh Tandojam, Livestock and Fisheries Department Government of Sindh, Karachi 70050, Pakistan; drshahroodsiddiqui@yahoo.com
7 Department of Animal Production, College of Food and Agriculture Sciences, King Saud University, Riyadh 11451, Saudi Arabia
* Correspondence: shoaibwzr@gmail.com (M.S.); aswelum@ksu.edu.sa (A.A.S.)

Citation: Shuaib, M.; Hafeez, A.; Tahir, M.; Sufyan, A.; Ullah, O.; Shams, M.A.; Siddiqui, S.A.; Swelum, A.A. Effects of β-Mannanase Supplementation and Soyhull Inclusion on Production Performance, Economics, Egg Quality, Blood Biochemicals, Nutrient Digestibility, and Intestinal Morphology in Golden Brown Hens (RIR × Fayoumi) during Late Peak Production. *Animals* **2024**, *14*, 2047. https://doi.org/10.3390/ani14142047

Academic Editor: Janghan Choi

Received: 13 June 2024
Revised: 5 July 2024
Accepted: 8 July 2024
Published: 12 July 2024

Simple Summary: This study explored the effects of the different dietary combinations of soyhulls and the β-mannanase enzyme on production performance, economics, egg quality, blood biochemicals, nutrient digestibility, and gut health in laying hens during the late peak production phase. Golden brown hens were fed different diets for four weeks, and the combination of 3% soyhulls and 30 mg/kg β-mannanase showed potential benefits. It improved production performance and increased egg weight, albumen weight, and height while significantly lowering total cholesterol, LDL, and VLDL levels. This combination also improved gut morphology and enhanced nutrient digestibility. Overall, the inclusion of 3% soyhulls and 30 mg/kg β-mannanase in the diet may have positive effects on production performance, nutrient digestibility, and gut health and potentially lower serum cholesterol levels in laying hens while maintaining acceptable egg quality.

Abstract: This study investigated the effects of the β-mannanase enzyme and soyhulls on production performance, economics, egg quality, hematology and serum biochemistry, nutrient digestibility, gut morphology, digesta viscosity, and excreta consistency in laying hens during the late peak production phase (37 to 40 weeks of age). Golden brown hens (RIR × Fayoumi; $n = 200$) were fed a control diet (no soyhulls or enzymes) and diets containing four combinations, i.e., 3% soyhulls with 20 mg/kg β-mannanase (D1), 3% soyhulls with 30 mg/kg β-mannanase (D2), 9% soyhulls with 20 mg/kg β-mannanase (D3), and 9% soyhulls with 30 mg/kg β-mannanase (D4), for four weeks in four replicates of 10 birds each. Overall, a significantly higher ($p < 0.05$) feed intake, weight gain, feed conversion ratio, and water intake were calculated in the D2 group as compared to the control and remaining combinations of soyhulls and β-mannanase. No mortality was recorded during the entire experiment. Economically, the D1 and D2 groups showed the best results as compared to the D3 and D4 groups. Egg quality parameters like egg weight, shell weight and shell thickness, yolk weight, albumen weight and height, and the Haugh unit remained unchanged ($p > 0.05$). Similarly, the D2 group showed significantly lower total cholesterol, LDL, and VLDL levels and enhanced gut morphology with greater villus width, height, crypt depth, and surface area across intestinal segments. Crude protein (CP), crude fiber (CF), crude fat, and ash digestibility were higher ($p < 0.05$) in the D1 and D2 groups compared to the control. Digesta viscosity, excreta consistency, and other egg quality parameters remained unaffected. In conclusion, the dietary inclusion of a combination of 3% soyhulls and 30 mg/kg β-mannanase may have potential benefits for laying hens by improving some production performance and egg quality indicators and economics, lowering blood cholesterol, LDL, and VLDL levels, enhancing nutrient digestibility, and improving gut morphology without affecting egg quality.

Keywords: feed intake; feces consistency; viscosity; FCR; Haugh unit

1. Introduction

The poultry industry faces a dual challenge: feeding the growing global population with protein while minimizing its environmental impact and production costs. A sustainable and cost-effective poultry production approach involves adopting practices that promote long-term ecological balance, resource conservation, and financial efficiency [1]. Formulating diets that cover the nutrient requirements of animals, including alternative feed ingredients such as soyhulls and corn distillers grains, presents ideas for producing animal-based food for human use in an efficient manner, with a substantial decrease in emissions of greenhouse gases [2]. In this context, dietary fiber represents a paradigm shift in sustainable poultry production. However, dietary fiber plays a complex role in the health and performance of poultry [3]. While insoluble fibers like cellulose and lignin are generally considered safe, their high levels can increase digesta retention time, potentially affecting nutrient absorption [4]. Soluble fibers like pectin and certain hemicelluloses, although beneficial for gut health, can lead to enteric problems at higher levels due to increased digesta viscosity [5].

Soyhulls, a by-product of oil extraction, are rich in insoluble fibers like lignocellulose but also contain varying levels of other fiber fractions [6]. Their rapid fermentation and the presence of beneficial sugars like galactomannan offer potential nutritional benefits [7–9]. However, soyhulls also contain high amounts of non-starch polysaccharides (NSPs), which can increase digesta viscosity, reduce the digestibility of nutrients, and depress growth performance [10]. Among NSPs, β-mannans are a group of related heat-resistant compounds that survive the drying–toasting phase of processing soybeans and make up 1.3 to 1.6% of dehulled or non-dehulled soybeans [11–13]. β-Mannans are mainly found in the hull and fiber fractions of soybean and are highly viscous and anti-nutritional [14]. Due to the lack of enzymes needed to efficiently digest the NSPs present in soyhulls, their utilization is limited in poultry [15]. Supplementation with exogenous enzymes like β-mannanase can address this challenge. These enzymes break down NSPs, reducing digesta viscosity and improving nutrient digestibility [16]. Therefore, β-mannanase can enhance the performance and production efficiency of poultry, making soyhulls more viable and cost-effective feed ingredients.

In the crucial peak production phase of laying hens, utilizing soyhulls and β-mannanase together in their diets holds the potential for both economic and performance benefits. Soyhulls, a cost-effective by-product of soybean processing, can partially replace soybean meal, lowering feed costs but also introducing indigestible fibers, and β-mannanase can break down β-mannans, improving nutrient digestibility and potentially increasing egg production without sacrificing egg quality. However, finding the optimal balance of soyhulls and β-mannanase requires further research, as higher soyhull levels might require more enzymes for optimal outcomes. This intriguing strategy could pave the way for sustainable and cost-effective laying-hen feed, increasing the farm profit while maintaining or even enhancing egg production during the peak period. It is therefore assumed that the addition of the enzyme β-mannanase (Hemicell^{TD}) to a soybean-hull-based diet may break down β-mannans, compensate for the negative effect of the high levels of fiber in the soyhulls, and improve the nutrient digestibility without effecting the egg quality. This study aims to investigate the effects of different combinations of soyhulls and β-mannanase on various aspects of laying hens' health and productivity during late peak egg production (33–36 weeks). By exploring the synergy between enzymes and soyhulls, this study seeks to contribute to the development of more sustainable and cost-effective feeding strategies for laying hens while ensuring optimal health and productivity.

2. Materials and Methods

2.1. Birds Housing, Experimental Diet, and Environment

A total of 200 golden brown (RIR × Fayoumi) laying hens at 36 weeks of age were purchased from a commercial market and weighed individually. The birds were randomly allocated to five groups of 40 birds each, with four replicates of 10 birds per group (five cages per replicate with 2 birds per cage). The experimental diets consisted of a control diet (no soyhulls or enzymes) and four treatment diets containing combinations of 3% or 9% soyhulls with 20 or 30 mg/kg β-mannanase (Hemicell™, Elanco Animal Health, Greenfield, IN, USA) (Table 1). All of the birds were housed under uniform environmental and management conditions. The room temperature was maintained at 23.8 °C, and the light period was 17 h/day. The flock was vaccinated (ND and IB) according to a routine schedule.

2.2. Production Performance Parameters

The formula for calculating feed intake (FI) was as follows: (FI = total feed offered − total feed used). Egg production was recorded on a daily basis. The formula for calculating hen-day egg production (HDEP) was as follows: (HDEP = total number of eggs in a particular time ÷ number of days × number of alive hens on each of these days). The weekly body weight gain (BWG) was calculated as (BWG = final body weight − starting body weight). Mortality and its causes were reported daily. Daily water intake was calculated by subtracting utilized water from given water. The formula used to calculate the feed conversion ratio (FCR) was (FCR = feed intake (kg) ÷ number of eggs × 12). [17].

2.3. Economics

The formula used to calculate total revenue (TR) was TR = total number of eggs × price per egg. The profit was found by subtracting the total cost from the total revenue. The cost–benefit ratio (CBR) was calculated by dividing the total revenue by the total cost. [18].

2.4. Egg Quality Traits

Four eggs were randomly selected from each replicate every week for four weeks to assess the internal and external quality. Individual eggs were weighed using a calibrated digital balance. The albumin and yolk were carefully separated from the eggshell, and the empty eggshell was dried overnight at 105 °C in a forced-air oven to ensure complete desiccation. Eggshell thickness, including the membranes, was measured at three locations on each egg using a micrometer screw gauge and averaged to provide a representative value. The egg contents were carefully transferred to a Petri dish, and albumin height was measured using a transparent plastic rod. The egg yolk was removed from the Petri dish using the suction technique, and the yolk weight was recorded. Finally, the separated albumin was weighed separately on the digital balance. Haugh units (HUs) were calculated using the equation outlined by [18,19]:

$$HU = 100\log10\,[H + 7.57 - 1.7W^{0.37}] \quad (1)$$

where H is the height of the albumen, and W is the weight of the egg.

Table 1. Ingredient composition of experimental diets.

Nutrient %	Experimental Diets				
	CON	D1	D2	D3	D4
Soyhulls (%)	0	3	3	9	9
β-Mannanase (mg/kg)	0	20	30	20	30
Corn	53.1	52.1	52.1	50.5	50.5
Canola meal (34%)	4.15	3.85	3.67	2.16	2.14
Soybean meal (44%)	24.3	23.6	23.6	22.2	22.2
Guar meal	0	1	1	1	1
PBM Hi fat	2.00	1.02	1.00	1.02	0.85
Poultry Oil	2.79	2.79	2.72	2.67	2.67
Common Salt	0.32	0.32	0.41	0.26	0.41
Sodium bicarbonate	0.10	0.10	0.10	0.10	0.10
Limestone	11.1	10.1	10.3	8.98	9.157
Celite	1	1	1	1	1
Dicalcium phosphate	0.77	0.77	0.75	0.77	0.62
DL-Methionine	0.08	0.08	0.08	0.07	0.08
Choline Chloride (70%)	0.10	0.10	0.10	0.10	0.10
Vitamin Premix *	0.07	0.07	0.07	0.07	0.07
Mineral Premix *	0.06	0.06	0.06	0.06	0.06
Phytase	0.01	0.01	0.01	0.01	0.01
Enramycin	0.02	0.02	0.02	0.02	0.02
Ethoxyquin	0.01	0.01	0.01	0.01	0.01
NSPs	0.02	0.00	0.00	0.00	0.00
Total	100	100	100	100	100
Analyzed value					
Dry matter	89.4	90.5	90.6	91.0	91.1
Crude protein	17.7	17.5	17.5	17.0	17.0
Crude fiber	2.85	2.87	2.87	2.90	2.90
Crude fat	4.82	4.76	4.77	4.72	4.72
Ash	13.5	13.7	13.7	13.3	13.4
Moisture	10.5	9.48	9.38	9.00	8.84
Nitrogen-Free Extract	51.6	49.0	49.0	45.8	45.8

* To provide 1 kg of diet: Retinyl acetate, 4400 IU; DL-α-tocopheryl acetate, 12 IU; Cholecalciferol, 118 μg; Thiamine, 2.5 mg; Menadione sodium bisulfite, 2.40 mg; Niacin, 30 mg; Vit. B_2, 4.8 mg; D-pantothenic acid, 10 mg; Vit. B_6, 5 mg; Vit. B_7, 130 μg; Cyanocobalamine, 19 μg; Vit. B_9, 2.5 mg; Mn, 85 mg; Zinc, 75 mg; Fe, 80 mg; Iodine, 1 mg; Selenium, 130 μg; Copper, 6 mg. PBM, Poultry by-product meal; DCP, Dicalcium phosphate; DLM, DL-Methionine; NSPs, non-starch polysaccharides. CON, control; D1 = 3%SH + 20 mg/kg β-mannanase; D2 = 3%SH + 30 mg/kg; D3 = 9%SH + 20 mg/kg; D4 = 9%SH + 30 mg/kg.

2.5. Hematology and Serum Biochemistry

On the fourth week of the study, blood samples were collected from four birds in each replicate for the analysis of white blood cells (WBCs), red blood cells (RBCs), hemoglobin (Hb), and packed cell volume (PCV). Wright–Giemsa-stained blood smears were prepared and examined under a microscope to identify and count different types of WBCs. Manual counting using a hemocytometer was performed to determine the total number of WBCs and RBCs per unit of blood volume. The microhematocrit technique, using capillary tubes and centrifugation, was employed to measure the PCV occupied by RBCs. The cyanmethemoglobin method, based on the conversion of hemoglobin to cyanmethemoglobin and subsequent spectrophotometric measurement, was used to quantify the Hb concentration in blood. Serum lipid profiles were determined using a commercial kit developed by Cromatest® Cholesterol MR (Linear Chemicals S.L., Barcelona, Spain). Total cholesterol (TC), high-density lipoprotein (HDL), and low-density lipoprotein (LDL) were analyzed in the blood serum. Very-low-density lipoprotein (VLDL) was calculated using the equation [20]

$$\mathrm{VLDL} = \mathrm{TC} - \mathrm{HDL} - \mathrm{LDL} \tag{2}$$

2.6. Digesta Viscosity and Excreta Consistency

At the end of the experiment, three birds/replicate were euthanized by cervical dislocation. To measure the digesta viscosity, fresh digesta from the gizzard to Meckel's diverticulum (proximal samples) and from Meckel's diverticulum to the junction of the ileum, cecum, and colon (distal samples) were collected from two birds per replicate. Two samples were thoroughly mixed, and approximately 1.5 g of each sample was transferred to microcentrifuge tubes. After centrifuging the tubes for 5 min at 12,700× *g*, the supernatant was collected, and viscosity was measured using a digital viscometer (Brookfield, Brookfield, MA, USA) at a constant shear rate of 42.5 s^{-1} and a controlled temperature of 40 °C following established protocols [21]. The excreta consistency was visually assessed and scored in the fourth week of the experiment using a modified scoring system (Table 2) [20,22].

Table 2. Scoring of excreta consistency.

Score	Description
1	Typical dry droppings with coning
2	No free water, loose droppings, and some coning.
3	Loose droppings, minor coning, and some free water
4	Very loose, non-coning droppings and copious quantities of free water

2.7. Apparent Total Tract Digestibility

The total collection method was used to determine nutrient digestibility. During the last three days of the experiment, three birds with similar average weights from each group were individually housed in metabolism cages equipped with separate feces collection trays. A known weight of feed was provided, and 1% Celite (Celite Corp., Lompoc, CA, USA) was added to all diets as an indigestible marker. Total excreta (including feces and urine) were collected twice daily (morning and evening) from each cage. Feathers and feed particles were carefully removed from the collected excreta, which were then weighed and stored at 4 °C for subsequent analysis. Feed and excreta samples were analyzed for dry matter (DM), crude protein (CP), crude fiber (CF), and ether extract (EE) following established protocols from AOAC (2000). The apparent digestibility of the nutrients was assessed using the index method according to the following equation, as outlined by [23,24]:

$$ATTD_X, \% = 100 - [(AIA_I/AIA_O) \times (N_O/N_I) \times 100] \quad (3)$$

where $ATTD_X$ is the apparent total tract digestibility of nutrients, AIA_I is the acid-insoluble ash concentration from dietary intake, AIA_O is the acid-insoluble ash concentration of fecal output, N_O is the nutrient concentration of fecal output, and N_I is the nutrient concentration of dietary intake.

2.8. Intestinal Histomorphology

Samples collected from the mid-regions of the duodenum, jejunum, and ileum were rinsed with 10× PBS (phosphate-buffered saline) and subsequently fixed in 10% buffered formalin for 24 h. After being preserved in 70% ethanol for 24 h, the samples were subjected to a progressive dehydration process that involved the use of increasing ethanol concentrations. This was followed by a clarifying process in xylene. The samples were immersed in paraffin wax and then cut into sections with a thickness of 5 µm. Two slices obtained at various depths were subjected to hematoxylin and eosin staining. Subsequently, images were captured at a magnification of 200× using a compound microscope (Nikon Eclipse 50, Nikon Corporation, Minato City, Tokyo, Japan). Five villi and crypts perpendicular to the muscularis mucosae and with a distinct border with the surrounding structure were chosen for further investigation. ImageJ software (version: 1.53u) was used to measure the width and height of five randomly chosen villi, as well as the depths of the crypts in each replication. The calculation of the villus surface area was performed using the formula provided by [20,25]:

$$\text{Villus surface area} = 2\pi \times (\text{Average villus width}/2) \times \text{villus height} \quad (4)$$

2.9. Statistical Analysis

Using a Completely Randomized Design (CRD), the collected data were analyzed by the GLM procedure in SPSS 21.0 (IBM Corp, Armonk, NY, USA) using a one-way ANOVA model. Replicate pens were considered experimental units. Tukey's test was applied to compare differences among the treatment groups. A p-value < 0.05 was considered statistically significant, and $0.05 \leq p\text{-value} \leq 0.1$ indicated a trend toward statistical significance.

3. Results

3.1. Production Performance Parameters

The results of the effect of soyhulls and β-mannanase on the production performance of laying hens are shown in Table 3. Feed intake (FI), weight gain (WG), the feed conversion ratio (FCR), and water intake (WI) were significantly affected ($p < 0.05$) weekly as well as overall. FI during weeks 37, 39, and 40 had significantly higher ($p < 0.05$) values in the D2 group than in the remaining groups, while at week 38, FI was significantly higher ($p < 0.05$) in the D1, D2, and D4 groups compared to the control and D3 groups. Overall, a significantly higher ($p < 0.05$) FI was calculated in the D2 group as compared to the control and remaining combinations of soyhulls and β-mannanase. WG at weeks 37 and 40 was significantly higher ($p < 0.05$) in the D1, D2, and D4 groups compared to the control and D3 groups. At week 38, WG had a significantly higher ($p < 0.05$) value in the D1 and D2 groups than in the control, D3, and D4 groups. At week 39 and overall, WG was significantly higher ($p < 0.05$) in the D2 group as compared to the control and other combinations of soyhulls and β-mannanase. The FCR was significantly lower (better) ($p < 0.05$) in the D1 and D2 groups at weeks 37 and 39 but in the D1 group at week 38 than in the remaining groups. At week 40, the FCR was significantly lower (better) ($p < 0.05$) in the D2 and D4 groups as compared to the control and other combinations of soyhull and β-mannanase. Overall, a significantly lower (better) ($p < 0.05$) FCR was calculated in the D1 and D2 groups compared to the control and D3 and D4 groups. WI recorded at weeks 37, 38, 39, and 40 was significantly higher ($p < 0.05$) in the D1 and D2 groups than in the control, D3, and D4 groups. Overall, WI had a significantly higher ($p < 0.05$) value in the D2 group in comparison to the control and other soyhull and β-mannanase combinations. Egg production (EP), hen-day egg production (HDEP), and mortality were not different ($p > 0.05$) among the groups.

3.2. Economics

Table 4 indicates the effect of the soyhull and β-mannanase combinations on the economics of laying hens. Total revenue (TR) was calculated to be significantly higher ($p < 0.05$) in the D2 group compared to the control and other combinations of soyhulls and β-mannanase. Profit and cost–benefit ratio (CBR) was calculated significantly higher ($p < 0.05$) in the control than in the D2, D3, and D4 groups. Overall, economically, the D1 and D2 groups showed the best results as compared to the D3 and D4 groups.

3.3. Egg Quality Traits

Table 5 shows the effect of the soyhull and β-mannanase combinations on external egg quality (egg weight, shell weight, and shell thickness). No significant differences in external egg quality traits were observed among the control and treatment groups ($p > 0.05$). However, egg weights were numerically higher in the D2 group and shell weights in the D1 group compared to the control (no soyhulls, no enzyme) and other soyhull–enzyme combinations. Similarly, eggshell thickness exhibited numerically higher overall values ($p > 0.05$) in the control and D1 groups compared to the other groups. No significant differences ($p > 0.05$) were observed in egg yolk weight, albumen weight, albumen height, or Haugh unit among any groups throughout the study (Table 6). However, during the second week of the experiment (i.e., 38 weeks of age), albumen weight showed a significantly higher increase in the D2 group compared to the D3 group.

Table 3. Effects of the dietary inclusion of different combinations of soyhulls and β-mannanase on the production performance of the laying hens.

Diets [1]	Weeks	CON	D1	D2	D3	D4	SEM	*p*-Value
Soyhulls (%)	-----	0	3	3	9	9		
β-Mannanase (mg/kg)	-----	0	20	30	20	30		
Feed intake (g)	37	770 ^{c}	778 ^{b}	787 ^{a}	770 ^{c}	778 ^{b}	0.88	0.047
	38	783 ^{c}	798 ^{a}	801 ^{a}	794 ^{b}	797 ^{a}	1.06	0.042
	39	765 ^{d}	786 ^{b}	793 ^{a}	780 ^{c}	788 ^{b}	1.12	0.031
	40	790 ^{c}	804 ^{b}	813 ^{a}	792 ^{c}	806 ^{b}	0.96	0.011
	Overall	3108 ^{d}	3166 ^{b}	3194 ^{a}	3136 ^{c}	3169 ^{b}	7.02	0.021
Weight gain (g)	37	13.0 ^{b}	14.0 ^{ab}	17.2 ^{a}	13.0 ^{b}	14.7 ^{ab}	1.04	0.019
	38	22.0 ^{b}	23.0 ^{ab}	24.5 ^{a}	18.0 ^{c}	21.0 ^{bc}	1.41	0.034
	39	20.0 ^{b}	18.0 ^{c}	23.0 ^{a}	21.0 ^{b}	21.2 ^{b}	1.48	0.046
	40	12.0 ^{c}	14.0 ^{ab}	15.2 ^{a}	13.0 ^{bc}	14.0 ^{ab}	1.29	0.041
	Overall	65.4 ^{c}	69.0 ^{b}	80.0 ^{a}	64.2 ^{c}	71.0 ^{b}	3.42	0.039
Egg production/bird	37	5.00	5.15	5.18	5.00	5.04	0.82	0.720
	38	5.00	5.20	5.17	5.00	5.04	0.76	0.611
	39	4.88	5.06	5.08	4.93	5.02	0.92	0.454
	40	4.87	4.98	5.10	4.89	5.02	0.58	0.150
	Overall	19.7	20.3	20.5	19.8	20.1	0.30	0.062
FCR	37	1.84 ^{a}	1.81 ^{b}	1.82 ^{b}	1.84 ^{a}	1.85 ^{a}	0.03	0.033
	38	1.87 ^{b}	1.84 ^{c}	1.86 ^{b}	1.90 ^{a}	1.89 ^{a}	0.08	0.015
	39	1.88 ^{a}	1.86 ^{b}	1.87 ^{ab}	1.89 ^{a}	1.88 ^{a}	0.06	0.046
	40	1.94 ^{a}	1.93 ^{ab}	1.91 ^{c}	1.94 ^{a}	1.92 ^{bc}	0.12	0.038
	Overall	1.88 ^{a}	1.86 ^{b}	1.86 ^{b}	1.89 ^{a}	1.89 ^{a}	0.14	0.049
Water intake (liter/bird)	37	1.23 ^{c}	1.27 ^{ab}	1.28 ^{a}	1.24 ^{c}	1.26 ^{b}	0.34	0.001
	38	1.24 ^{c}	1.27 ^{ab}	1.28 ^{a}	1.26 ^{b}	1.26 ^{b}	0.93	0.001
	39	1.26 ^{c}	1.29 ^{ab}	1.31 ^{a}	1.28 ^{b}	1.28 ^{b}	0.61	0.006
	40	1.25 ^{c}	1.28 ^{ab}	1.29 ^{a}	1.25 ^{c}	1.27 ^{b}	2.31	0.012
	Overall	5.02 ^{c}	5.12 ^{b}	5.18 ^{a}	5.05 ^{c}	5.07 ^{bc}	0.49	0.002
HDEP %	37	71.4	73.4	74.0	71.4	72.0	1.11	0.143
	38	71.4	74.2	74.0	71.4	72.0	1.19	0.239
	39	70.4	72.2	72.4	70.6	72.0	1.14	0.146
	40	70.2	71.4	73.0	70.5	72.0	1.10	0.117
	Overall	71.0	73.0	73.3	71.4	72.0	0.39	0.212
Mortality %	37	0.00	0.00	0.00	0.00	0.00	----	----
	38	0.00	0.00	0.00	0.00	0.00	----	----
	39	0.00	0.00	0.00	0.00	0.00	----	----
	40	0.00	0.00	0.00	0.00	0.00	----	----
	Overall	0.00	0.00	0.00	0.00	0.00	----	----

The means within a row that have different superscripts show a significant difference ($p < 0.05$). [1] CON, control; D1 = 3%SH + 20 mg/kg β-mannanase; D2 = 3%SH + 30 mg/kg; D3 = 9%SH + 20 mg/kg; D4 = 9%SH + 30 mg/kg. FCR = feed conversion ratio; HDEP = hen-day egg production.

Table 4. Effects of the dietary inclusion of different combinations of soyhulls and β-mannanase on the economics of laying hens.

Diets [1]	CON	D1	D2	D3	D4	SEM	*p*-Value
Soyhulls (%)	0	3	3	9	9		
β-Mannanase (mg/kg)	0	20	30	20	30		
Total revenue (R.s) [2]	324 [d]	334 [b]	337 [a]	326 [d]	331 [c]	1.18	0.002
Profit (R.s) [3]	132 [a]	130 [ab]	127 [b]	124 [c]	125 [bc]	1.44	0.016
Cost–benefit ratio	1.59 [a]	1.57 [ab]	1.55 [b]	1.52 [c]	1.51 [c]	0.03	0.002

The means within a row that have different superscripts show a significant difference ($p < 0.05$). [1] CON = control; D1 = 3%SH + 20 mg/kg β-mannanase; D2 = 3%SH + 30 mg/kg; D3 = 9%SH + 20 mg/kg; D4 = 9%SH + 30 mg/kg. [2] R.s = Rupees. [3] Profit was calculated as the difference between total revenue and total cost (feed, vaccination and medication, labor, electricity, and other miscellaneous costs).

Table 5. Effects of the dietary inclusion of different combinations of soyhulls and β-mannanase on the external egg quality.

Diets [1]	Week	CON	D1	D2	D3	D4	SEM	*p*-Value
Soyhulls (%)	-----	0	3	3	9	9		
β-Mannanase (mg/kg)	-----	0	20	30	20	30		
Egg weight (g)	37	56.5	57.0	57.7	56.8	57.3	0.64	0.706
	38	56.7	57.1	57.6	56.8	57.5	0.75	0.738
	39	57.0	57.6	58.0	57.4	58.2	0.78	0.765
	40	57.2	57.8	58.6	57.8	58.0	1.08	0.852
	Overall	56.8	57.4	58.0	57.2	57.7	0.57	0.441
Shell weight (g)	37	5.72	6.03	6.16	5.89	6.08	0.22	0.298
	38	5.22	5.73	5.81	5.39	5.58	0.18	0.100
	39	5.33	5.63	5.74	5.44	5.53	0.26	0.440
	40	5.54 [b]	5.73 [a]	5.90 [a]	5.57 [b]	5.64 [ab]	0.14	0.004
	Overall	5.50	5.82	5.74	5.50	5.70	0.09	0.220
Shell thickness (mm)	37	0.32	0.33	0.31	0.32	0.33	0.05	0.933
	38	0.35	0.33	0.34	0.33	0.31	0.20	0.368
	39	0.35	0.36	0.34	0.34	0.35	0.07	0.655
	40	0.34	0.34	0.34	0.33	0.35	0.05	0.451
	Overall	0.34	0.34	0.33	0.33	0.33	0.04	0.477

The means within a row that have different superscripts show a significant difference ($p < 0.05$). [1] CON = control; D1 = 3%SH + 20 mg/kg β-mannanase; D2 = 3%SH + 30 mg/kg; D3 = 9%SH + 20 mg/kg; D4 = 9%SH + 30 mg/kg.

3.4. Hematology and Serum Biochemistry

Most hematological and serum biochemistry parameters were within normal ranges and showed no significant differences ($p > 0.05$) across the various soyhull–enzyme combinations (Table 7). These included red blood cell (RBC) count, white blood cell (WBC) count, heterophil percentage, lymphocyte percentage, packed cell volume (PCV), hemoglobin (Hb), mean corpuscular hemoglobin concentration (MCHC), mean corpuscular volume (MCV), high-density lipoprotein (HDL), heterophile/lymphocyte ratio (H:L), and total protein (TP). However, significant reductions were observed in total cholesterol (TC), low-density lipoprotein (LDL), and very-low-density lipoprotein (VLDL) levels for several soyhull–enzyme combinations compared to the control group. Notably, the D4 group displayed the lowest TC, LDL, and VLDL levels. Moreover, the D3 and D4 groups exhibited lower TC, LDL, and VLDL levels compared to the control, D1, and D2 groups.

Table 6. Effects of the dietary inclusion of different combinations of soyhulls and β-mannanase on the internal egg quality.

Diets [1]	Week	CON	D1	D2	D3	D4	SEM	*p*-Value
Soyhulls (%)	-----	0	3	3	9	9		
β-Mannanase (mg/kg)	-----	0	20	30	20	30		
Yolk weight (g)	37	17.6	17.0	17.2	18.2	17.0	0.58	0.776
	38	17.0	16.6	16.5	17.7	16.7	0.82	0.873
	39	17.3	16.8	17.4	17.4	16.6	0.64	0.897
	40	17.0	17.2	17.6	18.1	17.2	0.49	0.675
	Overall	17.2	16.9	17.2	17.8	16.9	0.70	0.803
Albumen weight (g)	37	32.7	33.1	33.6	31.97	33.5	0.46	0.186
	38	33.9 [ab]	34.6 [a]	35.0 [a]	32.8 [b]	34.7 [a]	0.39	0.004
	39	34.1	34.9	34.6	34.0	35.0	0.60	0.710
	40	34.2	33.9	35.0	33.8	34.8	0.51	0.293
	Overall	33.7	34.1	34.5	33.5	34.4	0.27	0.061
Albumen height (mm)	37	6.73	7.00	7.08	6.88	6.95	0.24	0.881
	38	6.83	7.10	7.18	6.98	7.05	0.19	0.884
	39	6.65	6.90	7.08	6.75	6.80	0.22	0.660
	40	6.35	6.60	6.78	6.45	6.50	0.28	0.663
	Overall	6.64	6.90	7.03	6.76	6.83	0.17	0.689
Haugh unit	37	76.5	77.5	78.4	75.6	76.7	1.09	0.706
	38	76.7	77.7	78.4	77.4	78.6	1.84	0.946
	39	75.4	76.6	76.9	75.5	76.0	1.84	0.945
	40	76.2	76.9	77.2	76.4	77.0	1.21	0.991
	Overall	76.2	77.2	77.7	76.2	77.0	0.74	0.338

The means within a row that have different superscripts show a significant difference ($p < 0.05$). [1] CON = control; D1 = 3%SH + 20 mg/kg β-mannanase; D2 = 3%SH + 30 mg/kg; D3 = 9%SH + 20 mg/kg; D4 = 9%SH + 30 mg/kg.

Table 7. Effects of the dietary inclusion of different combinations of soyhulls and β-mannanase on hematological and serum biochemical indices in laying hens at 40 weeks of age.

Diets [1]	CON	D1	D2	D3	D4	SEM	*p*-Value
Soyhulls (%)	0	3	3	9	9		
β-Mannanase (mg/kg)	0	20	30	20	30		
Hematological Indices [2]							
RBCs (10^6μL)	3.17	3.21	3.23	3.18	3.20	0.02	0.070
WBCs (10^3μL)	3.10	3.06	3.05	3.07	3.08	0.01	0.061
Heterophile (%)	22.0	23.0	24.0	22.0	23.1	1.50	0.675
Lymphocyte (%)	58.0	59.1	59.2	58.2	58.7	1.19	0.765
H:L	0.38	0.38	0.40	0.37	0.39	0.03	0.857
PCV (%)	27.4	30.0	30.1	27.8	28.9	0.64	0.063
HB (g/dL)	10.3	11.0	11.2	10.4	10.5	0.12	0.061
MCHC (g/dL)	38.0	37.0	37.2	37.4	36.3	0.42	0.062
MCV (fL)	88.8	91.4	91.1	89.4	90.3	1.95	0.074
Serum Biochemical Indices [3]							
TC (mg/dL)	132 [a]	125 [b]	124 [b]	122 [c]	121 [c]	1.62	0.007
HDL (mg/dL)	67.0	67.0	68.0	70.0	70.0	1.41	0.129
LDL (mg/dL)	39.2 [a]	37.0 [ab]	35.3 [b]	33.4 [c]	32.1 [c]	1.06	0.002
VLDL (mg/dL)	25.4 [a]	21.0 [b]	21.3 [b]	18.3 [c]	18.0 [c]	2.02	0.042
TP (mg/dL)	5.00	5.07	5.12	5.04	5.10	0.02	0.074

The means within a row that have different superscripts show a significant difference ($p < 0.05$). [1] CON, control; D1 = 3%SH + 20 mg/kg β-mannanase; D2 = 3%SH + 30 mg/kg; D3 = 9%SH + 20 mg/kg; D4 = 9%SH + 30 mg/kg. [2] Red blood cells (RBCs); white blood cells (WBCs); heterophil-to-lymphocyte ratio (H:L); packed cell volume (PCV); hemoglobin (Hb); mean corpuscular hemoglobin concentrations (MCHC); mean corpuscular volume (MCV). [3] Total cholesterol (TC); high-density lipoprotein (HDL); low-density lipoprotein (LDL); very-low-density lipoprotein (VLDL); total protein (TP).

3.5. Total Tract Digestibility, Digesta Viscosity, and Excreta Consistency

Table 8 summarizes the effects of different soyhull and β-mannanase combinations on total tract digestibility, digesta viscosity, and excreta consistency in laying hens. Dry matter (DM) digestibility was non-significant ($p > 0.05$) among the groups. Crude protein digestibility was significantly higher ($p < 0.05$) in the D2 group compared to the control and other combinations. Crude fiber, crude fat, and ash digestibility were significantly higher ($p < 0.05$) in the D1 and D2 groups compared to the D3, D4, and control groups. Digesta viscosity was numerically the highest in the D3 group, but no significant differences ($p > 0.05$) were found between the groups. Excreta consistency was similar in all groups, typical dry droppings with coning. In summary, the D1 and D2 groups seem the most beneficial for digestibility, particularly for dry matter, crude protein, fiber, fat, and ash. Higher soyhull or lower β-mannanase levels appear less effective in improving digestibility. Digesta viscosity, while not statistically significant, showed a trend toward higher values with higher soyhull and lower β-mannanase levels. All groups maintained normal excreta consistency, indicating no adverse effects on gut health.

Table 8. Effects of the dietary inclusion of soybean hulls and enzyme on the total tract digestibility, digesta viscosity, and excreta consistency.

Diets [1]	CON	D1	D2	D3	D4	SEM	p-Value
Soyhulls (%)	0	3	3	9	9		
β-Mannanase (mg/kg)	0	20	30	20	30		
Dry matter %	79.0	81.4	82.0	80.0	82.0	1.02	0.062
Crude protein %	68.0 ^{c}	71.0 ^{b}	74.0 ^{a}	68.0 ^{c}	70.0 ^{b}	1.38	0.029
Crude fiber %	68.0 ^{c}	71.0 ^{ab}	72.0 ^{a}	69.0 ^{bc}	69.0 ^{bc}	1.03	0.005
Crude fat %	75.0 ^{c}	80.0 ^{a}	80.4 ^{a}	77.0 ^{b}	77.4 ^{b}	0.88	0.003
Ash %	56.2 ^{c}	59.0 ^{ab}	60.2 ^{a}	58.0 ^{b}	58.0 ^{b}	0.87	0.007
Viscosity (cP)	4.76	4.73	4.70	5.02	4.78	0.16	0.521
Excreta consistency [2]	1	1	1	1	1		------

The means within a row that have different superscripts show a significant difference ($p < 0.05$). [1] CON, control; D1 = 3%SH + 20 mg/kg β-mannanase; D2 = 3%SH + 30 mg/kg; D3 = 9%SH + 20 mg/kg; D4 = 9%SH + 30 mg/kg. [2] For excreta consistency, '1' shows normal dry droppings and cone formation. Centipoise, cP.

3.6. Intestinal Histomorphology

The effects of different dietary combinations of soyhulls and β-mannanase on the villus morphology of the duodenum, jejunum, and ileum in laying hens are shown in Table 9. In the duodenum, the villus width (Vw) and height (Vh), crypt depth (Cd), and surface area (VSA) were significantly higher in the D2 group compared to the control group and other combinations. The villus surface area (VSA) was significantly higher in the D2 group compared to all other groups, further solidifying its improved absorptive capacity. In the jejunum, Vw was significantly higher in the D2 and D4 groups compared to the control and other combinations of soyhulls and β-mannanase. Vh was significantly higher in the D2 group than in the control, D1, D3, and D4 groups. Cd was significantly higher in the D1 and D2 groups compared to the control, D3, and D4 groups. The VSA followed the same pattern as Vh, with the D2 group showing the highest value and significant differences compared to the control, D3, and D4 groups. In the ileum, Vw followed a similar pattern to that in the jejunum, with the D2 and D4 groups having the highest values and significant differences compared to the control, D1, and D3 groups. Similar to the jejunum Vh, the ileum Vh was significantly higher in the D2 group compared to the control and other combinations. Cd was again significantly higher in the D1, D2, and D4 groups and lower in the control and D3 groups. The VSA followed the same pattern as Vh, with the D2 group showing the highest value and significant differences compared to the control and other combinations of soyhulls and β-mannanase. Overall, the D2 group appears to be the most beneficial for intestinal villus morphology in laying hens. This diet promotes increased villus height and width while maintaining the highest overall absorptive surface area.

Table 9. Effects of dietary inclusion of soyhulls and enzyme (β-mannanase) on intestinal histomorphology.

Diets [1]	**CON**	**D1**	**D2**	**D3**	**D4**	**SEM**	***p*-Value**
Soyhulls (%)	0	3	3	9	9		
β-Mannanase (mg/kg)	0	20	30	20	30		
Duodenum [2]							
Vw (um)	75.0 [d]	83.0 [b]	89.0 [a]	79.0 [c]	83.0 [b]	2.65	0.022
Vh (um)	595 [d]	613 [b]	621 [a]	604 [c]	611 [b]	6.64	0.028
Cd (um)	100 [d]	107 [b]	111 [a]	102 [cd]	103 [d]	4.10	0.036
VSA (mm^2)	0.139 [d]	0.160 [b]	0.172 [a]	0.149 [c]	0.158 [b]	0.04	0.006
Jejunum							
Vw (um)	69.0 [c]	72.0 [b]	76.0 [a]	70.0 [c]	74.0 [ab]	2.39	0.039
Vh (um)	447 [d]	459 [c]	470 [a]	457 [c]	466 [b]	5.44	0.021
Cd (um)	53.0 [d]	69.0 [ab]	71.0 [a]	61.0 [c]	67.0 [b]	2.37	0.009
VSA (mm^2)	0.096 [d]	0.103 [b]	0.111 [a]	0.100 [b]	0.107 [c]	0.04	0.007
Ileum							
Vw (um)	62.0 [c]	65.0 [b]	68 [a]	63.0 [bc]	67.0 [a]	2.19	0.043
Vh (um)	397 [d]	407 [b]	412 [a]	403 [c]	401 [c]	4.88	0.032
Cd (um)	59.0 [bc]	60 [ab]	62.0 [a]	58.0 [c]	60.0 [ab]	2.12	0.032
VSA (mm^2)	0.078 [c]	0.083 [b]	0.086 [a]	0.079 [c]	0.082 [b]	0.03	0.038

The means within a row that have different superscripts show a significant difference ($p < 0.05$). [1] CON, control; D1 = 3%SH + 20 mg/kg β-mannanase; D2 = 3%SH + 30 mg/kg; D3 = 9%SH + 20 mg/kg; D4 = 9%SH + 30 mg/kg. [2] Vw = villus width; Vh = villus height; Cd = crypt depth; Vh: Cd = villus height to crypt depth ratio; VSA = villus surface area.

4. Discussion

The late phase of peak egg production in laying hens (around 32–44 weeks of age) presents unique challenges. As hens age, their metabolic rate slows down, and their appetite naturally decreases [26]. Hens prioritize egg production, channeling a notable portion of their dietary intake toward yolk and albumen synthesis. This creates an increased demand for essential nutrients like protein, calcium, vitamins, and minerals, which further exacerbates the nutrient deficiency issue, creating a vicious cycle. Dietary fiber, often seen as an antagonist to productivity, can play a surprisingly valuable role in this crucial period. Including moderate fiber (2–4%) boosts beneficial gut bacteria, enhancing nutrient absorption, digestion, and immunity [27,28], while insoluble fiber can increase the feeding time and induce a feeling of fullness, potentially helping hens regulate their feed intake and preventing overconsumption [5,27]. Studies suggest that certain fiber sources (e.g., inulin and lignocellulose) might improve eggshell thickness and reduce the incidence of shell cracks and breaks [29,30]. The physical stimulation provided by some fiber sources might redirect foraging behavior and reduce feather pecking, a common welfare concern in laying hens [31]. Different fiber sources have varying digestibility and fermentability properties. Highly fermentable fibers, while beneficial for gut health, may reduce the energy available for egg production at high inclusion levels [32]. Recent research suggests that combining soyhulls with β-mannanase supplementation in their diets presents a promising avenue for achieving this balance [33–35]. Soyhulls, a readily available and cost-effective by-product of soybean processing, offer the potential to partially replace soybean meal in poultry diets, thereby reducing feed costs [36]. However, their high fiber content, primarily composed of non-starch polysaccharides like beta mannans, limits nutrient digestibility and might negatively impact egg production [15]. This is where B-mannanase, an enzyme specific to mannan degradation, plays a vital role. By breaking down these complex fibers, B-mannanase increases intestinal nutrient digestibility and utilization [37]. Several studies have reported its efficacy in laying hens, demonstrating improvements in egg production, the feed conversion ratio, and egg weight without compromising egg quality [38–40]. However, determining the optimal synergistic combination of soyhulls and B-mannanase is critical for realizing their full potential. While higher soyhull inclusion can further reduce feed costs, it might also necessitate a higher B-mannanase dosage to maintain efficient nutrient utilization.

Our study investigated the effects of adding different combinations of soyhulls and the β-mannanase enzyme to the diets of laying hens. The inclusion of 3% soyhulls and 30 mg/kg β-mannanase enzyme resulted in significantly higher overall feed intake, weight gain, and water intake and a better FCR than in the remaining groups, while egg production and mortality were not different among the groups. These results are in agreement with [41], who found that broilers fed 20% soya bean hull meal with 1% Safzyme had significantly higher feed intake and non-significant mortality. The inclusion of 100g/ton enzyme complex increased hen feed consumption compared to the control [42]. The current research findings align with those of [43], who observed increased feed intake in laying hens when enzymes were included in the feed at a concentration of 0.1-0.5 percent. The quantities of distiller's dried grains with solubles (DDGS) and enzymes had a notable effect on feed efficiency (FCR) [44]. Similar to our findings, laying hens fed 10, 20, and 30% soybean husks with 2% cellulitic enzyme gained more weight than those fed them without the enzyme [45]. The results are consistent with [42], who found that 100 g/t of an enzyme complex (xylanase, ß-glucanase, and phytase-based) in feed formulations increased laying hens' performance and egg production. Adding xylanase and phytase alone or in combination with wheat-based laying-hen diets with low phosphorus and corn–soya-based layer diets did not alter egg production [46–48]. Similar to the current investigation, feeding the treatment groups 0.1-0.5% enzyme increased the feed conversion ratio (from 2.15 to 2.03) [43]. Similarly, the provision of the enzyme Quatrazyme (20 mg/kg) in the feed of broilers improved the FCR from 2.11 to 1.99 [44]. The better feed intake in the soybean hull and enzyme diet groups is due to the beneficial effect of the enzyme on the gastrointestinal tract and its ability to break down the cell wall of the soybean hull into easily digestible components, and similarly, [49] determined that an increase in feed intake occurs solely after the viscosity is reduced by enzymatic supplementation, which degrades the NSP components of the diet. The increased body weight gain in the soybean hull and enzyme diet group is due to the improved feed intake. The better FCR in the 3% soyhull and 30 mg/kg β-mannanase enzyme diet group than in the remaining groups is a result of the comparatively high egg output in this group. The type of dietary fiber determines whether water intake rises or decreases, although ambient temperature, feed composition, and the physicochemical characteristics of diet constituents and components may influence this connection [50]. The findings of the current investigation are consistent with the significant correlation between feed intake and water intake [51], whereby the higher feed intake in the enzyme and SH groups led to a corresponding rise in water consumption.

The total revenue was higher ($p < 0.05$) in the 3% soyhull and 30 mg/kg β-mannanase enzyme diet group as compared to all other groups, while the profit and cost–benefit ratio (CBR) were higher ($p < 0.05$) in the control and 3% soyhull and 20 mg/kg β-mannanase groups than in the remaining groups, and similarly, negative feed cost savings and higher feed costs were calculated when using 10 and 20% soybean hull meal and 0.1% cellulitic enzyme (Safzyme) in the broiler finisher diet [41]. Similarly, adding alternative fiber sources (coffee husks and soybean hulls) and 0.075 g/kg xylanase to the diet of laying hens resulted in increased feed costs per egg carton [52]. Numerous factors, such as the type and quantity of cereal consumed, the amount of anti-nutritive ingredients in a specific cereal, the amounts of enzymes used, the animal's age and type, the physiology of the bird, and the type of gut microflora, all affect the inclusion of enzymes in the diet and the improvement they achieve [53]. Due to increased egg production, the 3% soyhull and 30 mg/kg β-mannanase enzyme group had better overall income than the other groups, and the lower feed intake in the control group led to greater profits than in the other treatment groups. Feed, birds, eggs, and enzymes are all subject to market fluctuations, which primarily affect how dietary soybean hulls and enzymes are used in layer feed.

While statistically insignificant, we observed numerically higher egg weight in hens receiving 3% soyhulls and 30mg/kg enzyme. This finding aligns with previous research that also reported increased egg weight with enzyme supplementation in the diet [54–56]. However, other egg quality parameters, like the eggshell weight and thickness, yolk and albumen weight, albumen height, and Haugh unit, remained unaffected by our treatment. This suggests that while the enzyme may have slightly increased nutrient utilization and availability, it was not sufficient to significantly alter the proportions of different components within the egg.

Our study revealed a significant reduction in total cholesterol, LDL, and VLDL values with the combination of 9% soyhulls and 20 and 30 mg/kg enzyme compared to the control. These findings align with previous studies with the inclusion of higher levels of dietary fiber and a reduction in serum cholesterol levels in poultry [57–59]. Soyhulls contain both soluble and insoluble fiber portions, and the soluble portion of fiber lowers cholesterol by binding to it in the small intestine and preventing it from entering the bloodstream and exiting the body through the excreta [60,61]. The hematology and serum biochemistry parameters measured in our study all fell within the normal ranges [62]. This indicates that the dietary treatments did not negatively impact the overall health or internal physiology of the laying hens. Our findings align with previous studies reporting that the inclusion of up to 9% dietary fiber did not induce any significant changes in the internal physiology of the poultry [63,64]. This observation suggests that moderate levels of soyhulls with enzymes, as implemented in this study, are well tolerated and do not adversely affect the internal health of laying hens.

Our study indicated significantly higher total tract digestibility of crude protein, crude fiber, crude fat, and ash in the groups receiving 3% soyhulls and 20 or 30 mg/kg β-mannanase. This finding aligns with previous studies reporting improved digestibility with β-mannanase supplementation in poultry [65,66]. This may be explained by the fact that β-mannanase can hydrolyze the β-mannans in soyhulls, which are known to reduce nutrient digestibility by increasing the viscosity of intestinal digesta and inhibiting the activity of digestive enzymes [66,67]. By breaking down β-mannans, β-mannanase may reduce the viscosity of digesta, improve the mixing of enzymes and substrates, and increase the absorption of nutrients. The optimal level of β-mannanase for digestibility may depend on the content of β-mannans in the diet, as higher levels of β-mannans may require higher levels of β-mannanase to be effectively degraded. The results also indicate that higher levels of soyhulls or lower levels of β-mannanase may have negative effects on digestibility, as they may increase the viscosity of digesta and reduce the availability of nutrients. This is consistent with previous studies that reported the lower digestibility of dry matter, crude protein, fiber, and fat in pigs fed diets containing high levels of soyhulls or low levels of β-mannanase [67]. The increased viscosity of digesta may also affect intestinal morphology and health, as it may impair the mucosal barrier function and increase the susceptibility to pathogens [68]. However, in this study, no significant differences were observed in digesta viscosity or excreta consistency among the groups, suggesting that the levels of soyhulls and β-mannanase used in this study did not cause any adverse effects on gut health.

Our findings reveal a synergistic interaction between the combination of soyhulls and β-mannanase, significantly enhancing the intestinal villus structure and potentially improving nutrient absorption. In all three intestinal segments, hens fed the 3% soyhull and 30mg/kg β-mannanase diet exhibited the highest villus width and height. This observation aligns with previous studies demonstrating similar positive effects of moderate fiber inclusion and enzyme supplementation on gut morphology in poultry [68,69]. β-Mannanase specifically targets and degrades non-starch polysaccharides (NSPs) present in soyhulls, releasing trapped nutrients and potentially stimulating intestinal epithelial cell proliferation [70].

5. Conclusions

The findings of this study suggest that the combination of 3% soyhulls and 30mg/kg β-mannanase may offer a promising approach to improving the production performance, digestibility, gut health, cholesterol levels, and, potentially, economics of laying hens while maintaining acceptable egg quality.

Author Contributions: Conceptualization, M.S. and A.H.; methodology, M.S. and A.H.; validation, M.S., A.H. and M.T.; formal analysis, M.S.; investigation, M.S. and A.H.; resources, M.S.; data curation, A.H., M.T., A.S., O.U., M.A.S., S.A.S. and A.A.S.; writing—original draft preparation, M.S.; writing—review and editing, A.H., M.T., A.S., O.U., M.A.S., S.A.S. and A.A.S.; supervision, A.H. and M.T. All authors have read and agreed to the published version of the manuscript.

Funding: The authors extend their appreciation to the Researchers Supporting Project number (RSPD2024R971), King Saud University, Riyadh, Saudi Arabia, for funding this research.

Institutional Review Board Statement: The animal study protocol was approved by the Ethics Committee of The University of Agriculture Peshawar, Pakistan (protocol code: 6780-A/LM, B&G/UOA and date of approval: 31 December 2020).

Informed Consent Statement: Not applicable.

Data Availability Statement: The data that support the findings of this study are available from the corresponding author upon reasonable request.

Acknowledgments: We acknowledge the staff of the Department of Poultry Science and Faculty of Animal Husbandry and Veterinary Science (FAHVS), The University of Agriculture, Peshawar, Pakistan, which provided technical and laboratory facilities, and the Higher Education Commission (HEC) of Pakistan, which provided an HEC Indigenous Scholarship (grant number 518-74444-2AV5-110) for conducting the experiment in Pakistan. Also, the authors extend their appreciation to the Researchers Supporting Project number (RSPD2024R971), King Saud University, Riyadh, Saudi Arabia, for funding this research.

Conflicts of Interest: The authors declare no conflicts of interest.

References

1. Pretty, J. Agricultural Sustainability: Concepts, Principles and Evidence. *Philos. Trans. R. Soc. B Biol. Sci.* **2008**, *363*, 447–465. [CrossRef] [PubMed]
2. Niemann, H.; Kuhla, B.; Flachowsky, G. Perspectives for Feed-Efficient Animal Production1. *J. Anim. Sci.* **2011**, *89*, 4344–4363. [CrossRef]
3. Tejeda, O.J.; Kim, W.K. Role of Dietary Fiber in Poultry Nutrition. *Animals* **2021**, *11*, 461. [CrossRef]
4. Röhe, I.; Zentek, J. Lignocellulose as an Insoluble Fiber Source in Poultry Nutrition: A Review. *J. Anim. Sci. Biotechnol.* **2021**, *12*, 82. [CrossRef]
5. Mateos, G.G.; Jiménez-Moreno, E.; Serrano, M.P.; Lázaro, R.P. Poultry Response to High Levels of Dietary Fiber Sources Varying in Physical and Chemical Characteristics. *J. Appl. Poult. Res.* **2012**, *21*, 156–174. [CrossRef]
6. Liu, H.-M.; Li, H.-Y. Application and Conversion of Soybean Hulls. In *Soybean—The Basis of Yield, Biomass and Productivity*; Kasai, M., Ed.; InTech: Hirosaki, Japan, 2017; ISBN 978-953-51-3117-5.
7. Sun, X.; Debeni Devi, N.; Urriola, P.E.; Tiffany, D.G.; Jang, J.-C.; Shurson, G.G.; Hu, B. Feeding Value Improvement of Corn-Ethanol Co-Product and Soybean Hull by Fungal Fermentation: Fiber Degradation and Digestibility Improvement. *Food Bioprod. Process.* **2021**, *130*, 143–153. [CrossRef]
8. Whistler, R.L.; Saarnio, J. Galactomannan from Soy Bean Hulls. *J. Am. Chem. Soc.* **1957**, *79*, 6055–6057. [CrossRef]
9. Porfiri, M.C.; Wagner, J.R. Extraction and Characterization of Soy Hull Polysaccharide-Protein Fractions. Analysis of Aggregation and Surface Rheology. *Food Hydrocoll.* **2018**, *79*, 40–47. [CrossRef]
10. Choct, M.; Dersjant-Li, Y.; McLeish, J.; Peisker, M. Soy Oligosaccharides and Soluble Non-Starch Polysaccharides: A Review of Digestion, Nutritive and Anti-Nutritive Effects in Pigs and Poultry. *Asian-Australas. J. Anim. Sci.* **2010**, *23*, 1386–1398. [CrossRef]
11. Odetallah, N.; Ferket, P.; Grimes, J.; McNaughton, J. Effect of Mannan-Endo-1,4-Beta-Mannosidase on the Growth Performance of Turkeys Fed Diets Containing 44 and 48% Crude Protein Soybean Meal. *Poult. Sci.* **2002**, *81*, 1322–1331. [CrossRef]
12. Dierick, N.A. Biotechnology Aids to Improve Feed and Feed Digestion: Enzymes and Fermentation. *Arch. Für Tierernaehrung* **1989**, *39*, 241–261. [CrossRef] [PubMed]
13. Hsiao, H.-Y.; Anderson, D.M.; Dale, N.M. Levels of β-Mannan in Soybean Meal. *Poult. Sci.* **2006**, *85*, 1430–1432. [CrossRef] [PubMed]

14. Barros, V.R.S.M.D.; Lana, G.R.Q.; Lana, S.R.V.; Lana, Â.M.Q.; Cunha, F.S.A.; Neto, J.V.E. β-Mannanase and Mannan Oligosaccharides in Broiler Chicken Feed. *Ciênc. Rural* **2014**, *45*, 111–117. [CrossRef]
15. Nguyen, H.T.; Bedford, M.R.; Morgan, N.K. Importance of Considering Non-Starch Polysaccharide Content of Poultry Diets. *Worlds Poult. Sci. J.* **2021**, *77*, 619–637. [CrossRef]
16. Balasubramanian, B.; Ingale, S.L.; Park, J.H.; Rathi, P.C.; Shanmugam, S.; Kim, I.H. Inclusion of Dietary β-Mannanase Improves Performance and Ileal Digestibility and Reduces Ileal Digesta Viscosity of Broilers Fed Corn-Soybean Meal Based Diet. *Poult. Sci.* **2018**, *97*, 3097–3101. [CrossRef] [PubMed]
17. Shuaib, M.; Hafeez, A.; Chand, N.; Tahir, M. Effect of dietary inclusion of soybean hull on production performance and nutrient digestibility during peak egg production period with different phases in laying hens. *Pak. J. Zool.* **2022**, *55*, 397–405. [CrossRef]
18. Shuaib, M.; Hafeez, A.; Islam, Z.; Shah, A.A.; Ullah, S. Effect of dietary inclusion of soybean hulls in the diet on feed proximate analysis, egg quality parameters and economics during peak egg production stages in laying hens. *Pak. J. Zool.* **2022**, *55*, 1309–1315. [CrossRef]
19. Haugh, R. The Haugh Unit for Measuring Egg Quality. *U. S. Egg Poult. Mag.* **1937**, *43*, 552–555.
20. Shuaib, M.; Hafeez, A.; Kim, W.K.; Khan, A.; Sufyan, A. Effect of dietary inclusion of soybean hulls in basal diet on digesta viscosity, fecal consistency, hematology, serum biochemistry, and intestinal morphometric parameters in the laying hens during peak egg production stages. *Pak. J. Zool.* **2024**, *56*, 1079–1087. [CrossRef]
21. Bedford, M.R.; Classen, H.L. An In Vitro Assay for Prediction of Broiler Intestinal Viscosity and Growth When Fed Rye-Based Diets in the Presence of Exogenous Enzymes. *Poult. Sci.* **1993**, *72*, 137–143. [CrossRef]
22. Roland, D.A.; Caldwell, D. Relationship of Calcium to Wet Droppings in Laying Hens. *Poult. Sci.* **1985**, *64*, 1809–1812. [CrossRef]
23. Huang, K.H.; Li, X.; Ravindran, V.; Bryden, W.L. Comparison of Apparent Ileal Amino Acid Digestibility of Feed Ingredients Measured with Broilers, Layers, and Roosters. *Poult. Sci.* **2006**, *85*, 625–634. [CrossRef] [PubMed]
24. Shuaib, M.; Paneru, D.; Hafeez, A.; Tahir, T.; Kim, W.K. The chemical composition of soyhulls and their effect on amino acid and nutrient digestibility in laying hens during the peak of production. *Animals* **2023**, *13*, 2808. [CrossRef]
25. Solis De Los Santos, F.; Tellez, G.; Farnell, M.B.; Balog, J.M.; Anthony, N.B.; Pavlidis, H.O.; Donoghue, A.M. Hypobaric Hypoxia in Ascites Resistant and Susceptible Broiler Genetic Lines Influences Gut Morphology. *Poult. Sci.* **2005**, *84*, 1495–1498. [CrossRef] [PubMed]
26. Joyner, C.J.; Peddie, M.J.; Taylor, T.G. The Effect of Age on Egg Production in the Domestic Hen. *Gen. Comp. Endocrinol.* **1987**, *65*, 331–336. [CrossRef] [PubMed]
27. Desbruslais, A.; Wealleans, A.; Gonzalez-Sanchez, D.; Di Benedetto, M. Dietary Fibre in Laying Hens:A Review of Effects on Performance, Gut Health and Feather Pecking. *Worlds Poult. Sci. J.* **2021**, *77*, 797–823. [CrossRef]
28. Hou, L.; Sun, B.; Yang, Y. Effects of Added Dietary Fiber and Rearing System on the Gut Microbial Diversity and Gut Health of Chickens. *Animals* **2020**, *10*, 107. [CrossRef] [PubMed]
29. Shang, H.; Zhao, J.; Dong, X.; Guo, Y.; Zhang, H.; Cheng, J.; Zhou, H. Inulin Improves the Egg Production Performance and Affects the Cecum Microbiota of Laying Hens. *Int. J. Biol. Macromol.* **2020**, *155*, 1599–1609. [CrossRef] [PubMed]
30. Sozcu, A.; Ipek, A. The Effects of Lignocellulose Supplementation on Laying Performance, Egg Quality Parameters, Aerobic Bacterial Load of Eggshell, Serum Biochemical Parameters, and Jejunal Histomorphological Traits of Laying Hens. *Poult. Sci.* **2020**, *99*, 3179–3187. [CrossRef]
31. Mens, A.J.W.; Van Krimpen, M.M.; Kwakkel, R.P. Nutritional Approaches to Reduce or Prevent Feather Pecking in Laying Hens: Any Potential to Intervene during Rearing? *Worlds Poult. Sci. J.* **2020**, *76*, 591–610. [CrossRef]
32. Jha, R.; Mishra, P. Dietary Fiber in Poultry Nutrition and Their Effects on Nutrient Utilization, Performance, Gut Health, and on the Environment: A Review. *J. Anim. Sci. Biotechnol.* **2021**, *12*, 51. [CrossRef]
33. Scapini, L.; Rorig, A.; Ferrarini, A.; Fülber, L.; Canavese, M.; Silva, A.; Fernandes, J. Nutritional Evaluation of Soybean Hulls with or without β-Mannanase Supplement on Performance, Intestinal Morphometric and Carcass Yield of Broilers Chickens. *Braz. J. Poult. Sci.* **2018**, *20*, 633–642. [CrossRef]
34. Kiarie, E.G.; Steelman, S.; Martinez, M.; Livingston, K. Significance of Single β-Mannanase Supplementation on Performance and Energy Utilization in Broiler Chickens, Laying Hens, Turkeys, Sows, and Nursery-Finish Pigs: A Meta-Analysis and Systematic Review. *Transl. Anim. Sci.* **2021**, *5*, txab160. [CrossRef] [PubMed]
35. Kong, C.; Lee, J.H.; Adeola, O. Supplementation of β-Mannanase to Starter and Grower Diets for Broilers. *Can. J. Anim. Sci.* **2011**, *91*, 389–397. [CrossRef]
36. Loy, D.D.; Lundy, E.L. Nutritional Properties and Feeding Value of Corn and Its Coproducts. In *Corn*; Elsevier: Amsterdam, The Netherlands, 2019; pp. 633–659. ISBN 978-0-12-811971-6.
37. Dekker, R.F. Biodegradation of the Hemicelluloses. In *Biosynthesis and Biodegradation of Wood Components*; Academic Press: Cambridge, MA, USA, 1985; pp. 505–533.
38. Carvalho, C.L.; Andretta, I.; Galli, G.M.; Bastos Stefanello, T.; Camargo, N.D.O.T.; Mendes, R.E.; Pelisser, G.; Balamuralikrishnan, B.; Melchior, R.; Kipper, M. Dietary Supplementation with β-Mannanase and Probiotics as a Strategy to Improve Laying Hen Performance and Egg Quality. *Front. Vet. Sci.* **2023**, *10*, 1229485. [CrossRef]
39. White, D.; Adhikari, R.; Wang, J.; Chen, C.; Lee, J.H.; Kim, W.K. Effects of Dietary Protein, Energy and β-Mannanase on Laying Performance, Egg Quality, and Ileal Amino Acid Digestibility in Laying Hens. *Poult. Sci.* **2021**, *100*, 101312. [CrossRef]

40. Ryu, M.H.; Hosseindoust, A.; Kim, J.S.; Choi, Y.H.; Lee, S.H.; Kim, M.J.; Lee, J.H.; Chae, B.J. *β*-Mannanase Derived from *Bacillus Subtilis WL-7* Improves the Performance of Commercial Laying Hens Fed Low or High Mannan-Based Diets. *J. Poult. Sci.* **2017**, *54*, 212–217. [CrossRef] [PubMed]
41. Esonu, B.O.; Izukanne, R.; Emenalom, O.O.; Etuk, E.B.; Samuel, S.; Ezeoke, F.; Inyang, O.A. Evaluation and economics of enzyme supplementation on the performance of broiler finishers fed soybean hull meal based diets. *Nigerian J. Anim. Prod.* **2006**, *33*, 216–221. [CrossRef]
42. Abreu, M.T.; Fassani, E.J.; Silveira, M.M.B.M.; Viveiros, M.P. Enzymatic complex based on xylanase, beta-glucanase and phytase on feeds for light commercial layers at peak production. *Anim. Indust. Bullet.* **2018**, *75*, 17–24. [CrossRef]
43. Danang, B.; Rostini, T. The effect of protease enzyme supplementation to productivity eggs of alabio duck. *Int. J. Biosc.* **2016**, *8*, 203–208. Available online: https://repo-dosen.ulm.ac.id//handle/123456789/9816 (accessed on 15 June 2020).
44. Javer Alves, V.F.; Geraldo, A.; Machado, L.C.; Brito, J.A.D.; Bertechini, A.G.; Murakami, E.S.F. Effect of protease supplementation on the production performance of laying hens. *Anim. Sci.* **2015**, *37*, 29–33. [CrossRef]
45. Esonu, B.O.; Izukanne, R.O.; Inyang, O.A. Evaluation of cellulolytic enzyme supplementation on production indices and nutrient utilization of laying hens fed soybean hull based diets. *Int. J. Poult. Sci.* **2005**, *4*, 213–216. [CrossRef]
46. Mathlouthi, N.; Mohammed, M.A.; Larbier, M. Effect of enzyme preparation containing xylanase and β-glucanase on the performance of laying hens fed wheat/ barley or maize/ soybean meal-based diets. *Br. Poult. Sci.* **2003**, *44*, 60–66. [CrossRef]
47. Silversides, F.G.; Scott, T.A.; Korver, D.R.; Afsharmanesh, M.; Hruby, M. A study on the interaction of xylanase and phytase enzymes in wheat-based diets fed to commercial white and brown egg-laying hens. *Poult. Sci.* **2006**, *85*, 297–305. [CrossRef]
48. Jalal, M.A.; Scheideler, S.E. Effect of supplementation of two different sources of phytase on egg production parameters in laying hens and nutrient digestibility. *Poult. Sci.* **2001**, *80*, 1463–1471. [CrossRef]
49. Almirall, M.; Brufau, J.; Esteve-Garcia, E. Effects of intestinal viscosity on digestive enzyme activities at different ages supplemented with β-glucanases. In Proceedings of the 1st Symposium of Enzymes in Animal Nutrition, Kartause Ittingen, Switzerland, 13–16 October 1993; pp. 69–72.
50. Carre, B.; Lessire, M.; Juin, H. Prediction of metabolisable energy value of broiler diets and water excretion from dietary chemical analyses. *J. Anim.* **2013**, *7*, 1246–1258. [CrossRef]
51. Jiménez-Moreno, E.; Coca-Sinova, A.D.; González-Alvarado, J. Inclusion of insoluble fiber sources in mash or pellet diets for young broilers. Effects on growth performance and water intake. *Poult. Sci.* **2016**, *95*, 41–52. [CrossRef]
52. Sousa, L.S.D.; Carvalho, T.S.M.; Nogueira, F.A.; Saldanha, M.M.; Vaz, D.P.; Bertechini, A.G.; Lara, L.J.C. Fiber source and xylanase on performance, egg quality, and gastrointestinal tract of laying hens. *Rev. Bras. Zootec.* **2019**, *48*, e20170286. [CrossRef]
53. Bedford, M.R. The effect of enzymes on digestion. *J. App. Poult. Res.* **1996**, *5*, 370–378. [CrossRef]
54. Jackson, M.; Fodge, D.; Hsiao, H. Effects of Beta-Mannanase in Corn-Soybean Meal Diets on Laying Hen Performance. *Poult. Sci.* **1999**, *78*, 1737–1741. [CrossRef]
55. Lee, J.Y.; Kim, S.Y.; Lee, J.H.; Lee, J.H.; Ohh, S.J. Effect of Dietary β-Mannanase Supplementation and Palm Kernel Meal Inclusion on Laying Performance and Egg Quality in 73 Weeks Old Hens. *J. Anim. Sci. Technol.* **2013**, *55*, 115–122. [CrossRef]
56. Mangi, M.H.; Hussain, T.; Shahid, M.S.; Sabir, N.; Kalhoro, M.S.; Zhou, X.; Yuan, J. Effects of Flaxseed and Multi-Carbohydrase Enzymes on the Cecal Microbiota and Liver Inflammation of Laying Hens. *Animals* **2021**, *11*, 600. [CrossRef]
57. Weiss, F.G.; Scott, M.L. Effects of Dietary Fiber, Fat and Total Energy Upon Plasma Cholesterol and Other Parameters in Chickens. *J. Nutr.* **1979**, *109*, 693–701. [CrossRef]
58. Viveros, A.; Ortiz, L.T.; Rodríguez, M.L.; Rebolé, A.; Alzueta, C.; Arija, I.; Centeno, C.; Brenes, A. Interaction of Dietary High-Oleic-Acid Sunflower Hulls and Different Fat Sources in Broiler Chickens. *Poult. Sci.* **2009**, *88*, 141–151. [CrossRef]
59. McNaughton, J.L. Effect of Dietary Fiber on Egg Yolk, Liver, and Plasma Cholesterol Concentrations of the Laying Hen. *J. Nutr.* **1978**, *108*, 1842–1848. [CrossRef]
60. Gunness, P.; Gidley, M.J. Mechanisms Underlying the Cholesterol-Lowering Properties of Soluble Dietary Fibre Polysaccharides. *Food Funct.* **2010**, *1*, 149. [CrossRef]
61. Chawla, R.; Patil, G.R. Soluble Dietary Fiber. *Compr. Rev. Food Sci. Food Saf.* **2010**, *9*, 178–196. [CrossRef]
62. Evans, G. Removal of Blood from Laboratory Mammals and Birds. *Lab. Anim.* **1994**, *28*, 178–179. [CrossRef]
63. Adewole, D. Effect of Dietary Supplementation with Coarse or Extruded Oat Hulls on Growth Performance, Blood Biochemical Parameters, Ceca Microbiota and Short Chain Fatty Acids in Broiler Chickens. *Animals* **2020**, *10*, 1429. [CrossRef]
64. Martínez, Y.; Carrión, Y.; Rodríguez, R.; Valdivié, M.; Olmo, C.; Betancur, C.; Liu, G.; Al-Dhabi, N.; Duraipandiyan, V. Growth Performance, Organ Weights and Some Blood Parameters of Replacement Laying Pullets Fed with Increasing Levels of Wheat Bran. *Rev. Bras. Ciênc. Avícola* **2015**, *17*, 347–354. [CrossRef]
65. Mohammadigheisar, M.; Shouldice, V.L.; Balasubramanian, B.; Kim, I.H. Effect of Dietary Supplementation of β-Mannanase on Growth Performance, Carcass Characteristics, Excreta Microflora, Blood Constituents, and Nutrient Ileal Digestibility in Broiler Chickens. *Anim. Biosci.* **2021**, *34*, 1342–1349. [CrossRef] [PubMed]
66. Yaqoob, M.U.; Yousaf, M.; Khan, M.I.; Wang, M. Effect of β-Mannanase Supplementation on Growth Performance, Ileal Digestibility, Carcass Traits, Intestinal Morphology, and Meat Quality in Broilers Fed Low-ME Diets. *Animals* **2022**, *12*, 1126. [CrossRef] [PubMed]

67. Kim, J.S.; Ingale, S.L.; Hosseindoust, A.R.; Lee, S.H.; Lee, J.H.; Chae, B.J. Effects of Mannan Level and β-Mannanase Supplementation on Growth Performance, Apparent Total Tract Digestibility and Blood Metabolites of Growing Pigs. *Animal* **2017**, *11*, 202–208. [CrossRef] [PubMed]
68. Montagne, L.; Pluske, J.R.; Hampson, D.J. A Review of Interactions between Dietary Fibre and the Intestinal Mucosa, and Their Consequences on Digestive Health in Young Non-Ruminant Animals. *Anim. Feed Sci. Technol.* **2003**, *108*, 95–117. [CrossRef]
69. Singh, A.K.; Kim, W.K. Effects of Dietary Fiber on Nutrients Utilization and Gut Health of Poultry: A Review of Challenges and Opportunities. *Animals* **2021**, *11*, 181. [CrossRef]
70. Habte-Tsion, H.-M.; Kumar, V. Nonstarch Polysaccharide Enzymes—General Aspects. In *Enzymes in Human and Animal Nutrition*; Elsevier: Amsterdam, The Netherlands, 2018; pp. 183–209. ISBN 978-0-12-805419-2.

animals

MDPI

Article

Effect of Tannic Acid on Antioxidant Function, Immunity, and Intestinal Barrier of Broilers Co-Infected with Coccidia and *Clostridium perfringens*

Zhengfan Zhang [1,†], Pengtao Xu [1,†], Chengao Liu [1], Jing Chen [1], Bingbing Ren [1], Encun Du [2], Shuangshuang Guo [1], Peng Li [1], Lanlan Li [1] and Binying Ding [1,*]

1 Engineering Research Center of Feed Protein Resources on Agricultural By-Products, Ministry of Education, Hubei Key Laboratory of Animal Nutrition and Feed Science, Wuhan Polytechnic University, Wuhan 430023, China; zhang850820@hotmail.com (Z.Z.); xupengtao113@126.com (P.X.); lllca0010@163.com (C.L.); 15139433336@163.com (J.C.); qaz123mlp4562022@163.com (B.R.); guo1shuangshuang@163.com (S.G.); pengli@whpu.edu.cn (P.L.); lan_lanli@163.com (L.L.)

2 Hubei Key Laboratory of Animal Embryo and Molecular Breeding, Institute of Animal Science and Veterinary Medicine, Hubei Academy of Agricultural Sciences, Wuhan 430064, China; qiaowan77@126.com

* Correspondence: dbying7471@126.com

† The authors contributed equally to this work.

Simple Summary: Necrotic enteritis (NE) is mainly caused by coccidia and *Clostridium perfringens* (CCP), which can induce intestine injury and oxidative stress in broilers. Tannic acids (TA) are natural polyphenolic compounds with anti-bacterial, anti-inflammatory, and anti-oxidation functions. It has been demonstrated that dietary supplementation with hydrolyzable TA has beneficial effects on the growth and antioxidant capacity of broilers. However, the effects of TA on intestinal health and antioxidative function in broilers with NE conditions still need to be clarified. Thus, this study aimed to evaluate the effects of TA on the antioxidant function, immunity, and intestinal barrier in broilers co-infected with CCP. The results showed that the addition of 1000 mg/kg TA to the diet could improve the jejunal barrier, attenuate the inflammatory response of the jejunum, and increase the antioxidant capacity of the liver and jejunum through the activation of the transcription factor Nrf2 downstream of the Nrf2-Keap1 pathway in CCP infected broilers.

Citation: Zhang, Z.; Xu, P.; Liu, C.; Chen, J.; Ren, B.; Du, E.; Guo, S.; Li, P.; Li, L.; Ding, B. Effect of Tannic Acid on Antioxidant Function, Immunity, and Intestinal Barrier of Broilers Co-Infected with Coccidia and *Clostridium perfringens*. *Animals* **2024**, *14*, 955. https://doi.org/10.3390/ani14060955

Academic Editor: Janghan Choi

Received: 29 January 2024
Revised: 15 March 2024
Accepted: 17 March 2024
Published: 19 March 2024

Abstract: The purpose of this study was to determine the efficacy of tannic acid on the antioxidative function, immunity, and intestinal barrier of broilers co-infected with coccidia and *Clostridium perfringens* (CCP). A total of 294 1-day-old arbor acres(AA) broilers were divided into three groups: control group (CON), CCP co-infected group (CCP), and 1000 mg/kg TA + CCP co-infected group (CTA). This trial lasted for 28 days. The results showed that the CCP group decreased the activity of glutathione peroxidase (GSH-Px), total superoxide dismutase (T-SOD), catalase (CAT), and total antioxidant capacity (T-AOC) levels and increased the contents of hydrogen peroxide (H_2O_2) and malondialdehyde (MDA) in the jejunum ($p < 0.05$). The mRNA levels of GSH-Px3 and CAT in the liver and jejunum, and the mRNA levels of GSH-Px3, SOD, HO-1, and NAD(P)H quinone oxidoreductase I (NQO1) in the liver were down-regulated by CCP challenge ($p < 0.05$). In addition, the Keap1 and Nrf2 mRNA levels in the liver and jejunum, jejunal glutathione S-transferase (GST), and heme-oxygenase-1 (HO-1) were upregulated in the CCP group compared with CON ($p < 0.05$). The mRNA levels of interleukin 8 (IL-8), IL-1β, inducible nitric oxide synthase (iNOS), and interferon γ (IFN-γ) in the jejunum were elevated, and jejunal mRNA levels of IL-10, zonula occludens protein1 (ZO-1), claudin-1, claudin-2, and occludin were decreased in the CCP treatment ($p < 0.05$). Dietary supplementation with 1000 mg/kg TA increased the activity of GSH-Px, T-SOD, CAT, and T-AOC and decreased the contents of H_2O_2 and MDA in the jejunum ($p < 0.05$). Compared with the CCP group, TA decreased the mRNA level of Keap1 and Nrf2 in the liver and jejunum, increased the GSH-Px3, SOD, and CAT mRNA in the liver, and alleviated the rise of IL-8, IL-1β, iNOS, and IFN-γ and decrease in IL-10, occludin gene expression in the jejunum ($p < 0.05$). In conclusion, the addition of 1000 mg/kg TA to the diet improved the jejunal barrier, mitigated the jejunal inflammation, and increased the antioxidant capacity of the liver and jejunum through the activation of the transcription factor Nrf2 downstream of the Nrf2-Keap1 pathway in broilers with NE condition.

Keywords: antioxidant; broiler; intestinal barrier; necrotic enteritis; tannic acid

1. Introduction

Necrotic enteritis (NE) is a prevalent infectious gastrointestinal illness in broiler production, predominantly caused by *Clostridium perfringen*. *C. perfringens* is secondary to coccidia (coccidia is primary), or they may be co-infected, resulting in an inflammatory reaction of the intestinal mucosa and oxidative stress, which can cause a tremendous economic loss in the poultry industry [1–4]. Oxidative stress and inflammation disrupt mucosal cells and tight junction proteins, impeding the self-repair mechanisms of the intestinal barrier [5,6]. Antibiotics protect broilers from intestinal diseases and increase productivity. However, antibiotics in animal feed have been limited or outlawed in many countries due to bacteria and antibiotic resistance. Therefore, it is urgent to find natural and environmentally friendly feed additives to control the growing prevalence of NE in poultry effectively.

Tannins are natural polyphenolic substances in numerous plants with anti-bacterial, anti-viral, anti-inflammatory, and anti-oxidation functions [7,8]. Research has demonstrated that dietary supplementation with hydrolyzed tannins can reduce oxidative stress and improve antioxidant capacities in broilers [9,10]. Dietary TA can also alleviate tissue inflammation and protect intestinal health by lowering the mRNA levels of tissue inflammation elements and increasing the mRNA levels of intestinal intact proteins in broilers [11–13]. It has been shown that TA can alleviate the inflammation of broilers with NE conditions and enhance the antioxidative ability by activating the Nrf2-Keap1 signaling pathway in rats [14–17]. However, whether TA could improve antioxidant capacity and alleviate oxidative stress through the Nrf2-Keap1 signaling pathway in broilers with NE condition has yet to be reported.

Therefore, this study aimed to investigate the efficacy of TA on antioxidant functions, immune, and gut barrier of broilers co-infected with CCP to provide some data references and theoretical backup for protecting TA against poultry with NE condition.

2. Materials and Methods

2.1. Animal Care and Diet

The study was performed for 28 days in a one-way, fully randomized design. A total of 294 1-day-old AA broilers, with an initial BW of 46.40 ± 0.40 g, were randomized to 3 treatment groups with 7 replicates of 14 broilers in each pen (7 males and 7 females). Treatments were as follows: the control group without CCP challenge (CON), the co-infected group with CCP challenge (CCP), and the 1000 mg/kg TA (≥80% the hydrolysable TA from Chinese gallnut, Wufeng Chicheng Biotechnology Company Limited, Yichang, China) + co-infected group with CCP challenge (CTA). The diets were fed in Phase 1 (Day 1 to Day 14) and Phase 2 (Day 15 to Day 28). The basal diets are formulated concerning and in conjunction with the Arbor Acres Broiler Nutrition Specifications (2019); the compositions and nutrition contents of the basal diets are summarized in Table 1. On days 7 and 10 of the study, broilers in the CCP and CTA groups were orally injected with 1 mL of quadrivalent anti-coccidial vaccine (purchased from Foshan Zhengdian Biotechnology Co., Ltd., Foshan, Guangdong, China). The quadrivalent vaccine consisted of 1×10^5 oocysts of the *E. acervuline* strain PAHY, plus the 5×10^4 oocysts of *E. necatrix* strain PNHZ, *E. tenella* strain PTMZ, and *E. maxima* strain PMHY. The recommended inoculation dose for each bird was 1100 ± 110 sporulated oocysts, and the dose used in the present study was 30 times the recommended dose. Meanwhile, the CON group received an equivalent amount of saline. From day 16 to day 20, birds in the CCP and CTA groups per bird per day were orally given 1 mL of fresh *C. perfringens* (1×10^8 CFU/mL, gavage using a 1 mL pipette gun, the *C. perfringens*, CVCC2030, was obtained from the China Veterinary Microbial Culture Collection and Management Center, Beijing, China), while birds in the CON group were orally administered an equivalent amount of sterile broth. All broilers were kept in cages with free access to feed and water throughout the trial. The ambient temperature was maintained at 33 ± 2 °C for the first week and then progressively lowered to 22 °C until the experiment ended. The broilers were exposed to 23 h of light and 1 h of darkness per day throughout the experimental period.

Table 1. Compositions and nutrient content of the basal diet (%).

Items	1 to 14 Days	15 to 28 Days
Ingredients		
Corn	10.00	10.00
Wheat	55.42	58.90
Soybean meal	23.50	19.37
Fish meal	5.00	5.00
Soybean oil	2.50	3.80
$CaHPO_4$	1.05	0.75
Limestone	1.00	0.80
NaCl	0.35	0.35
L-lysine•HCl	0.35	0.25
DL-methionine	0.20	0.15
D-threonine	0.15	0.15
Choline chloride	0.25	0.25
Premix [a]	0.23	0.23
Total	100.00	100.00
Nutrient levels [b]		
ME/(MJ/kg)	12.55	12.95
Crude protein	21.53	20.14
Ca	0.96	0.86
Non-phytate phosphorus	0.45	0.40
Digestible lysine	1.25	1.05
Digestible methionine	0.53	0.46
Digestible threonine	0.81	0.76
Digestible tryptophan	0.23	0.22

[a] The premix provided the following per kg of diets: Cu 10 mg, Zn 100 mg, Fe 80 mg, Mn 100 mg, Se 0.3 mg, I 0.7 mg, VA 12,000 IU, VD_3 3000 IU, VK_3 3.2 mg, VB_1 3 mg, VB_2 8.0 mg, VB_{12} 0.025 mg, VE 44 IU, biotin 0.0325 mg, folic acid 2.00 mg, pantothenic acid 15 mg, and nicotinic acid 15 mg. [b] Non-phytate phosphorus, ME and digestible amino acids were calculated values, while Ca and CP were measured values.

2.2. Sample Collection

On days 14, 21, and 28, two broilers of average weight were chosen in each replicate, which was euthanized by cervical dislocation and then promptly slaughtered for sampling. Liver and jejunum(mid-jejunum) were collected and frozen in liquid nitrogen, then transferred to a freezer at −80 °C for storage.

2.3. Antioxidant Indexes

Liver and jejunal samples were placed in a mortar, to which liquid nitrogen was added and then crushed with a pestle. A total of 0.1 g of liver and jejunum samples were placed in 1.5 mL centrifuge tubes, respectively, and 0.9 mL of saline was injected into the sample centrifuge tubes using a pipette, followed by centrifugation in a centrifuge (Fresco 21, Thermo Scientific, Wilmington, DE, USA) for 15 min at 3500× *g* at 4 °C, and the supernatant was collected. Off-the-shelf test kits were purchased from Nanjing Jianjian Bioengineering Institute, Nanjing, China. The activity of GSH-Px, T-SOD, and CAT, the concentration of H_2O_2 and MDA, and T-AOC in the liver and jejunum, were tested following the manufacturer's directions.

2.4. RNA Isolation and Quantitative Real-Time PCR

RNA isolation and quantitative real-time PCR of liver and jejunal samples were performed, as reported by Guo and Li [18,19]. The SYBR Premix Ex Taq kit (Takara Biotechnology (Dalian) Co., Ltd., Dalian, China) used cDNA from each sample and primers. The qPCR was performed on an Applied Biosystems 7500 Fast Real-Time PCR System (Foster City, CA, USA). The relative expression of each gene was analyzed using the $2^{-\Delta\Delta Ct}$ method (Relative quantification) [20]. β-actin was used as the endogenous reference gene for all the genes tested. Table 2 lists the sequences of the relevant assay genes and internal reference primers used in this study.

Table 2. List of gene primer sequences [a].

Gene	Primer Sequence (5′ to 3′)	Product Length	NCBI Number
β-actin	F-ACTCTGGTGATGGTGTTAC R-GGCTGTGATCTCCTTCTG	497	NM 205518.2
Nrf-2	F-ATCACCTCTTCTGCACCGAA R-GCTTTCTCCCGCTCTTTCTG	229	NM 205117.2
Keap1	F-CAACTTCGCCGAGCAGA R-CGTGGAACACCTCCGACT	179	KU 321503179
GST	F-AGAGTCGAAGCCTGATGCAC R-CACTCCGCTTATCAGCAAACA	220	NM_001001777.2
HO-1	F-ACGAGTTCAAGCTGGTCACG R-GGATGCTTCTTGCCAACGAC	244	NM_205344.2
NQO1	F-TCGCCGAGCAGAAGAAGATTGAAG R-GGTGGTGAGTGACAGCATGGC	191	NM_001277621.1
GSH-px1	F-GACCAACCCGCAGTACATCA R-GAGGTGCGGGCTTTCCTTTA	204	NM_001277853.3
GSH-px3	F-AAGTGCCAGGTGAACGGGAAGG R-AGGGCTGTAGCGGCGGAAAG	204	NM 001163232.3
SOD	F-GGTGCTCACTTTAATCCTG R-CTACTTCTGCCACTCCTCC	109	NM 205064.2
CAT	F-GGTTCGGTGGGGTTGTCTTT R-CACCAGTGGTCAAGGCATCT	213	NM_001031215.2
iNOS	F-CCTGGAGGTCCTGGAAGAGT R-CCTGGGTTTCAGAAGTGGC	82	NM_204961.2
TNF-α	F-GAGCGTTGACTTGGCTGTC R-AAGCAACAACCAGCTATGCAC	64	NM_204267.2
IFN-γ	F-AGCTGACGGTGGACCTATTATT R-GGCTTTGCGCTGGATTC	259	Y07922.1
IL-1β	F-ACTGGGCATCAAGGGCTA R-GGTAGAAGATGAAGCGGGTC	131	NM_204524.2
IL-8	F-GCAAGGTAGGACGCTGGTAA R-GCGTCAGCTTCACATCTTGA	107	NM_205498.2
TGF-β4	F-CGGGACGGATGAGAAGAAC R-CGGCCCACGTAGTAAATGAT	258	M31160.1
ZO-1	F-CTTCAGGTGTTTCTCTTCCTCCTC R-CTGTGGTTTCATGGCTGGATC	131	XM_413773.4
Claudin-1	F-CATACTCCTGGGTCTGGTTGGT R-GACAGCCATCCGCATCTTCT	100	AY750897.1
Claudin-2	F-CAACTGGAAGATCAGCTCCT R-TGTAGATGTCGCACTGAGTG	119	NM_001277622.1
Occludin	F-ACGGCAGCACCTACCTCAA R-GGGCGAAGAAGCAGATGAG	123	D21837.1
Mucin-2	F-TTCATGATGCCTGCTCTTGTG R-CCTGAGCCTTGGTACATTCTTGT	93	XM_421035.2

F Upstream primer, R Downstream primers. [a] The primers were synthesized by Shanghai Shenggong Biotechnology Co., Ltd. (Shanghai, China)

2.5. Statistical Analysis

Data from this trial were analyzed by one-way ANOVA conducted using least significant difference (LSD) multiple comparisons using SPSS Statistics 25 (SPSS Inc., Chicago, IL, USA). Data presented as mean ± standard deviation (SD). Statistically significant at $p < 0.05$. The graphs were created using GraphPad Prism 10.0 software (GraphPad Software, LLC, San Diego, CA, USA).

3. Results

3.1. Antioxidant Indexes

In our previous study [21], we found that the hepatic CAT and T-SOD activity was decreased, and the MDA and H_2O_2 contents in the liver were decreased in the CCP group than those in the CON and CTA groups ($p < 0.05$).

Jejunal antioxidant capacity results are shown in Table 3. Jejunal GSH-Px and CAT activities were lower in the CCP group than in the CON and CTA groups on day 14 ($p < 0.05$). The GSH-Px level of jejunum was decreased in the CCP group compared with the CON and CTA, and T-SOD activity in CCP was lower than the CTA on day 21 ($p < 0.05$). Jejunal T-AOC, T-SOD, and CAT levels were lower in the CCP group than in the CON and CTA groups on day 28 ($p < 0.05$). On days 14 and 21, the content of jejunal H_2O_2 was increased in the CCP compared with the CON and CTA groups, and MDA content was more significant than the CON, and MDA content on day 21 was markedly decreased in the CTA group ($p < 0.05$).

Table 3. Effects of TA on jejunal antioxidant capacity of broilers co-infected with CCP.

Item	CON	CCP	CTA	*p*-Value
Day 14				
GSH-Px(U/mg prot)	21.97 ± 4.40^{b}	16.90 ± 4.45^{c}	26.96 ± 2.03^{a}	<0.001
T-SOD (U/mg prot)	192.04 ± 19.56	181.88 ± 13.81	180.72 ± 28.72	0.322
CAT (U/mg prot)	1.31 ± 0.12^{a}	0.82 ± 0.23^{c}	1.13 ± 0.10^{b}	<0.001
H_2O_2 (mmol/g prot)	1.88 ± 0.22^{b}	3.03 ± 0.29^{a}	2.01 ± 0.40^{b}	<0.001
MDA (nmol/mg prot)	7.15 ± 1.42^{b}	10.52 ± 1.11^{a}	9.37 ± 3.06^{a}	<0.001
T-AOC (mmol/g prot)	0.11 ± 0.01	0.10 ± 0.01	0.10 ± 0.01	0.052
Day 21				
GSH-Px (U/mg prot)	28.82 ± 3.75^{b}	24.12 ± 2.17^{c}	31.18 ± 2.76^{a}	<0.001
T-SOD (U/mg prot)	178.16 ± 12.25^{ab}	167.95 ± 11.94^{b}	184.79 ± 18.14^{a}	0.013
CAT (U/mg prot)	2.27 ± 0.39	2.30 ± 0.32	2.37 ± 0.35	0.747
H_2O_2 (mmol/g prot)	6.16 ± 0.50^{b}	7.62 ± 1.94^{a}	5.38 ± 1.35^{b}	0.001
MDA (nmol/mg prot)	6.09 ± 0.91^{b}	7.03 ± 1.39^{a}	5.83 ± 1.14^{b}	0.024
T-AOC (mmol/g prot)	0.15 ± 0.01	0.15 ± 0.02	0.16 ± 0.02	0.084
Day 28				
GSH-Px (U/mg prot)	68.42 ± 9.34	63.18 ± 9.80	64.33 ± 13.58	0.428
T-SOD (U/mg prot)	364.21 ± 21.01^{a}	313.83 ± 32.65^{c}	339.47 ± 28.04^{b}	<0.001
CAT (U/mg prot)	6.32 ± 0.99^{a}	5.07 ± 0.38^{b}	5.85 ± 0.81^{a}	<0.001
H_2O_2 (mmol/g prot)	2.58 ± 0.60	3.08 ± 0.76	2.64 ± 0.50	0.086
MDA (nmol/mg prot)	5.85 ± 1.63	7.58 ± 1.56	7.12 ± 1.56	0.076
T-AOC (mmol/g prot)	0.17 ± 0.05^{a}	0.12 ± 0.01^{b}	0.17 ± 0.06^{a}	0.028

Numbers are expressed as mean and SD, n = 14. [a,b,c] Values not sharing a superscript in the same row are markedly different.

3.2. The mRNA Levels of Antioxidant Enzymes in the Liver

As shown in Figure 1, hepatic GSH-Px1, GSH-Px3, CAT, and SOD mRNA levels on days 14 and 21 were reduced by CCP challenge (Figure 1A,B, $p < 0.05$); GSH-Px3 mRNA levels in the liver on day 28 were markedly greater in the CON group than in the CTA and CCP groups (Figure 1C, $p < 0.05$). The CTA could alleviate the downregulation of hepatic GSH-Px1, CAT, and SOD mRNA levels of broilers challenged by CCP ($p < 0.05$).

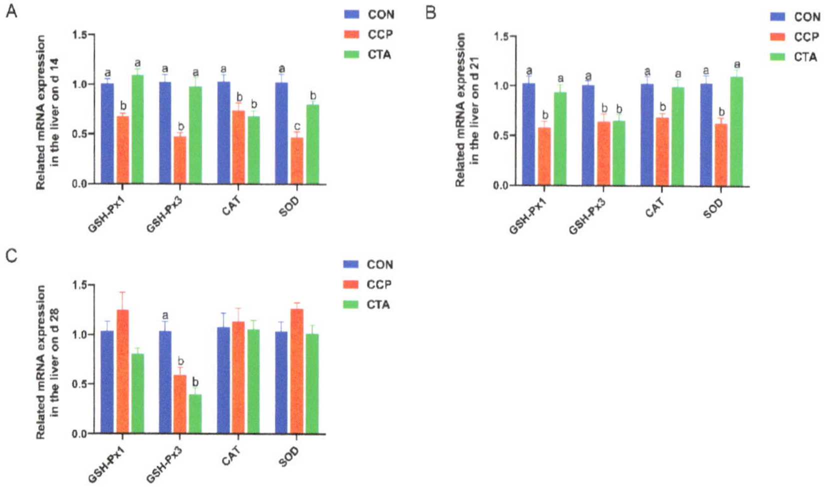

Figure 1. Antioxidant enzymes mRNA expression in the liver ((**A**) on day 14, (**B**) on day 21, (**C**) on day 28). Numbers are expressed as mean and SD, n = 14. [a,b,c] Values not sharing a superscript in the same row are markedly different.

3.3. The mRNA Levels of Antioxidant Enzymes in the Jejunum

The results of jejunal antioxidant enzyme mRNA levels are shown in Figure 2. Compared to the CON and CTA groups, the CCP significantly lowered jejunal GSH-Px3 and CAT mRNA levels on day 14 (Figure 2A, $p < 0.05$). The SOD mRNA level in the CTA group on day 21 was markedly higher than those in the CON and CCP groups. Additionally, the CCP significantly downregulated the CAT mRNA level in the jejunum on day 21 compared with the CON and CTA (Figure 2B, $p < 0.05$). There was no significant difference in relevant antioxidant enzymes in the jejunum among treatments on day 28 (Figure 2C, $p > 0.05$).

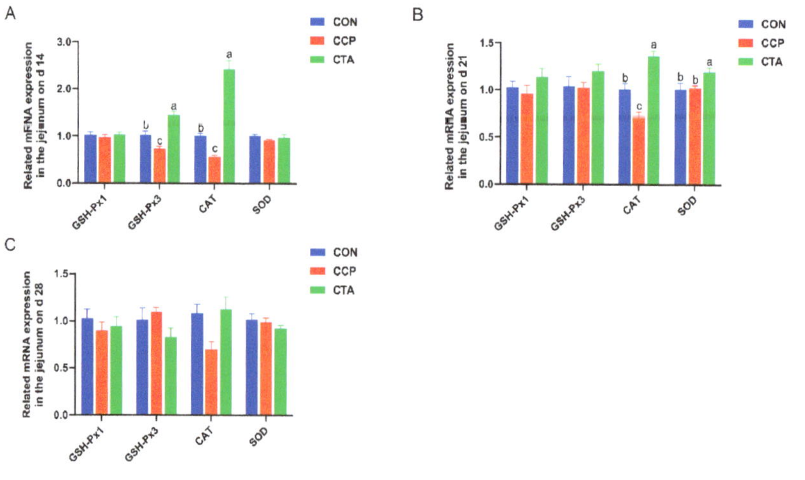

Figure 2. Jejunal antioxidant enzyme mRNA expression ((**A**) on day 14, (**B**) on day 21, (**C**) on day 28). Numbers are expressed as mean and SD, n = 14. [a,b,c] Values not sharing a superscript in the same row are markedly different.

3.4. The mRNA Levels of the Antioxidant Pathway in the Liver

As presented in Figure 3, on day 14, the Nrf2 mRNA level was lower in the CON in the liver than in the CTA and CCP groups, and the GST mRNA level was markedly higher in CTA than in CON and CCP (Figure 3A, $p < 0.05$). On day 21, the CCP significantly increased the Keap1 and Nrf2 mRNA levels and significantly reduced the mRNA levels of HO-1 and NQO1 compared to the CON; compared to the CCP, the CTA markedly reduced the mRNA levels of Keap1 and Nrf2 and significantly increased the mRNA levels of GST and NQO1(Figure 3B, $p < 0.05$). On day 28, the CCP significantly enhanced the mRNA levels of Keap1 and Nrf2 compared to the CON; the CTA markedly decreased the mRNA levels of Keap1, Nrf2, and GST compared to the CCP (Figure 3C, $p < 0.05$).

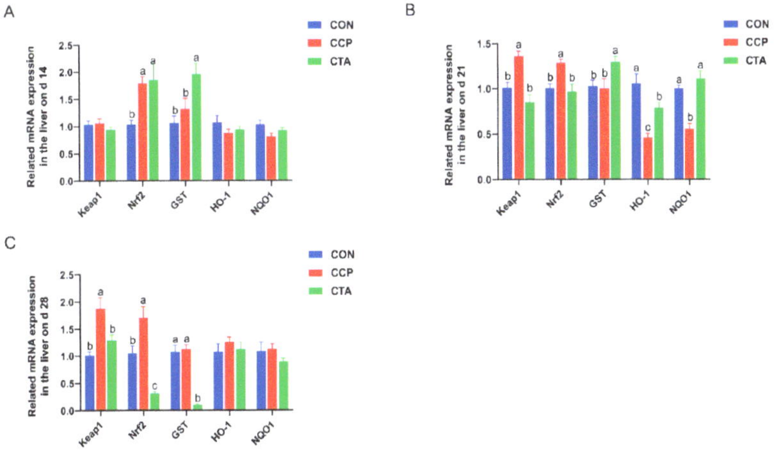

Figure 3. Antioxidant pathway mRNA expression in the liver ((**A**) on day 14, (**B**) on day 21, (**C**) on day 28). Numbers are expressed as mean and SD, n = 14. [a,b,c] Values not sharing a superscript in the same row are markedly different.

3.5. The mRNA Levels of the Antioxidant Pathway in the Jejunum

The results of jejunal antioxidant pathways mRNA levels are shown in Figure 4. On day 14, the mRNA level of Keap1 was markedly more significant in the CCP in the jejunum than in the CON and CTA; the mRNA level of Nrf2 was markedly lower in the CTA in the jejunum than in COT and CCP (Figure 4A, $p < 0.05$). The mRNA levels of Keap1, Nrf2, GST, and HO-1 in the jejunum on day 21 were upregulated by the CCP challenge (Figure 4B, $p < 0.05$); the mRNA levels of Keap1 and GST in the jejunum on day 28 were increased with CCP challenge (Figure 4C, $p < 0.05$); The CTA could alleviate the upregulation of Keap1, Nrf2, GST and HO-1 in the jejunum on days 21 and 28 in the CCP-challenged broilers (Figure 4B,C, $p < 0.05$).

3.6. The mRNA Levels of the Cytokines in the Jejunum

As shown in Figure 5, on day 14, the CCP markedly elevated the mRNA levels of IL-8, IL-1β, iNOS, and IFN-γ in the jejunum compared with the CON and CTA groups (Figure 5A, $p < 0.05$). The CTA markedly elevated the mRNA levels of IL-10 in the jejunum on day 21 compared with the CON and CCP groups (Figure 5B, $p < 0.05$). The mRNA levels of IL-8, iNOS, and IFN-γ in the jejunum on days 21 and 28 were elevated by CCP challenge (Figure 5A,B, $p < 0.05$). Alternatively, the mRNA level of IL-1β was elevated, and the mRNA level of IL-10 was reduced in the jejunum on day 28 with CCP challenge (Figure 5C, $p < 0.05$). The CTA could alleviate the increase in IL-8, IL-1β, iNOS, and IFN-γ mRNA levels and the decrease in IL-10 mRNA level in the jejunum in the CCP-challenged broilers ($p < 0.05$).

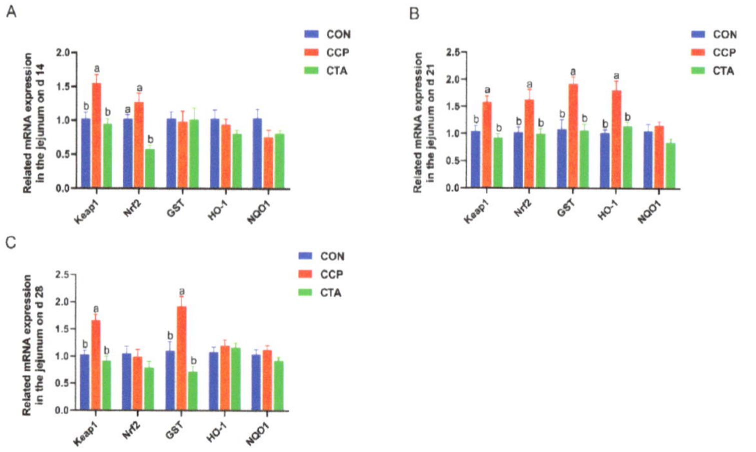

Figure 4. Antioxidant pathway mRNA expression in the jejunum ((**A**) on day 14, (**B**) on day 21, (**C**) on day 28). Numbers are expressed as mean and SD, n = 14. [a,b] Values not sharing a superscript in the same row are markedly different.

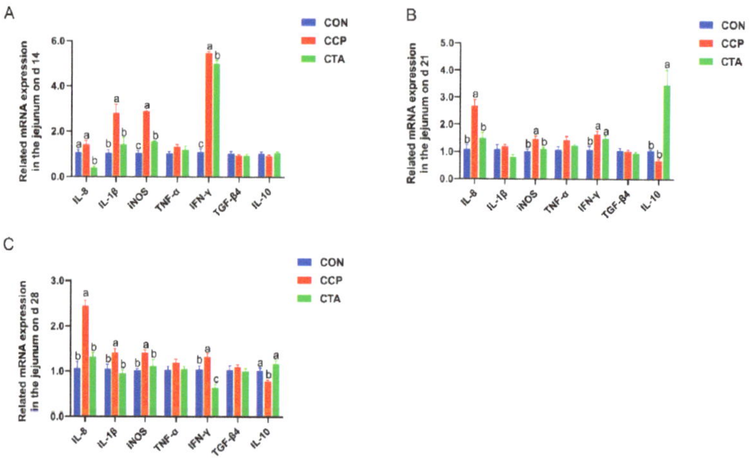

Figure 5. The cytokines mRNA expression in the jejunum ((**A**) on day 14, (**B**) on day 21, (**C**) on day 28). Numbers are expressed as mean and SD, n = 14. [a,b,c] Values not sharing a superscript in the same row are markedly different.

3.7. The mRNA Levels of Barrier Factors in the Jejunum

According to Figure 6. The ZO-1 mRNA level in the jejunum on day 14 was reduced by CCP challenge in comparison with the CON group; the CCP markedly reduced the mRNA levels of Claudin-2 in the jejunum on day 14 compared to the CON and CTA groups (Figure 6A, $p < 0.05$). Jejunal mRNA levels of ZO-1 and Occludin were markedly reduced by CCP challenge on day 21 compared to the CON group (Figure 6B, $p < 0.05$). The mRNA levels of Claudin-1, Claudin-2, and Occludin in the jejunum on day 28 were significantly reduced by CCP challenge (Figure 6C, $p < 0.05$). The CTA could alleviate the decrease in Claudin-2 and Occludin mRNA in the jejunum in the CCP-challenged broilers ($p < 0.05$).

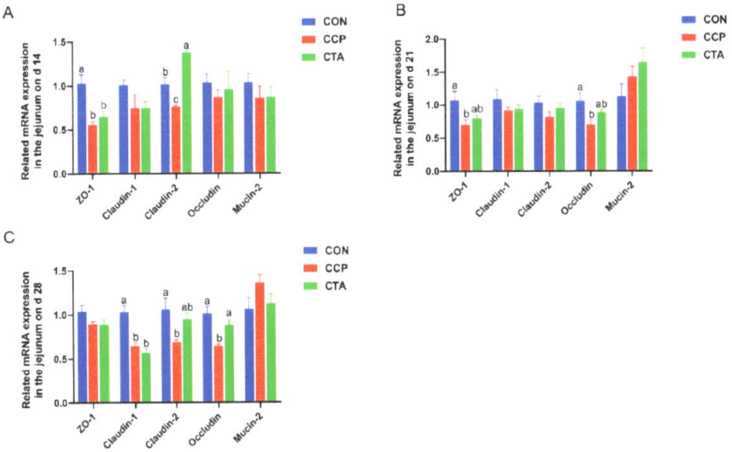

Figure 6. Barrier factors mRNA expression in the jejunum ((**A**) on day 14, (**B**) on day 21, (**C**) on day 28). Numbers are expressed as mean and SD, n = 14. [a,b,c] Values not sharing a superscript in the same row are markedly different.

4. Discussion

The NE is a prevalent infectious gastrointestinal illness in broiler production, predominantly associated with *C. perfringens* types A and G. *C. perfringens* is secondary to coccidia (coccidia is primary), or they may be co-infected. This study used the NE model constructed by coccidia and *C. perfringens* type A. In our previous study, we found that the infection of coccidia and *C. perfringens* reduced the growth performance and caused intestinal morphological structure damage in broilers [21]. Therefore, the infection model is well-established. We also found significantly decreased levels of high-density lipoprotein (HDL), gamma-glutamyl transferase (GGT), and lactate dehydrogenase (LDH) in serum in broilers under NE conditions compared with the control group. HDL is associated with lipid metabolism, and GGT and LDH are associated with liver function [21]. Therefore, in this study, we investigated the effects of TA on hepatic antioxidant function and jejunal function in broilers with NE conditions.

The antioxidant system consists of several components, including enzymatic species such as SOD, CAT, and GSH-Px, as well as non-enzymatic species such as T-AOC [22,23]. When animals are exposed to pathogens, viruses, and harsh environments, the body produces large amounts of reactive oxygen species (ROS) and MDA, which may cause oxidative stress [24]. It has been shown that broilers with NE condition reduced antioxidant function, induced inflammatory response, and impaired intestinal barrier function [19,25,26]. Consistent with previous findings, we found that Jejunal H_2O_2 and MDA contents were increased, and T-AOC, SOD, CAT, and GSH-Px levels were decreased by CCP challenge, which indicated that CCP co-infection caused oxidative stress in the broiler. Dietary supplementation with TA has been shown to enhance the activities of intestinal GSH-Px, SOD, and CAT, and to decrease H_2O_2 and MDA contents, thereby increasing intestinal antioxidant capacity and protecting the intestinal tract from damage caused by oxidative stress [9,10,27]. In the present study, we found that dietary supplementation with 1000 mg/kg TA reduced the increase in peroxides caused by CCP challenge and enhanced the activities of jejunal antioxidant enzymes. In our previous study, dietary supplementation with 1000 mg/kg TA could alleviate the decrease in antioxidant enzyme levels and increase H_2O_2 and MDA contents in the liver by CCP challenge [21]. These results indicated that dietary supplementation with 1000 mg/kg TA could alleviate the oxidative stress in the liver and jejunum of broilers caused by CCP challenge.

The Nrf2-Keap1 system is a significant regulatory pathway against oxidative stress [28]. The activation of the Nrf2-Keap1 pathway enhances antioxidant defense factors-associated genes, which include the GSH-Px1, GSH-Px3, SOD, CAT, HO-1, NQO1, and GST [29]. GSH-Px1 and GSH-Px3 belong to selenium-containing GSH-Px, which have the function of cleaning up reactive oxygen species, with the difference that GSH-Px1 exists in the cytoplasm and mitochondria, and GSH-Px3 exists in the extracellular [30–33]. Invasion by pathogens can upregulate the expression of the Nrf2 and keap1 genes, which causes oxidative stress in the organism and promotes the entry of Nrf2 into the nucleus to participate in the synthesis of several antioxidant enzymes in reaction to oxidative stress [19,34,35]. In our study, Keap1 and Nrf2 mRNA levels were elevated in the liver and jejunum, and jejunal GST and HO-1 mRNA levels were increased in the CCP-challenged broilers, suggesting that CCP infection induces oxidative stress in broilers. However, we found that the mRNA levels of GSH-Px1, CAT, and SOD in the liver and jejunal CAT were decreased, and jejunal H_2O_2 and MDA content in broilers with the CCP challenge were increased. It showed that CCP-infected broilers responded but could not successfully increase antioxidant enzymes to scavenge free radicals, causing oxidative damage to the liver and jejunum. Studies have shown that tannins can enhance the mRNA levels of antioxidant enzymes and reduce oxidative stress to protect the health of the body [36–38]. Consistent with these studies, we found that dietary supplementation with 1000 mg/kg TA could alleviate the up-regulation of Keap1 and Nrf2 mRNA levels and enhance the mRNA levels of GSH-Px1, SOD, CAT in the liver and jejunal CAT in the CCP-challenged broilers, which suggested that dietary supplementation with 1000 mg/kg TA could enhance antioxidant enzyme activities through the activation of the transcription factor Nrf2 downstream of the Nrf2-Keap1 pathway to alleviate oxidative damage in the liver and jejunum caused in the CCP-challenged broilers.

Intestinal immune barrier homeostasis is maintained by releasing inflammation and anti-inflammation factors by bowel-related lymphoid tissues composed of various cells that prevent pathogen invasion [39]. This study aimed to determine the expression of specific immune genes in the intestine, mainly belonging to pro-inflammatory (IL-8, IL-1β, iNOS, IFN-γ, TNF-α) and anti-inflammatory (TGF-β4, IL-10) mediators. Pro- and anti-inflammatory cytokines are essential to immune response homeostasis and inflammation [40]. It has been found that when inflammation occurs in the gut upon invasion by pathogens, there is a significant elevation in the expression of the pro-inflammatory genes, which may or may not coincide with a reduction in the anti-inflammatory gene expression [18,41,42]. This study found that jejunal pro-inflammatory mediators (IL-8, IL-1β, iNOS, IFN-γ) were significantly upregulated, the anti-inflammatory mediator (IL-10) was significantly downregulated in broilers with the CCP challenge, and these results suggested that CCP-infected broilers elicited an intestinal inflammatory response in the jejunum. In our study, dietary supplementation with 1000 mg/kg TA significantly diminished pro-inflammatory mediators and elevated anti-inflammatory mediators with CCP challenge, consistent with previous studies [43–45]. These results suggest that dietary supplementation with 1000 mg/kg TA is beneficial in alleviating intestinal inflammatory responses in broilers co-infected with CCP.

Tight junction proteins are essential for maintaining the epithelial barrier, which maintains the diffusion barrier and closes the cell gap. ZO-1, occludin, and claudins are the most critical proteins of the intestinal barrier [42]. It was found that both inflammatory mediators and cytokines result in abnormal expression of tight junction proteins such as ZO-1, claudin-1, and occludin, elevating intestinal mucosal permeability and impairing intestinal barrier function [46–50]. It was shown that CCP infection of broilers resulted in significant downregulation of tight junction proteins, causing intestinal damage in broilers [51,52]. In the present study, the mRNA levels of ZO-1, claudin-1, claudin-2, and occludin in the jejunum were significantly downregulated in CCP-infected broilers, which suggested that the jejunal barrier is impaired in CCP-infected broilers. Tannins significantly upregulate ZO-1 levels to protect against jejunal barrier damage. Research has shown that dietary TA supplementation can increase the mRNA levels of ZO-1, claudin-1, claudin-2, and occludin to mitigate intestinal barrier damage [53,54]. In line with prior research, we found that dietary supplementation with 1000 mg/kg TA raised the mRNA levels of claudin-2 and occludin, which suggested that dietary supplementation with TA is beneficial in ameliorating jejunal barrier damage in CCP-challenged broilers.

In this study, dietary supplementation with 1000 mg/kg TA could improve liver and jejunal function in broilers with NE conditions. However, we also found that TA increases GST mRNA levels, which may cause some adverse effects on the liver in broilers. The reason may be related to TA's origin, dose, and degree of polymerization of TA, which are closely related to its bioavailability. Studies have shown that highly polymerized tannins, with high molecular weight, are more poorly absorbed in the small intestine [55], and hydrolyzed tannins are oligomers of gallic acid, ellagic acid, and glucose that are partially hydrolyzed by digestive acids in the small intestine for easier absorption [56,57]. However, adding high doses of TA to diets may negatively affect performance, lymphoid organ weights, and ileal digestibility of amino acids in broilers and cause liver damage in mice [58,59]. Therefore, the increase in the GST mRNA level in the liver in this trial may be related to the dose of TA. We will further investigate the optimal dose of TA added to the ration to improve the health of broilers with NE conditions.

5. Conclusions

The co-infection of CCP reduced antioxidant capacity, induced intestinal inflammatory response, and impaired the intestinal barrier functions of broilers. Dietary supplementation with 1000 mg/kg TA could protect the intestinal barrier, mitigate the inflammatory response, and enhance antioxidant capacity by the activation of the transcription factor Nrf2 downstream of the Nrf2-Keap1 pathway, thereby protecting the intestinal health of co-infected broilers with CCP.

Author Contributions: B.D.: conceptualization, study design, methodology, and supervision; Z.Z. and P.X.: formal analysis and writing the original draft; C.L. and J.C.: validation and formal analysis; B.R., E.D. and S.G.: investigation and data curation; P.L. and L.L.: statistical analysis and manuscript review and editing. All authors have read and agreed to the published version of the manuscript.

Funding: This study was funded by the Natural Science Foundation of Hubei Province (2022CFB396), the Innovative Group Program of the Natural Science Foundation of Hubei Province (2023AFA018), the Scientific and Technological Projects of Hubei Province (2023EHA041), and the PhD Start-up Program of Wuhan Polytechnic University (2022RZ068).

Institutional Review Board Statement: This study was approved by the Animal Care and Use Committee at Wuhan Polytechnic University (NO.WPU202204152). All animal experiments were conducted following the guidelines of the Research Ethics Committee of the College of Animal Science and Nutritional Engineering, Wuhan Polytechnic University.

Informed Consent Statement: Informed consent was obtained from all subjects involved in the study.

Data Availability Statement: The datasets used and/or analyzed during the current study are available from the corresponding author upon reasonable request.

Acknowledgments: We thank our students and technicians for contributing to this research.

Conflicts of Interest: The authors declare that there is no conflict of interest regarding the publication of this paper.

Abbreviations

TA: tannic acid; NE: Necrotic enteritis; CCP: coccidia and *Clostridium perfringens*; MDA: malondialdehyde; H_2O_2: hydrogen peroxide; T-SOD: total superoxide dismutase; CAT: catalase; T-AOC: total antioxidant capacity; GSH-Px: glutathione peroxidase; *Keap1*: Kelch-like-ECH-associated protein1; *Nrf2*: nuclear factor erythroid 2-related factor 2; *GSH-Px1*: glutathione peroxidase1; *GSH-Px3*: glutathione peroxidase3; *SOD*: superoxide dismutase; *GST*: glutathione S-transferase; *HO-1*: heme-oxygenase-1; *NQO1*: NAD(P)H quinone oxidoreductase I; *iNOS*: inducible nitric oxide synthase;

TNF-α: tumor necrosis factor α; *IL-1β* interleukin 1β; *IL-8*: interleukin 8; *IL-10*: interleukin 10; *TGF-β4*: ransforming growth factorβ4; *IFN-γ*: Interferon γ; *ZO-1*: zonula occludens protein1.

References

1. Emami, N.K.; Dalloul, R.A. Centennial Review: Recent developments in host-pathogen interactions during necrotic enteritis in poultry. *Poult. Sci.* **2021**, *100*, 101330. [CrossRef]
2. Gu, C.; Lillehoj, H.S.; Sun, Z.; Lee, Y.; Zhao, H.; Xianyu, Z.; Yan, X.; Wang, Y.; Lin, S.; Liu, L.; et al. Characterization of virulent $netB^+/tpeL^+$ *Clostridium perfringens* strains from necrotic enteritis-affected broiler chicken farms. *Avian Dis.* **2019**, *63*, 461–467. [CrossRef] [PubMed]
3. Mehdizadeh Gohari, I.; A Navarro, M.; Li, J.; Shrestha, A.; Uzal, F.; A McClane, B. Pathogenicity, and virulence of *Clostridium perfringens*. *Virulence* **2021**, *12*, 723–753. [CrossRef] [PubMed]
4. Yu, C.; Tong, Y.; Li, Q.; Wang, T.; Yang, Z. Trans-anethole ameliorates intestinal injury through activation of Nrf2 signaling pathway in subclinical necrotic enteritis-induced broilers. *Front. Vet. Sci.* **2022**, *9*, 877066. [CrossRef] [PubMed]
5. Forman, H.J.; Zhang, H. Targeting oxidative stress in disease: Promise and limitations of antioxidant therapy. *Nat. Rev. Drug Discov.* **2021**, *20*, 689–709. [CrossRef]
6. Di Tommaso, N.; Gasbarrini, A.; Ponziani, F.R. Intestinal barrier in human health and disease. *Int. J. Environ. Res. Public. Health* **2021**, *18*, 12836. [CrossRef] [PubMed]
7. Jing, W.; Xiaolan, C.; Yu, C.; Feng, Q.; Haifeng, Y. Pharmacological effects and mechanisms of tannic acid. *Biomed. Pharmacother.* **2022**, *154*, 113561. [CrossRef] [PubMed]
8. Baldwin, A.; Booth, B.W. Biomedical applications of tannic acid. *J. Biomater. Appl.* **2022**, *36*, 1503–1523. [CrossRef] [PubMed]
9. Zhang, Z.F.; Xi, Y.; Wang, S.T.; Zheng, L.Y.; Qi, Y.; Guo, S.S.; Ding, B.Y. Effects of Chinese gallnut tannic acid on growth performance, blood parameters, antioxidative status, intestinal histomorphology, and cecal microbial shedding in broilers challenged with aflatoxin B1. *J. Anim. Sci.* **2022**, *100*, skac099. [CrossRef]
10. Xi, Y.; Chen, J.; Guo, S.; Wang, S.; Liu, Z.; Zheng, L.; Qi, Y.; Xu, P.; Li, L.; Zhang, Z.; et al. Effects of tannic acid on growth performance, relative organ weight, antioxidative status, and intestinal histomorphology in broilers exposed to aflatoxin B1. *Front. Vet. Sci.* **2022**, *9*, 1037046. [CrossRef]
11. Liu, H.W.; Li, K.; Zhao, J.S.; Deng, W. Effects of chestnut tannins on intestinal morphology, barrier function, pro-inflammatory cytokine expression, microflora and antioxidant capacity in heat-stressed broilers. *J Anim. Physiol. Anim. Nutr.* **2018**, *102*, 717–726. [CrossRef]
12. Guo, P.; Tong, Y.; Yang, R.; Zhang, M.; Lin, Q.; Lin, S.; Wang, C. Effects of hydrolyzed gallotannin on intestinal physical barrier, immune function, and microbiota structure of yellow-feather broilers. *Poult. Sci.* **2023**, *102*, 103010. [CrossRef]
13. Cui, X.; Zhang, S.; Jiang, S.; Gou, Z.; Wang, Y. Dietary protocatechuic acid ameliorates ileal mucosal barrier injury and inflammatory response and improves intestinal microbiota composition in Yellow chickens challenged with Salmonella typhimurium. *Poult. Sci.* **2023**, *102*, 102496. [CrossRef]
14. Xu, H.; Fu, J.; Luo, Y.; Li, P.; Song, B.; Lv, Z.; Guo, Y. Effects of tannic acid on the immunity and intestinal health of broiler chickens with necrotic enteritis infection. *J. Anim. Sci. Biotechnol.* **2023**, *14*, 72. [CrossRef]
15. Xu, H.; Zhang, X.; Li, P.; Luo, Y.; Fu, J.; Gong, L.; Lv, Z.; Guo, Y. Effects of tannic acid supplementation on the intestinal health, immunity, and antioxidant function of broilers challenged with necrotic enteritis. *Antioxidants* **2023**, *12*, 1476. [CrossRef]
16. Li, M.; Liu, P.; Xue, Y.; Liang, Y.; Shi, J.; Han, X.; Zhang, J.; Chu, X.; Chu, L. Tannic acid attenuates hepatic oxidative stress, apoptosis and inflammation by activating the Keap1-Nrf2/ARE signaling pathway in arsenic trioxide-toxicated rats. *Oncol. Rep.* **2020**, *44*, 2306–2316. [CrossRef]
17. Jin, W.; Xue, Y.; Xue, Y.; Han, X.; Song, Q.; Zhang, J.; Li, Z.; Cheng, J.; Guan, S.; Sun, S.; et al. Tannic acid ameliorates arsenic trioxide-induced nephrotoxicity, contribution of NF-κB and Nrf2 pathways. *Biomed. Pharmacother.* **2020**, *126*, 110047. [CrossRef]
18. Guo, S.; Xi, Y.; Xia, Y.; Wu, T.; Zhao, D.; Zhang, Z.; Ding, B. Dietary Lactobacillus fermentum and Bacillus coagulans supplementation modulates intestinal immunity and microbiota of broiler chickens challenged by *Clostridium perfringens*. *Front. Vet. Sci.* **2021**, *8*, 680742. [CrossRef] [PubMed]
19. Li, P.; Liu, C.; Niu, J.; Zhang, Y.; Li, C.; Zhang, Z.; Guo, S.; Ding, B. Effects of dietary supplementation with vitamin A on antioxidant and intestinal barrier function of broilers co-Infected with coccidia and *Clostridium perfringens*. *Animals* **2022**, *12*, 3431. [CrossRef] [PubMed]
20. Livak, K.J.; Schmittgen, T.D. Analysis of relative gene expression data using real-time quantitative PCR and the 2(-Delta Delta C(T)) Method. *Methods* **2001**, *25*, 402–408. [CrossRef] [PubMed]
21. Xu, P.; Chen, Y.; Chen, J.; He, L.; Liu, C.; Guo, S.; Zhang, Z.; Ding, B. Effects of tannic acid on growth performance, antioxidant function, and intestinal morphology of broilers infected with coccidia and *Clostridium perfringens*. *Chin. J. Anim. Nutr.* **2023**, *35*, 7036–7048. [CrossRef]
22. Demirci-Çekiç, S.; Özkan, G.; Avan, A.N.; Uzunboy, S.; Çapanoğlu, E.; Apak, R. Biomarkers of oxidative stress and antioxidant defense. *J. Pharm. Biomed. Anal.* **2022**, *209*, 114477. [CrossRef] [PubMed]

23. Zhang, S.; Wang, C.; Sun, Y.; Wang, G.; Chen, H.; Li, D.; Yu, X.; Chen, G. Xylanase and fermented polysaccharide of hericium caputmedusae reduce pathogenic infection of broilers by improving antioxidant and anti-Inflammatory properties. *Oxidative Med. Cell Longev.* **2018**, *2018*, 4296985. [CrossRef] [PubMed]
24. Jelic, M.D.; Mandic, A.D.; Maricic, S.M.; Srdjenovic, B.U. Oxidative stress and its role in cancer. *J. Cancer Res. Ther.* **2021**, *17*, 22–28. [CrossRef] [PubMed]
25. Daneshmand, A.; Kermanshahi, H.; Mohammed, J.; Sekhavati, M.H.; Javadmanesh, A.; Ahmadian, M.; Alizadeh, M.; Razmyar, J.; Kulkarni, R.R. Intestinal changes and immune responses during *Clostridium perfringens*-induced necrotic enteritis in broiler chickens. *Poult. Sci.* **2022**, *101*, 101652. [CrossRef] [PubMed]
26. Zhang, X.; Zhao, Q.; Ci, X.; Chen, S.; Xie, Z.; Li, H.; Zhang, H.; Chen, F.; Xie, Q. Evaluation of the efficacy of chlorogenic acid in reducing small intestine injury, oxidative stress, and inflammation in chickens challenged with *Clostridium perfringens* type A. *Poult. Sci.* **2020**, *99*, 6606–6618. [CrossRef]
27. Wang, M.; Huang, H.; Liu, S.; Zhuang, Y.; Yang, H.; Li, Y.; Chen, S.; Wang, L.; Yin, L.; Yao, Y.; et al. Tannic acid modulates intestinal barrier functions associated with intestinal morphology, antioxidative activity, and intestinal tight junction in a diquat-induced mouse model. *RSC Adv.* **2019**, *9*, 31988–31998. [CrossRef]
28. Hu, H.; Dai, S.; Li, J.; Wen, A.; Bai, X. Glutamine improves heat stress-induced oxidative damage in the broiler thigh muscle by activating the nuclear factor erythroid 2-related 2/Kelch-like ECH-associated protein 1 signaling pathway. *Poult. Sci.* **2020**, *99*, 1454–1461. [CrossRef]
29. Liu, H.; Li, X.; Shi, S.; Zhou, Y.; Zhang, K.; Wang, Y.; Zhao, J. Chlorogenic acid improves growth performance and intestinal health through autophagy-mediated nuclear factor erythroid 2-related factor 2 pathway in oxidatively stressed broilers induced by dexamethasone. *Poult. Sci.* **2022**, *101*, 102036. [CrossRef]
30. Flohe, L.; Günzler, W.A.; Schock, H.H. Glutathione peroxidase: A selenoenzyme. *FEBS Lett.* **1973**, *32*, 132–134. [CrossRef]
31. Takahashi, K.; Avissar, N.; Whitin, J.; Cohen, H. Purification and characterization of human plasma glutathione peroxidase: A selenoglycoprotein distinct from the known cellular enzyme. *Arch. Biochem. Biophys.* **1987**, *256*, 677–686. [CrossRef]
32. Brigelius-Flohé, R.; Maiorino, M. Glutathione peroxidases. *Biochim. Biophys. Acta.* **2013**, *1830*, 3289–3303. [CrossRef]
33. Handy, D.E.; Loscalzo, J. The role of glutathione peroxidase-1 in health and disease. *Free Radic. Biol. Med.* **2022**, *188*, 146–161. [CrossRef]
34. He, F.; Antonucci, L.; Karin, M. NRF2 as a regulator of cell metabolism and inflammation in cancer. *Carcinogenesis.* **2020**, *41*, 405–416. [CrossRef] [PubMed]
35. Liu, S.; Pi, J.; Zhang, Q. Signal amplification in the KEAP1-NRF2-ARE antioxidant response pathway. *Redox Biol.* **2022**, *54*, 102389. [CrossRef] [PubMed]
36. Wei, X.; Luo, C.; He, Y.; Huang, H.; Ran, F.; Liao, W.; Tan, P.; Fan, S.; Cheng, Y.; Zhang, D.; et al. Hepatoprotective effects of different extracts from triphala against CCl4-Induced acute liver injury in mice. *Front. Pharmacol.* **2021**, *12*, 664607. [CrossRef] [PubMed]
37. Kizir, D.; Karaman, M.; Ceylan, H. Tannic acid may ameliorate doxorubicin-induced changes in oxidative stress parameters in rat spleen. *Naunyn Schmiedebergs Arch. Pharmacol.* **2023**, *396*, 3605–3613. [CrossRef] [PubMed]
38. Wang, Z.; Guo, L.; Ding, X.; Li, F.; Xu, H.; Li, S.; Wang, X.; Li, K.; Yue, X. Supplementation of chestnut tannins in diets can improve meat quality and antioxidative capability in Hu lambs. *Meat Sci.* **2023**, *206*, 109342. [CrossRef]
39. Broom, L.J.; Kogut, M.H. Gut immunity: Its development and reasons and opportunities for modulation in monogastric production animals. *Anim. Health Res. Rev.* **2018**, *19*, 46–52. [CrossRef] [PubMed]
40. Kolehmainen, M.; Mykkänen, O.; Kirjavainen, P.V.; Leppänen, T.; Moilanen, E.; Adriaens, M.; Laaksonen, D.E.; Hallikainen, M.; Puupponen-Pimiä, R.; Pulkkinen, L.; et al. Bilberries reduce low-grade inflammation in individuals with features of metabolic syndrome. *Mol. Nutr. Food Res.* **2012**, *56*, 1501–1510. [CrossRef]
41. Fasina, Y.O.; Lillehoj, H.S. Characterization of intestinal immune response to *Clostridium perfringens* infection in broiler chickens. *Poult. Sci.* **2019**, *98*, 188–198. [CrossRef]
42. Tang, Y.; Zhang, X.; Wang, Y.; Guo, Y.; Zhu, P.; Li, G.; Zhang, J.; Ma, Q.; Zhao, L. Dietary ellagic acid ameliorated *Clostridium perfringens*-induced subclinical necrotic enteritis in broilers via regulating inflammation and cecal microbiota. *J. Anim. Sci. Biotechnol.* **2022**, *13*, 47. [CrossRef]
43. Jing, C.; Niu, J.; Liu, Y.; Jiao, N.; Huang, L.; Jiang, S.; Yan, L.; Yang, W.; Li, Y. Tannic acid extracted from galla chinensis supplementation in the diet improves intestinal development through suppressing inflammatory responses via blockage of NF-κB in broiler chickens. *Animals* **2022**, *12*, 2397. [CrossRef]
44. Liu, S.; Wang, K.; Lin, S.; Zhang, Z.; Cheng, M.; Hu, S.; Hu, H.; Xiang, J.; Chen, F.; Li, G.; et al. Comparison of the effects between tannins extracted from different natural plants on growth performance, antioxidant capacity, immunity, and intestinal flora of broiler chickens. *Antioxidants* **2023**, *12*, 441. [CrossRef]
45. Song, X.; Chen, Y.; Chen, X.; Zhao, X.; Zou, Y.; Li, L.; Zhou, X.; Li, M.; Zhang, D.; Ye, G.; et al. Exosomes from tannic acid-stimulated macrophages accelerate wound healing through miR-221-3p mediated fibroblasts migration by targeting CDKN1b. *Int. J. Biol. Macromol.* **2023**, *244*, 125088. [CrossRef] [PubMed]
46. Liao, P.; Liao, M.; Li, L.; Tan, B.; Yin, Y. Effect of deoxynivalenol on apoptosis, barrier function, and expression levels of genes involved in nutrient transport, mitochondrial biogenesis and function in IPEC-J2 cells. *Toxicol Res* **2017**, *6*, 866–877. [CrossRef] [PubMed]

47. Wu, Y.; Zhen, W.; Geng, Y.; Wang, Z.; Guo, Y. Pretreatment with probiotic Enterococcus faecium NCIMB 11181 ameliorates necrotic enteritis-induced intestinal barrier injury in broiler chickens. *Sci. Rep.* **2019**, *9*, 10256. [CrossRef]
48. Ma, J.; Wang, P.; Liu, Y.; Zhao, L.; Li, Z.; Xue, Y. Krüppel-like factor 4 regulates blood-tumor barrier permeability via ZO-1, occludin and claudin-5. *J. Cell. Physiol.* **2014**, *229*, 916–926. [CrossRef] [PubMed]
49. Qiu, K.; Li, C.L.; Wang, J.; Qi, G.H.; Gao, J.; Zhang, H.J.; Wu, S.G. Effects of dietary supplementation with Bacillus subtilis, as an alternative to antibiotics, on growth performance, serum immunity, and intestinal health in broiler chickens. *Front. Nutr.* **2021**, *8*, 786878. [CrossRef] [PubMed]
50. Rajasekaran, S.A.; Barwe, S.P.; Gopal, J.; Ryazantsev, S.; Schneeberger, E.E.; Rajasekaran, A.K. Na-K-ATPase regulates tight junction permeability through occludin phosphorylation in pancreatic epithelial cells. *Am. J. Physiol. Gastrointest. Liver Physiol.* **2007**, *292*, G124–G133. [CrossRef] [PubMed]
51. Xu, X.; Yang, S.; Olajide, J.S.; Qu, Z.; Gong, Z.; Wang, J.; Zhang, Y.; Wang, H.; Xiong, L.; Zhang, K.; et al. Clostridium butyricum supplement can ameliorate the intestinal barrier roles in broiler chickens experimentally infected with *Clostridium perfringens*. *Front. Physiol.* **2021**, *12*, 737481. [CrossRef]
52. Li, P.; Zheng, L.; Qi, Y.; Liu, Z.; Du, E.; Wei, J.; Zhang, Z.; Guo, S.; Ding, B. Dietary *Lactobacillus fermentum* and *Lactobacillus paracasei* improve the intestinal health of broilers challenged with coccidia and *Clostridium perfringens*. *Front. Vet. Sci.* **2022**, *9*, 1025677. [CrossRef] [PubMed]
53. Li, X.; Sun, R.; Liu, Q.; Gong, Y.; Ou, Y.; Qi, Q.; Xie, Y.; Wang, X.; Hu, C.; Jiang, S.; et al. Effects of dietary supplementation with dandelion tannins or soybean isoflavones on growth performance, antioxidant function, intestinal morphology, and microbiota composition in Wenchang chickens. *Front. Vet. Sci.* **2023**, *9*, 1073659. [CrossRef]
54. Yu, J.; Song, Y.; Yu, B.; He, J.; Zheng, P.; Mao, X.; Huang, Z.; Luo, Y.; Luo, J.; Yan, H.; et al. Tannic acid prevents post-weaning diarrhea by improving intestinal barrier integrity and function in weaned piglets. *J. Anim. Sci. Biotechnol.* **2020**, *11*, 87. [CrossRef]
55. Landete, J.M. Ellagitannins, ellagic acid and their derived metabolites: A review about source, metabolism, functions and health. *Food Res. Int.* **2011**, *44*, 1150–1160. [CrossRef]
56. Fraga-Corral, M.; Otero, P.; Cassani, L.; Echave, J.; Garcia-Oliveira, P.; Carpena, M.; Chamorro, F.; Lourenço-Lopes, C.; Prieto, M.A.; Simal-Gandara, J. Traditional Applications of Tannin Rich Extracts Supported by Scientific Data: Chemical Composition, Bioavailability and Bioaccessibility. *Foods* **2021**, *10*, 251. [CrossRef]
57. Kawabata, K.; Yoshioka, Y.; Terao, J. Role of Intestinal Microbiota in the Bioavailability and Physiological Functions of Dietary Polyphenols. *Molecules* **2019**, *24*, 370. [CrossRef] [PubMed]
58. Hidayat, C.; Irawan, A.; Jayanegara, A.; Sholikin, M.M.; Prihambodo, T.R.; Yanza, Y.R.; Wina, E.; Sadarman, S.; Krisnan, R.; Isbandi, I. Effect of dietary tannins on the performance, lymphoid organ weight, and amino acid ileal digestibility of broiler chickens: A meta-analysis. *Vet. World.* **2021**, *14*, 1405–1411. [CrossRef]
59. Qin, L.S.; Zhao, H.P.; Zhao, Y.L.; Ma, Z.J.; Zeng, L.N.; Zhang, Y.M.; Zhang, P.; Yan, D.; Bai, Z.F.; Li, Y.; et al. Protection and bidirectional effect of rhubarb anthraquinone and tannins for rats' liver. *Zhongguo Zhong Xi Yi Jie He Za Zhi* **2014**, *34*, 698–703. (In Chinese)

animals

MDPI

Article

Efficacy and Equivalency of Phytase for Available Phosphorus in Broilers Fed an Available Phosphorus-Deficient Diet

Myunghwan Yu [1], Elijah Ogola Oketch [1], Jun-Seon Hong [1], Nuwan Chamara Chathuranga [1], Eunsoo Seo [1], Haeeun Park [1], Bindhu Lakshmibai Vasanthakumari [2], Hans Lee [3] and Jung-Min Heo [1,*]

1 Department of Animal Science and Biotechnology, Chungnam National University, Daejeon 34134, Republic of Korea; tomymh@cnu.ac.kr (M.Y.); elijahogolah@gmail.com (E.O.O.); gospel0342@naver.com (J.-S.H.); nuonechathuranga@gmail.com (N.C.C.); sksjang1@naver.com (E.S.); qkrgodms2@naver.com (H.P.)

2 Kemin Industries Inc., Des Moines, IA 50317, USA; b.l.vasanthakumari@kemin.com

3 Kemin Animal Nutrition and Health, Asia Pacific, Seongnam 13631, Republic of Korea; hans.lee@kemin.com

* Correspondence: jmheo@cnu.ac.kr; Tel.: +82-42-821-5777

Simple Summary: Continuous concerns persist regarding skeletal disorders and their associated welfare issues in modern fast-growing broiler chickens. As an essential and the third most expensive nutrient in the poultry diet, phosphorus plays a crucial role in bone growth, and the strength and rigidity of the skeleton. About 60% of dietary phosphorus is utilized by poultry, with the remaining portion being excreted and potentially contributing to pollution when released into the environment. Hence, it was postulated that a multi-phased approach involving the reduction in phosphorus content in the feed, coupled with increased phosphorus utilization through phytase supplementation, could alleviate the economic burden associated with both phosphorus excretion and feed costs. Therefore, this study was conducted to assess the efficacy of phytase on the performance, carcass traits, nutrient digestibility, tibia characteristics, and inositol phosphorus degradation in broiler chickens fed an available phosphorus-deficient diet. The results indicate that a reduction in the available phosphorus to 0.20% with phytase inclusion initiated phytate degradation and, as a result, improved the productive performance, nutrient digestibility, and tibia traits of the broilers. These findings support the application of low-phosphorus diets with phytase supplementation in the poultry industry.

Abstract: This study was conducted to assess the effectiveness of phytase on the performance, carcass traits, nutrient digestibility, tibia characteristics, and inositol phosphorus (IP) degradation in broiler chickens. Additionally, the available phosphorus (AP) equivalency of phytase in AP-deficient diets was estimated for 35 days after hatching. A total of 336 one-day-old Ross 308 broiler chicks were allocated to one of seven dietary treatments with six replications with eight birds per cage. The dietary treatments were as follows: (1) positive control containing 0.45% AP of the starter and 0.42% AP of the grower diet (PC), (2) 0.10% AP deficiency from the PC (NC-1), (3) 0.15% AP deficiency from the PC (NC-2), (4) 0.20% AP deficiency from the PC (NC-3), (5) NC-3 +phytase (500 FTU/kg; NC-3-500), (6) NC-3 + phytase (1000 FTU/kg; NC-3-1000), and (7) NC-3 + phytase (1500 FTU/kg; NC-3-1500). On d 35, the NC-3 diet exhibited lower tibia weight compared to the other treatments ($p < 0.001$). The NC-3-1500 group had higher calcium and phosphorus contents in the tibia than the other treatments on d 35 ($p < 0.01$). Phytase supplementation led to a reduced IP_6 concentration and increased IP_3 concentrations in different sections of the gastrointestinal tract on d 21 and 35 compared to the control diet ($p < 0.05$). In conclusion, based on the tibia phosphorus content, this study determined that 500 FTU/kg phytase was equivalent to 0.377% and 0.383% AP in the diet on d 21, and 0.317% and 0.307% AP in the diet on d 35, respectively. Likewise, 1000 FTU/kg was determined to be equivalent to 0.476% and 0.448% AP on d 21, and 0.437% and 0.403% AP on d 35, respectively. Furthermore, 1500 FTU/kg was determined to be equivalent to 0.574% and 0.504% AP on d 21, and 0.557% and 0.500 AP on d 35, respectively.

Keywords: available phosphorus; broiler; equivalency value; low-phosphorus diet; phytase; tibia

Citation: Yu, M.; Oketch, E.O.; Hong, J.-S.; Chathuranga, N.C.; Seo, E.; Park, H.; Vasanthakumari, B.L.; Lee, H.; Heo, J.-M. Efficacy and Equivalency of Phytase for Available Phosphorus in Broilers Fed an Available Phosphorus-Deficient Diet. *Animals* **2024**, *14*, 41. https://doi.org/10.3390/ani14010041

Academic Editor: Jeffrey Downing

Received: 14 November 2023
Revised: 17 December 2023
Accepted: 20 December 2023
Published: 21 December 2023

1. Introduction

Alongside energy and protein supply, the provision of phosphorus (P) is of critical importance to poultry diets, since it is the third most expensive nutrient. Phosphorus is involved in bone mineralization, energy metabolism, and nucleic acid synthesis and, subsequently, has an impact on the birds' overall growth performance [1,2]. According to [3], the plant seeds used in corn and soybean meal diets for poultry contain approximately 50–80% of their total phosphorus in the form of phytate [Myo-inositol (1, 2, 3, 4, 5, 6) hexakisphosphate; IP_6]; a compound that is largely ingestible by non-ruminant animals. Ref. [4] noted that phytate, a compound found in plant-based feedstuffs, can have anti-nutritional effects by increasing endogenous nutrient losses and forming insoluble complexes through nutrient binding and mineral chelation, which can reduce the bioavailability of important nutrients and minerals, such as calcium, phosphorus, and zinc.

Phytase, known as Myo-inositol hexakisphosphate phosphohydrolase, is an enzyme capable of decomposing nutrient-binding phytate into inositol penta-, tetra-, tri-, di-, and monophosphate (IP_5 through IP_1). This enzymatic process releases one phosphate molecule at a time in a sequential manner [5]. Because $IP_{6\text{-}3}$ have the anti-nutritional effect of reducing nutrient absorption in monogastric animals, it is important to reduce the presence of $IP_{6\text{-}3}$ by using phytase [6]. The overall target is the complete breakdown of IP_6, IP_4, and IP_3, and to release inositol. In this regard, supplementation with phytase generally increases P utilization in poultry, leading to improved poultry productivity through several specific metabolic pathways. Furthermore, it reduces the reliance on inorganic phosphates in the diet and minimizes P excretion into the environment [2,7]. Therefore, accurately estimating the efficacy of phytase is crucial for optimizing its application to modern poultry diets.

The precise estimation of the available phosphorus (AP) concentrations with the dietary inclusion of phytase is challenging due to various factors, such as phytase sources, varying supplemental levels, and the interactions between nutrient components [8]. Numerous studies have assessed the AP equivalency of different phytase sources in broiler diets [9–11]. Nevertheless, there is a lack of experiments investigating the AP equivalency of phytase in broiler diets using monocalcium phosphate (MCP) as the inorganic P source. Hence, the purpose of this study was to compare the effects of different AP concentrations and phytase levels in the diets on growth performance, carcass traits, nutrient digestibility, tibia traits, and phytate degradation. This study also investigated the AP equivalency of phytase by evaluating its relationship to the tibia phosphorus concentration of broilers fed a typical corn–soybean meal.

2. Materials and Methods

2.1. Birds and Housing

A total of 336 one-day-old male Ross 308 broiler chickens (41.16 ± 0.23 g) were obtained from a local hatchery (Dongsan hatchery, Cheonan, Republic of Korea) and used in the current 35 d study. The birds were individually weighed and allocated to cages in a completely randomized design. Each battery cage (76 × 61 × 46 cm^3) housed eight birds. The cages were equipped with two nipple drinkers and a metal trough to provide water and feed efficiently. The experimental diets were offered *ad libitum*, and the birds had continuous access to clean drinking water through the nipple drinkers. The birds were vaccinated against infectious bronchitis and Newcastle disease at the hatchery. The room temperature was maintained at 30 ± 1 °C from day 1 to 3, and then gradually decreased to 25 ± 1 °C until d 14 of age. Thereafter, a 24 ± 1 °C temperature was maintained throughout the experiment according to the Ross 308 broiler management guidelines [12]. The birds were allowed continuous lighting during the entire experimental period.

2.2. Experimental Design and Diets

The birds were assigned to one of seven dietary treatments in a completely randomized design, with six replicate pens per treatment. Phygest HT, an *Escherichia coli*-derived 6-phytase (Kemin, Industries Asia Pte, Senoko Drive, Singapore) was used as the phytase source in this study. The experimental diets were formulated to evaluate the effects of dietary phosphorus deficiency and phytase supplementation. The experimental diets consisted of various formulations to evaluate the effects of AP deficiency and phytase supplementation. The dietary treatments included a nutritionally adequate positive control diet (PC), with appropriate levels of starter (day 1–21) and grower (day 22–35) diets containing 0.45% and 0.42% AP, respectively. To induce phosphorus deficiencies, three negative control diets (NC-1, NC-2, NC-3) were formulated with decreasing levels of AP compared to the PC (0.10% deficiency, 0.15% deficiency, and 0.20% deficiency, respectively). Furthermore, the NC-3 was supplemented with phytase at three different inclusion levels: NC-3-500 (500 FTU/kg), NC-3-1000 (1000 FTU/kg), and NC-3-1500 (1500 FTU/kg). The NC diets were carefully formulated to contain 0.35%, 0.30%, and 0.25% AP in the starter phase, and 0.32%, 0.27%, and 0.22% AP in the grower phase, while ensuring that the basal PC diet, which included corn and soybean meal, met or exceeded the nutritional specifications set by [13] and the Ross 308 nutrient recommendations [14]. The composition and calculated analysis of the starter and grower diets are presented in Table 1. The diets were provided in a mash form on an *ad-libitum* basis. Furthermore, Cr_2O_3 (chromium oxide powder, >99.9% purity, Sigma-Aldrich, St. Louis, MO, USA) was added as an indigestible marker for digestibility analysis at a proportion of 0.3% to all the experimental diets.

Table 1. Composition (%, as-fed basis) of the experiment diets [1].

Ingredient (%)	Starter Phase (d 1–21)				Grower Phase (d 22–35)			
	PC	NC-1	NC-2	NC-3	PC	NC-1	NC-2	NC-3
Corn	50.94	51.52	51.79	52.06	55.02	55.61	55.88	56.19
Soybean meal, 45%	39.16	39.04	39.00	38.96	34.28	34.20	34.16	34.08
Limestone	1.42	1.60	1.69	1.78	1.46	1.63	1.72	1.81
Mono-calcium phosphate	1.48	1.04	0.84	0.60	1.40	0.96	0.72	0.52
Iodized salt	0.30	0.30	0.30	0.30	0.30	0.30	0.30	0.30
Beef tallow	5.36	5.16	5.04	4.96	6.32	6.08	6.00	5.88
DL-methionine, 98%	0.32	0.32	0.32	0.32	0.33	0.33	0.33	0.33
L-lysine-Sulfate, 65%	0.42	0.42	0.42	0.42	0.29	0.29	0.29	0.29
Vit-Min premix [2]	0.30	0.30	0.30	0.30	0.30	0.30	0.30	0.30
Cr_2O_3	0.30	0.30	0.30	0.30	0.30	0.30	0.30	0.30
Calculated values								
Dry matter	88.17	88.10	88.06	88.03	88.23	88.16	88.12	88.09
ME, kcal/kg	3050	3050	3050	3050	3150	3149	3150	3149
Crude protein	23.0	23.0	23.0	23.0	20.0	20.0	20.0	20.0
Crude fiber	3.83	3.84	3.84	3.85	3.64	3.66	3.66	3.66
Calcium	0.90	0.90	0.90	0.90	0.88	0.88	0.88	0.88
Total phosphorus	0.71	0.60	0.56	0.50	0.66	0.56	0.51	0.46
Available phosphorus	0.45	0.35	0.30	0.25	0.42	0.32	0.27	0.22
Lysine	1.45	1.45	1.45	1.45	1.25	1.25	1.25	1.25
Methionine + Cysteine	0.99	0.99	0.99	0.99	0.95	0.95	0.95	0.95
Analyzed values								
Dry matter	88.84	88.88	88.79	88.51	88.60	88.76	87.84	88.06
Crude protein (Nitrogen × 6.25)	21.15	22.79	21.14	20.97	20.16	19.90	20.22	20.24
Gross energy, kcal/kg	4100	4052	4017	4063	4113	4152	4194	4182
Crude ash	6.58	6.27	6.33	6.59	6.15	6.10	6.15	6.61
Calcium	1.09	1.15	1.07	1.08	0.98	1.02	0.94	0.94
Total phosphorus	0.61	0.55	0.51	0.46	0.64	0.62	0.60	0.58

[1] PC: Positive control diet contained the recommended calcium and available phosphorus; NC-1: negative control diet contained the recommended calcium and 0.10% available phosphorus deficiency; NC-2: negative control diet contained the recommended calcium and 0.15% available phosphorus deficiency; NC-3: negative control diet contained the recommended calcium and 0.20% available phosphorus deficiency. [2] Provided per kilogram of diet: vitamin A (trans-retinyl acetate), 14,000 IU; vitamin D3 (cholecalciferol), 3000 IU; vitamin E (DL-α Tocopherol acetate), 40 mg; vitamin K3, 2.4 mg; thiamin, 1.2 mg; riboflavin, 50 mg; pyridoxine, 3 mg; vitamin B12, 20 μg; niacin, 40 mg; pantothenic acid, 10 mg; folic acid, 0.5 mg; Fe (from iron sulfate), 17 mg; Cu (from copper sulfate), 13 mg; Zn (from zinc oxide), 92 mg; Mn (from manganese oxide), 100 mg; I (from potassium iodide), 1 mg; Co, 0.15 mg; Se (from sodium selenite), 0.25 mg.

2.3. Growth Performance Measurements

The initial body weights of the birds were recorded upon arrival, and subsequent body weights (BW) were measured weekly (d 7, 14, 21, 28, and 35) throughout this study. The average daily gain (ADG) was calculated using the BW data. Feed intake was measured weekly based on the feed consumption in the individual cages. The mortality-corrected average daily feed intake (ADFI) and feed conversion ratio (FCR) were calculated for each cage for each week of the experiment.

2.4. Post-Mortem Procedure and Sample Collection

On d 21 and 35, six birds per treatment (one bird per cage) that had body weights closest to the mean were selected and euthanized using carbon dioxide asphyxiation for sample collection. The dressing percentage of meat with giblets (heart, gizzard, and liver) was determined by dividing its weight by the live weight of the birds. The drumsticks (skinless), breast meat, and the right tibias were removed from the carcasses and weighed. The percentages of the breast meat, drumstick, and right tibia were calculated relative to the weight of the entire carcass. Subsequently, the right tibia was de-fleshed and dried for further analysis.

Abdominal incisions were made on each euthanized bird, and then the duodenum, jejunum, and ileum were separated from the gastrointestinal tract. The ileum was defined as the segment of the small intestine that extends from Meckel's diverticulum to the ileocecal junction. The digesta of the ileum, as well as the duodenum/jejunum from birds subjected to the same treatment, was gently flushed with distilled water into labeled plastic containers and stored at −80 °C until further analysis for inositol phosphate degradation. Furthermore, the ileal digesta of the birds was also obtained for the evaluation of nutrient digestibility.

2.5. Nutrient Digestibility

The apparent ileal digestibility (AID) of the dry matter (DM), crude protein (CP), gross energy, crude ash, calcium, and phosphorus contents were determined to estimate the rate of nutrient digestibility at the terminal ileum. Previously collected ileal digesta samples were thawed, dried at 55 °C for 24 h, ground, and strained through a 0.75 mm sieve (ZM 200 Ultra-Centrifugal Mill; Retsch GmbH & Co., KG, Haan, Germany). Standard procedures [15] were employed for the analysis of the nutrient fractions, while the concentration of chromic oxide was determined using the method of [16]. The AID was then calculated as follows:

$$\text{AID } (\%) = 100 - (100 \times \frac{M_{diet} \times N_{digest}}{M_{digest} \times N_{diet}})$$

M_{diet} is the marker concentration in the diet, N_{digest} is the nutrient concentration in the ileal digesta, M_{digest} is the marker concentration in the ileal digesta, and N_{diet} is the nutrient concentration in the diet.

2.6. Tibia Characteristics

The right tibia bone from the eviscerated birds was dehydrated in ethanol for 72 h and defatted in a diethyl ether: methanol mixture (9:1) for 72 h. Subsequently, the samples were dried overnight at 105 °C and then ashed at 600 °C overnight in porcelain crucibles (AOAC method 942.05). The resulting weight of the tibia ash was quantified as grams of ash per 100 g dry fat-free tibia. The analysis of calcium (AOAC method 927.02) and phosphorus (AOAC method 965.17) in the ashed tibia bones was carried out following standard procedures [15].

2.7. Inositol Phosphate Degradation Assays

The digesta collected from the duodenum/jejunum and ileum, amounting to 0.5 mL, was filtered in buffer, and 20 µL or 40 µL with or without added internal $IP_{2\text{-}6}$ standards at 1.5 mM were injected into the column. Myo-inositol phosphate esters $IP_{1\text{-}6}$ standards (Sichem and Sigma-Aldrich, St. Louis, MO, USA), prepared in buffer G were also analyzed under the same conditions, with an injection volume of 40 µL. The linear range for IP_6 was 10 to 250 µg in an injection volume of 40 µL. Filtrates were analyzed using high-performance ion chromatography and UV detection at 290 nm after post-column derivatization. IPs with different degrees of phosphorylation (IP_{3-6}) and their positional isomers were separated, without enantiomer differentiation, onto a Carbo Pac PA 200 column and corresponding guard column. An $Fe(NO_3)_3$ solution with 0.1% $Fe(NO_3)_3 \cdot 9H_2O$ in 2% $HClO_4$ was used as the reagent for derivatization according to [17]. The elution order of the IP isomers was established using commercial standards. Peaks were detected, and their retention times corresponded to the retention times of the IP standards. IP_6 was used for quantification, and the correction factors for differences in the detector responses for IP_{3-5} were used according to [18]. The limit of detection was defined for a signal:noise ratio of 3:1, and was 0.1 µmol/g of DM for IP_{3-4} isomers and 0.05 µmol/g of DM for IP_6. The limit of quantification was defined for a signal:noise ratio of 6:1.

2.8. Statistical Analysis

The data obtained from the experiment were analyzed using the general linear model (GLM) procedure for one-way ANOVA in Statgraphics Centurion Version 18.1.12 software. A completely randomized design was employed for the analysis. In terms of the growth performance measurements, the experimental unit was defined as the cage. For carcass trait weights, digestibility, tibia composition, and inositol degradation, selected individual birds were considered as the experimental unit. Statistical significance was determined at a significance level of $p < 0.05$. Whenever treatment effects were found to be significant ($p < 0.05$), the means were further analyzed and compared using Tukey's multiple range test procedures implemented in Statgraphics Centurion.

Linear and quadratic regression analyses were performed to test the effect of the increasing levels of phytase, and were also conducted to calculate the AP equivalency values of phytase, based on the concentrations and total amounts of phosphorus in the tibia as the response criteria, using the polynomial regressions described by [19]. The regression equations for the AP levels in the diet and supplemental phytase levels for particular response variables were equated and solved for x.

$$Y_s = a_s + b_s x_s \text{ (linear regression equation for AP)} \quad (1)$$

$$Y_P = a_p + b_p x_p \text{ (linear regression equation for phytase)} \quad (2)$$

$$\text{Equation (1) = Equation (2): } a_s + b_s x_s = a_p + b_p x_p$$

where Y is the response criterion (concentrations of tibia phosphorus), x_s is the AP level in the diet, x_p is the supplemented phytase, a_s is the intercept of the linear regression for AP levels in the diets, a_p is the intercept of the linear regression for the supplemented phytase, b_s is the slope of the response criterion to the dietary AP levels, and b_p is the slope of the response criterion to the supplemented phytase. The linear response equations for the AP in the diet, and that for supplemented phytase, were set to be equal and were solved for the AP equivalency values for their respective variables.

The quadratic regression equation was also calculated in the same manner as the linear equation, as follows:

$$Y_s = a_s + b_s x_s + c_s {x_s}^2 \text{ (quadratic regression equation for AP)} \quad (3)$$

$$Y_P = a_p + b_p x_p + c_p {x_p}^2 \text{ (quadratic regression equation for phytase)} \quad (4)$$

$$\text{Equation (3) = Equation (4): } a_s + b_s x_s + c_s {x_s}^2 = a_p + b_p x_p + c_p {x_p}^2$$

3. Results

3.1. Growth Performance

As presented in Table 2, the broilers fed the NC diets recorded lower ($p < 0.001$) BW compared to the PC and the phytase-supplemented NC-3-500, NC-3-1000, and NC-3-1500 diets for all the periods measured, except for day 1. Throughout the entire experimental period, the lowest BW values were recorded for the NC-3 diet ($p < 0.001$). No significant differences ($p > 0.05$) were observed in the BWs of the broilers, not only between the phytase-supplemented diets and the PC, but also among the different phytase levels for all periods. Likewise, lower ADG values were recorded for the NC diets compared to the PC and phytase-supplemented NC-3-500, NC-3-1000, and NC-3-1500 diets. Furthermore, no significant differences ($p > 0.05$) were noted in the ADG among the broilers fed the various phytase-supplemented diets compared to the PC during the starter (d 1–21) and over the entire period (d 1–35). Similarly, feeding the NC-3 diet resulted in lower ($p < 0.001$) average daily gains throughout the entire experimental period. The daily feed intake levels were unaffected by phosphorus reduction or phytase supplementation throughout the experiment. Considering the FCR, birds fed the NC-3 diet reported lower ($p < 0.01$) feed efficiency compared to the PC and phytase-supplemented diets during the entire period of the experiment (d 1 to 35). Moreover, there were no changes ($p > 0.05$) in the FCR compared to the PC and all the phytase diets during the starter (d 1 to 21) or the entire period of the experiment (d 1 to 35).

Table 2. Effect of phytase inclusion in diets on growth performance of broiler chickens [1].

Period	Dietary Treatment [2]							SEM [3]	*p*-Value	Polynomial Contrast [4]	
	PC	NC-1	NC-2	NC-3	NC-3-500	NC-3-1000	NC-3-1500			Linear	Quadratic
BW, g											
Day 1	41.23	41.03	41.12	40.83	41.33	40.98	41.58	0.101	0.552	0.161	0.376
Day 7	121.70 ^{cd}	113.17 ^{ab}	113.88 ^{abc}	108.07 ^{a}	119.63 ^{bcd}	122.67 ^{d}	122.60 ^{d}	1.302	0.004	0.003	0.002
Day 14	323.56 ^{e}	270.75 ^{abc}	263.01 ^{ab}	245.72 ^{a}	290.68 ^{bcd}	301.69 ^{cde}	304.35 ^{de}	5.503	<0.001	0.001	0.001
Day 21	745.74 ^{d}	620.50 ^{bc}	565.17 ^{ab}	537.61 ^{a}	670.74 ^{cd}	716.24 ^{d}	710.18 ^{d}	14.980	<0.001	<0.001	<0.001
Day 28	1336.75 ^{c}	1148.91 ^{ab}	1087.05 ^{a}	1046.89 ^{a}	1232.90 ^{bc}	1326.34 ^{c}	1302.83 ^{c}	22.814	<0.001	<0.001	<0.001
Day 35	1989.76 ^{c}	1825.97 ^{b}	1771.90 ^{ab}	1676.36 ^{a}	1975.83 ^{c}	2061.68 ^{c}	2028.07 ^{c}	27.831	<0.001	<0.001	<0.001
ADG, g/d											
Day 7	11.50 ^{cd}	10.31 ^{ab}	10.40 ^{abc}	9.61 ^{a}	11.19 ^{bcd}	11.67 ^{d}	11.57 ^{d}	0.183	0.005	0.004	0.003
Day 14	28.84 ^{d}	22.51 ^{abc}	21.30 ^{ab}	19.67 ^{a}	24.44 ^{bc}	25.57 ^{cd}	25.96 ^{cd}	0.647	<0.001	0.002	0.002
Day 21	60.31 ^{d}	49.96 ^{bc}	43.17 ^{ab}	41.70 ^{a}	54.29 ^{cd}	59.22 ^{d}	57.98 ^{d}	1.427	<0.001	<0.001	<0.001
Day 28	84.43 ^{bc}	75.49 ^{ab}	74.55 ^{ab}	72.75 ^{a}	80.31 ^{abc}	87.16 ^{c}	84.66 ^{bc}	1.524	0.047	0.010	0.013
Day 35	93.29 ^{a}	96.72 ^{ab}	97.84 ^{abc}	89.93 ^{a}	106.13 ^{c}	105.05 ^{cd}	103.61 ^{bcd}	1.337	0.001	0.008	<0.001
Day 1–21	33.55 ^{d}	27.59 ^{bc}	24.96 ^{ab}	23.66 ^{a}	29.97 ^{cd}	32.16 ^{d}	31.84 ^{d}	0.712	<0.001	<0.001	<0.001
Day 22–35	88.86 ^{abc}	86.11 ^{ab}	86.19 ^{ab}	81.34 ^{a}	93.22 ^{bc}	96.10 ^{c}	94.13 ^{c}	1.208	0.004	0.004	0.001
Day 1–35	55.67 ^{c}	51.00 ^{b}	49.45 ^{ab}	46.73 ^{a}	55.27 ^{b}	57.73 ^{c}	56.76 ^{c}	0.794	<0.001	<0.001	<0.001
ADFI, g/d											
Day 7	13.02	14.60	14.78	14.67	14.81	14.42	13.23	0.277	0.390	0.141	0.220
Day 14	33.92	36.02	36.09	37.86	33.84	34.52	33.48	1.173	0.961	0.475	0.725
Day 21	73.24	74.68	66.82	68.35	69.07	71.19	72.21	1.452	0.803	0.340	0.641
Day 28	109.44	113.83	110.70	116.09	109.63	123.32	119.79	2.370	0.653	0.401	0.692
Day 35	137.13	151.94	149.31	153.22	153.35	151.13	148.95	1.803	0.209	0.456	0.738
Day 1–21	40.06	41.77	39.23	40.29	39.24	40.04	39.64	0.782	0.987	0.913	0.985
Day 22–35	123.29	132.88	130.00	134.65	131.49	137.23	134.37	1.658	0.408	0.795	0.967
Day 1–35	73.35	78.21	75.54	78.04	76.14	78.92	77.53	0.985	0.795	0.913	0.989
FCR, g/g											
Day 7	1.14 ^{a}	1.42 ^{ab}	1.43 ^{ab}	1.52 ^{b}	1.36 ^{abc}	1.25 ^{ab}	1.14 ^{a}	0.036	0.009	0.002	0.008
Day 14	1.19 ^{a}	1.62 ^{abc}	1.69 ^{bc}	1.97 ^{c}	1.41 ^{ab}	1.36 ^{ab}	1.26 ^{ab}	0.065	0.010	0.013	0.022
Day 21	1.23 ^{a}	1.53 ^{ab}	1.64 ^{b}	1.67 ^{b}	1.27 ^{a}	1.21 ^{a}	1.25 ^{a}	0.052	0.027	0.005	0.001
Day 28	1.30	1.52	1.54	1.63	1.37	1.42	1.43	0.043	0.485	0.265	0.265
Day 35	1.47	1.58	1.54	1.72	1.45	1.44	1.44	0.028	0.054	0.015	0.009
Day 1–21	1.21 ^{a}	1.53 ^{bc}	1.60 ^{c}	1.72 ^{c}	1.31 ^{ab}	1.25 ^{a}	1.24 ^{a}	0.044	0.001	0.003	0.002
Day 22–35	1.39	1.55	1.53	1.68	1.41	1.43	1.43	0.030	0.108	0.046	0.030
Day 1–35	1.32 ^{a}	1.54 ^{bc}	1.54 ^{bc}	1.68 ^{c}	1.38 ^{ab}	1.37 ^{ab}	1.37 ^{ab}	0.030	0.006	0.008	0.003

[1] Values are mean of six replicates per treatment. [2] PC: Positive control diet contained the recommended Ca and non-phytate phosphorus; NC-1: negative control diet contained the recommended Ca and 0.10% available P deficiency; NC-2: negative control diet contained the recommended Ca and 0.15% available P deficiency; NC-3: negative control diet contained the recommended Ca and 0.20% available P deficiency; NC-3-500, NC-3-1000, and NC-3-1500 are NC-3 plus phytase 500, 1000, 1500 FTU/kg, respectively. [3] Pooled standard error of mean. [4] Orthogonal polynomial contrasts were conducted to assess the significance of the linear or quadratic effects of the supplementation with phytase in the broilers. $^{a–e}$ Values in a row with different superscripts differ significantly ($p < 0.05$).

3.2. Carcass Traits

No significant differences ($p > 0.05$) were observed in the dressing percentages, and the relative breast meat and drumstick weights among the various dietary treatments on both d 21 and 35 are shown in Table 3. However, the broilers fed the NC-3 diet exhibited lower tibia weight ($p < 0.001$) compared to the other dietary treatments on d 35. On the other hand, there were no differences ($p > 0.05$) in tibia weight between the PC diet and all the phytase-supplemented diets on both d 21 and 35.

Table 3. Effect of phytase inclusion in diets on carcass traits of broiler chickens [1].

Period	Dietary Treatment [2]							SEM [3]	*p*-Value	Polynomial Contrast [4]	
	PC	NC-1	NC-2	NC-3	NC-3-500	NC-3-1000	NC-3-1500			Linear	Quadratic
Dressing (%) [5]											
Day 21	90.48	91.02	90.21	90.42	91.04	91.36	90.85	0.130	0.186	0.306	0.150
Day 35	92.64	91.31	92.22	92.15	92.96	92.22	92.27	0.169	0.255	0.824	0.578
Breast (%) [6]											
Day 21	23.32	23.32	22.72	21.29	23.27	23.30	23.76	0.229	0.080	0.020	0.036
Day 35	27.44	26.42	25.86	24.29	25.68	27.02	26.42	0.284	0.067	0.030	0.041
Drumstick (%) [7]											
Day 21	9.89	9.77	9.52	9.21	9.90	10.06	9.92	0.094	0.215	0.068	0.059
Day 35	10.54	10.00	10.15	9.95	10.13	10.14	10.75	0.087	0.118	0.006	0.012
Tibia weight (g)											
Day 21	5.83	5.50	4.83	4.67	5.33	5.33	5.50	0.124	0.158	0.078	0.158
Day 35	16.67 [b]	15.17 [ab]	15.00 [ab]	12.50 [a]	16.33 [b]	16.67 [b]	16.67 [b]	0.307	<0.001	<0.001	<0.001

[1] Values are mean of six replicates per treatment. [2] PC: Positive control diet contained the recommended Ca and non-phytate phosphorus; NC-1: negative control diet contained the recommended Ca and 0.10% available P deficiency; NC-2: negative control diet contained the recommended Ca and 0.15% available P deficiency; NC-3: negative control diet contained the recommended Ca and 0.20% available P deficiency; NC-3-500, NC-3-1000, and NC-3-1500 are NC-3 plus phytase 500, 1000, 1500 FTU/kg, respectively. [3] Pooled standard error of mean. [4] Orthogonal polynomial contrasts were conducted to assess the significance of the linear or quadratic effects of the supplementation with phytase in the broilers. [5] (Carcass weight/live body weight) × 100. [6] (Breast meat weight/carcass weight) × 100. [7] (Drumstick weight/carcass weight) × 100. [a,b] Values in a row with different superscripts differ significantly ($p < 0.05$).

3.3. Nutrient Digestibility

The AID of energy was decreased ($p < 0.001$) in all the NC diets compared to the PC diet on d 21 and 35 (Table 4). The AID of energy in the phytase-supplemented diets was median to that of the PC and NC diets. The apparent CP digestibility was not affected ($p > 0.05$) for either the PC or NC-3 with phytase diets on d 21 and 35. However, the AID of energy on d 35 was improved ($p < 0.001$) in the birds fed the NC-3-1000 diet compared to all the treatments. Nonetheless, the broilers fed the NC-3 diet had a lower ($p < 0.05$) crude ash digestibility than those fed the PC or NC-3 with phytase diets on d 35. The NC-3-1500 diet increased ($p < 0.05$) the AID of phosphorus on d 21 and 35 compared to the NC diets.

3.4. Tibia Traits

Feeding NC-3 diets resulted in lower ($p < 0.001$) calcium and phosphorus contents in the tibia compared to the PC on d 21 (see Table 5). The tibia phosphorus content was higher ($p < 0.001$) in the broilers fed the NC-3-1500 compared to the other treatments on d 21 and 35. Furthermore, the birds fed the NC-3-1500 had higher ($p < 0.01$) contents of calcium on d 35 than those receiving the other treatments.

3.5. Inositol Phosphate Degradation

All the phytase-supplemented NC-3 diets recorded lower ($p < 0.01$) IP_6 concentrations, not only in the duodenum/jejunum but also in the ileum, on d 21 and 35 compared to the PC diets (see Table 6). Regardless of the concentration, phytase supplementation in the NC-3 resulted in a higher ($p < 0.001$) IP_4 concentration in the ileum on d 35 compared to the treatments without phytase. Birds fed the NC-3-1000 and NC-3-1500 diets had higher ($p < 0.001$) IP_3 concentrations in the duodenum/jejunum, as well as in the ileum, on d 21 and 35 compared to the PC diets.

Table 4. Effect of phytase inclusion in diets on nutrient digestibility in broiler chickens [1].

Period	Dietary Treatment [2]							SEM [3]	p-Value	Polynomial Contrast [4]	
	PC	NC-1	NC-2	NC-3	NC-3-500	NC-3-1000	NC-3-1500			Linear	Quadratic
Day 21 (%)											
Dry matter	64.20	64.84	64.73	64.83	64.36	64.71	64.51	0.397	0.999	0.711	0.906
Crude protein	80.72	81.05	79.33	79.50	79.95	80.87	79.95	0.214	0.180	0.452	0.417
Energy	79.32 e	75.22 c	72.70 b	70.03 a	76.03 cd	77.06 d	77.59 de	0.502	<0.001	<0.001	<0.001
Ash	53.37	48.43	47.50	46.82	48.69	51.92	53.46	1.060	0.434	0.007	0.020
Calcium	53.67	52.37	52.32	52.53	53.98	56.20	55.10	1.058	0.955	0.219	0.331
Phosphorus	60.56 ab	59.44 a	58.93 a	57.34 a	59.64 a	65.72 ab	73.29 b	1.342	0.011	0.001	0.004
Day 35 (%)											
Dry matter	76.03	76.22	75.96	75.36	76.68	76.08	75.67	0.382	0.988	0.950	0.775
Crude protein	89.09 cd	87.62 bc	86.95 ab	85.32 a	87.56 bc	90.93 d	88.93 c	0.344	<0.001	0.001	<0.001
Energy	82.82 d	79.07 b	80.76 c	77.43 a	81.42 c	84.51 e	82.79 d	0.368	<0.001	<0.001	<0.001
Ash	60.93 b	56.68 ab	56.63 ab	52.89 a	63.01 b	63.25 b	63.56 b	1.047	0.021	0.032	0.032
Calcium	53.30	52.76	51.74	50.18	52.57	53.29	53.31	1.223	0.995	0.522	0.774
Phosphorus	63.16 abc	60.15 a	61.95 ab	59.87 a	64.02 abc	65.16 bc	66.67 c	0.642	0.026	0.006	0.019

[1] Values are mean of six replicates per treatment. [2] PC: Positive control diet contained the recommended Ca and non-phytate phosphorus; NC-1: negative control diet contained the recommended Ca and 0.10% available P deficiency; NC-2: negative control diet contained the recommended Ca and 0.15% available P deficiency; NC-3: negative control diet contained the recommended Ca and 0.20% available P deficiency; NC-3-500, NC-3-1000, and NC-3-1500 are NC-3 plus phytase 500, 1000, 1500 FTU/kg, respectively. [3] Pooled standard error of mean. [4] Orthogonal polynomial contrasts were conducted to assess the significance of the linear or quadratic effects of the supplementation with phytase in the broilers. [a–e] Values in a row with different superscripts differ significantly ($p < 0.05$).

Table 5. Effect of phytase inclusion in diets on tibia traits of broiler chickens [1].

Period	Dietary Treatment [2]							SEM [3]	p-Value	Polynomial Contrast [4]	
	PC	NC-1	NC-2	NC-3	NC-3-500	NC-3-1000	NC-3-1500			Linear	Quadratic
Ash (%)											
Day 21	50.24	44.42	43.67	41.37	46.69	47.59	51.48	1.053	0.120	0.007	0.028
Day 35	50.04 bc	48.26 abc	46.46 ab	44.50 a	48.38 bc	48.73 bc	52.03 c	0.581	0.012	<0.001	0.003
Ca (%)											
Day 21	38.18 c	32.65 ab	30.28 a	28.20 a	37.18 bc	37.88 bc	41.83 c	1.500	<0.001	<0.001	<0.001
Day 35	35.33 b	33.08 ab	31.97 ab	30.28 a	33.37 ab	34.31 b	41.62 c	0.687	<0.001	<0.001	<0.001
P (%)											
Day 21	17.48 c	14.40 ab	12.90 a	12.66 a	16.12 bc	16.54 bc	20.46 d	0.470	<0.001	<0.001	<0.001
Day 35	16.80 a	15.44 a	15.32 a	15.15 a	15.60 a	16.35 a	18.60 b	0.272	0.003	<0.001	<0.001

[1] Values are mean of six replicates per treatment. [2] PC: Positive control diet contained the recommended Ca and non-phytate phosphorus; NC-1: negative control diet contained the recommended Ca and 0.10% available P deficiency; NC-2: negative control diet contained the recommended Ca and 0.15% available P deficiency; NC-3; negative control diet contained the recommended Ca and 0.20% available P deficiency; NC-3-500, NC-3-1000, and NC-3-1500 are NC-3 plus phytase 500, 1000, 1500 FTU/kg, respectively. [3] Pooled standard error of mean. [4] Orthogonal polynomial contrasts were conducted to assess the significance of the linear or quadratic effects of the supplementation with phytase in the broilers. [a–d] Values in a row with different superscripts differ significantly ($p < 0.05$).

Table 6. Effect of phytase inclusion in the diets on digesta IP_{6-3} concentrations in the duodenum/jejunum and ileum of the broiler chickens [1].

Period	Dietary Treatment [2]							SEM [3]	p-Value	Polynomial Contrast [4]	
	PC	NC-1	NC-2	NC-3	NC-3-500	NC-3-1000	NC-3-1500			Linear	Quadratic
	Digesta duodenum/jejunum on day 21 (μM/g)										
IP_6	0.791 c	0.702 c	0.726 c	0.687 c	0.503 b	0.359 a	0.404 ab	0.030	<0.001	0.001	0.001
IP_5	0.398	0.375	0.484	0.345	0.275	0.343	0.258	0.027	0.350	0.540	0.828
IP_4	0.097 a	0.123 ab	0.127 ab	0.130 ab	0.192 b	0.267 b	0.217 b	0.013	0.002	0.037	0.035
IP_3	0.026 ab	0.005 a	0.015 a	0.017 a	0.047 b	0.077 c	0.145 d	0.008	<0.001	<0.001	<0.001
	Digesta ileum on day 21 (μM/g)										
IP_6	0.661 b	0.667 b	0.668 b	0.651 b	0.322 a	0.255 a	0.229 a	0.036	<0.001	<0.001	<0.001
IP_5	0.388	0.316	0.362	0.290	0.217	0.209	0.195	0.025	0.229	0.115	0.228
IP_4	0.111	0.130	0.121	0.141	0.202	0.158	0.192	0.011	0.183	0.442	0.683
IP_3	0.0258 a	0.0218 a	0.0358 a	0.0320 a	0.119 b	0.088 b	0.114 b	0.009	<0.001	0.058	0.078
	Digesta duodenum/jejunum on day 35 (μM/g)										
IP_6	0.689 b	0.638 b	0.604 b	0.657 b	0.359 a	0.356 a	0.351 a	0.027	<0.001	<0.001	<0.001
IP_5	0.315	0.315	0.299	0.302	0.240	0.264	0.222	0.016	0.566	0.303	0.583
IP_4	0.124	0.110	0.118	0.128	0.167	0.164	0.182	0.009	0.187	0.116	0.268

Table 6. *Cont.*

Period	Dietary Treatment [2]							SEM [3]	*p*-Value	Polynomial Contrast [4]	
	PC	NC-1	NC-2	NC-3	NC-3-500	NC-3-1000	NC-3-1500			Linear	Quadratic
IP_3	0.0097 [a]	0.0083 [a]	0.0153 [a]	0.0093 [a]	0.214 [b]	0.176 [b]	0.167 [b]	0.0181	<0.001	0.058	0.013
Digesta ileum on day 35 (μM/g)											
IP_6	0.676 [c]	0.684 [c]	0.620 [c]	0.606 [bc]	0.448 [ab]	0.433 [a]	0.423 [a]	0.027	0.005	0.083	0.133
IP_5	0.332 [bc]	0.349 [c]	0.403 [c]	0.357 [c]	0.242 [ab]	0.224 [a]	0.199 [a]	0.016	0.001	<0.001	<0.001
IP_4	0.0913 [a]	0.103 [a]	0.0907 [a]	0.103 [a]	0.212 [b]	0.190 [b]	0.190 [b]	0.0098	<0.001	0.014	0.001
IP_3	0.0183 [a]	0.0277 [a]	0.0078 [a]	0.0441 [a]	0.133 [b]	0.147 [b]	0.106 [b]	0.0111	<0.001	0.152	0.029

[1] $IP_{6\text{-}3}$ (IP_6 = Myo-inositol hexa-phosphates, IP_5 = Myo-inositol penta-phosphates, IP_4 = Myo-inositol tetra-phosphates, and IP_3 = Myo-inositol tri-phosphates): the means calculated from six replicates per treatment. [2] PC: Positive control diet contained the recommended Ca and non-phytate phosphorus; NC-1: negative control diet contained the recommended Ca and 0.10% available P deficiency; NC-2: negative control diet contained the recommended Ca and 0.15% available P deficiency; NC-3: negative control diet contained the recommended Ca and 0.20% available P deficiency; NC-3-500, NC-3-1000, and NC-3-1500 are NC-3 plus phytase 500, 1000, 1500 FTU/kg, respectively. [3] Pooled standard error of mean. [4] Orthogonal polynomial contrasts were conducted to assess the significance of the linear or quadratic effects of the supplementation with phytase in the broilers. [a–d] Values in a row with different superscripts differ significantly ($p < 0.05$).

3.6. Regression Equations and Phosphorus Equivalency Values

The research findings regarding the relationship between increasing levels of dietary AP and supplemental phytase are summarized in Table 7. The calculated phosphorus equivalency values of the levels of phytase with NC-3 using linear and quadratic regressions are presented in Table 8. Based on the linear and quadratic regression models, 500 FTU/kg phytase was determined to be equivalent to 0.377% and 0.383% AP in the diet on d 21, and 0.317% and 0.307% AP in the diet on d 35, based on the concentrations of tibia phosphorus, respectively. Likewise, 1000 FTU/kg phytase was determined to be equivalent to 0.476% and 0.448% AP in the diet on d 21, and 0.437% and 0.403% AP in the diet on d 35, respectively. Finally, 1500 FTU/kg phytase was determined to be equivalent to 0.574% and 0.504% AP in the diet on d 21, and 0.557% and 0.500 AP in the diet on d 35, respectively.

Table 7. Regressions of the concentrations of tibia phosphorus on available phosphorus from the diets and phytase supplementation.

Items	Day 21				Day 35			
	Regression Equation							
	Linear	R^2	Quadratic	R^2	Linear	R^2	Quadratic	R^2
Available phosphorus (AP)								
Tibia P (%)	$Y = 5.80 + 25.37X_{AP}$	0.9542	$Y = 15.80 - 33.90X_{AP} + 83.82X_{AP}^2$	0.9916	$Y = 13.11 + 8.36X_{AP}$	0.8876	$Y = 17.83 - 22.53X_{AP} + 47.73X_{AP}^2$	0.9915
Phytase								
Tibia P (%)	$Y = 12.87 + 0.005X_{phy}$	0.9283	$Y = 12.99 + 0.004X + (4.6 \times 10^{-7})X_{phy}^2$	0.9300	$Y = 14.76 + 0.002X_{phy}$	0.8748	$Y = 15.21 - 0.0005X_{phy} + (1.800 \times 10^{-6})X_{phy}^2$	0.9898

Y = predicted performance for a given criteria; X_{AP} = percentage of AP from the diet; X_{phy} = FTU of phytase per kg.

Table 8. Equivalency of available phosphorus (AP) and equivalency values (%) of phytase.

Items		Equivalency of AP Values (FTU/kg) [1]		
		500	1000	1500
Calculated matrix value (%) [2]				
Linear equation	Day 21	0.377	0.476	0.574
	Day 35	0.317	0.437	0.557
Quadratic equation	Day 21	0.383	0.448	0.504
	Day 35	0.307	0.403	0.500

[1] Equivalency of AP values (%) relative to NC-3 diet. [2] Calculated values were calculated by substituting the phosphorus percent in the tibia to estimate the equivalency available phosphorus of phytase.

4. Discussion

Several studies have evaluated the AP equivalency of phytase for a couple of inorganic sources (i.e., monosodium phosphate and dicalcium phosphate) to determine their effectiveness in diet formulation [9,11,19]. Despite having the highest P digestibility among inorganic P sources in broiler diets [1,20], there is a scarcity of available data regarding the AP equivalency of phytase with mono-calcium phosphate (MCP). The present research was designed to investigate the efficacy and AP equivalency of phytase on growth performance, carcass traits, nutrient digestibility, tibia bone mineralization, and inositol degradation in broilers.

In the current study, the growth performance of the broilers, except for the ADFI, was found to be reduced in all periods as the content of available phosphorus decreased. However, the body weight on d 35 increased by 17.86, 22.99%, and 20.98%, respectively, when 500, 1000, and 1500 FTU/kg of phytase were added to the NC-3. Likewise, the ADG for the entire experimental period improved by 18.28%, 23.54%, and 21.46%, respectively, for the diets containing 500, 1000, and 1500 FTU/kg of phytase relative to NC-3. Furthermore, the NC-3-500, NC-3-1000, and NC-3-1500 diets improved feed efficiency for the whole period by 17.86%, 18.45%, and 18.45%, respectively, compared to NC-3. The results of this current study are collectively in agreement with the findings of previous studies that have shown that phytase improves the growth performance of broilers fed available phosphorus-deficient diets [11,19,21]. Moreover, the observed decline in growth performance and feed efficiency with decreasing levels of dietary available phosphorus substantiates the inability of these birds to effectively utilize phytate phosphorus. This partial disagreement with previous studies could be due to differences in the composition of the feed ingredients and the type of inorganic phosphates.

With the exception of the tibia weights, the carcass traits were not influenced by a P reduction or phytase supplementation among the different treatments. An available phosphorus reduction without dietary phytase supplementation resulted in lowered tibia weights. The outcome of this study suggests the capacity of phytase supplementation to improve ileal mineral absorption and bone mineralization, which ultimately improves the tibia weight. Phytase supplementation could also correct the negative impact of low available phosphorus contents in the diets by improving the ileal Ca and P digestibility for improved tibia weights, as was observed. Although these results are consistent with [22], who found that supplementation with phytase did not affect the carcass traits, the research regarding the available phosphorus content and carcass traits remains insufficient.

In the current study, as the AP levels were decreased in the adequate PC diet, which was sufficient to meet nutrient requirements, the digestibility of the energy and phosphorus in the diets was negatively affected. According to [23], elevated dietary phytate concentrations result in diminished energy utilization and ileal digestibility of protein due to the inhibitory effects of phytate. The inhibitory action of elevated phytate levels results in nutrient-binding complexes that lower the hydrolyzing of endogenous enzymes and, thus, the availability of minerals and energy is hampered. In this present study, the digestibility of energy, phosphorus, crude protein, and ash on d 35 was increased in response to phytase supplementation in the reduced AP of diets. Increased nutrient digestibility is evidence of phytase activity resulting in the considerable breakdown of the phytate structure and the reduction in anti-nutritional factors. The breakdown of phytate into lower esters leads to a subsequent improvement not only of P, but could exert "extra-phosphoric effects" to improve the digestibility of energy, crude protein, and ash, as has been previously reported [4,24].

Due to its involvement in improving Ca and P digestibility and, thus, improved bone mineralization, it was also observed that dietary phytase supplementation improved the tibia characteristics, especially the total ash content, as has been previously reported [11,19,25]. While the current study is consistent with previous studies [11,19,25] regarding the tibia ash content, a similar trend was observed for the concentrations of calcium and phosphorus in the tibia. Likewise, ref. [9] demonstrated improvements in the tibia phosphorus content when increasing the graded levels of phytase in low-phosphorus diets. In line with the findings of [26], phytase supplementation contributes to increased phosphorus digestibility, consequently leading to higher concentrations of substrate available for tibia mineral deposition. This effect is supported by the observed increase in tibia ash and phosphorus contents in this study. However, ref. [21] noted that there was no difference in tibia phosphorus and calcium level related to the non-phytate phosphorus and phytase contents of the diets. The discrepancy between these results could be due to differences in the composition of the feed ingredients, inorganic phosphate types, and the age of the broilers.

It is expected that phytase supplementation facilitates a process in which phosphate molecules are released through a step-by-step dephosphorylation pathway, breaking down IP_6 into lower esters, including IP4 and IP_3, which are also known to be antinutritive in nature [27]. The analysis of the inositol phosphate profiles in the duodenal/jejunal and ileal digesta from this experiment suggests that the highest proportion of IP_6 concentrations are found in diets without phytase supplementation, particularly in the PC and NC diets (i.e., NC-1, NC-2, and NC-3). Subsequently, phytase supplementation in the NC-3-500, NC-3-1000, and NC-3-1500 diets results in the hydrolysis of IP_6 into $IP_{5\text{-}3}$ and lower esters. It has been reported [28] that phytase readily targets higher molecular weight esters to release IP_5 more efficiently than the other esters. However, a focus should also be placed on improving the content of the lower esters, including IP_4 and IP_3. The superdosing of phytase as a concept is targeted at supplying higher phytase levels to result in as much IP_5, IP_4, and IP_3 as possible, as well as the eventual release of inositol. These findings are consistent with those of [6,29], who demonstrated a reduction in IP_6 and an increase in IP_3 in the ileum following dietary phytase supplementation. Going forward, the focus should be not only on the release of IP_4 and IP_3, but on their subsequent breakdown as well which results in the eventual release of inositol. Zinc, as a co-factor of pancreatic secretions, is known to be precipitated at higher pH levels in the ileum by IP_4 and IP_3, thus hindering protein degradation. This could be due to the pancreatic duct secreting zinc into the small intestine, which induces an increasing digesta pH and reduces the activity of phytase or, more plausibly, the impact of dietary phosphorus on the hydrolysis of phytate by exogenous phytase [30,31]. Unfortunately, we were not able to test this notion in the current study.

5. Conclusions

Phytase supplementation resulted in a reduction in the IP_6 concentration and a corresponding increase in IP_3 concentrations in various sections of the GI tract at 21 and 35 days of age compared to the control diet. Based on the linear regression model, supplementation with 500, 1000, and 1500 phytase units per kg of diet was determined to be equivalent to 0.377, 0.476, and 0.574% AP on d 21, and 0.317, 0.437, and 0.557% on d 35, respectively. Additionally, employing the quadratic equation, supplementation with 500, 1000, and 1500 phytase units per kg of diet corresponded to 0.383, 0.448, and 0.504% AP on d 21, and 0.307, 0.403, and 0.500% on d 35, respectively. Our data consistently suggest that phytase can be used from 1 to 35 d of age in available phosphorus-deficient broiler diets to improve nutrient digestibility (Ca, P, energy, ash, and crude protein) and to stimulate phytate degradation into lower esters. Improved nutrient digestibility and phytate degradation are directly responsible for the improved growth performance and bone mineralization observed in the tibia in the current study.

Author Contributions: Conceptualization, M.Y., B.L.V., H.L. and J.-M.H.; methodology, M.Y.; software, M.Y.; validation, M.Y., J.-S.H., H.L. and J.-M.H.; formal analysis, M.Y.; investigation, M.Y., E.O.O., J.-S.H., N.C.C., E.S. and H.P.; resources, B.L.V., H.L. and J.-M.H.; data curation, M.Y.; writing—original draft preparation, M.Y.; writing—review and editing, M.Y., E.O.O., B.L.V. and J.-M.H.; visualization, M.Y.; supervision, J.-M.H.; project administration, B.L.V., H.L. and J.-M.H. All authors have read and agreed to the published version of the manuscript.

Funding: The authors are grateful to Kemin Industries Asia Pte Ltd., Singapore, for their financial support and supply of the additives in this study.

Institutional Review Board Statement: The experimental procedures and protocol were reviewed and approved by the Animal Ethics Committee of Chungnam National University (protocol number 202206A-CNU-102) (approval date 28 June 2022).

Informed Consent Statement: Not applicable.

Data Availability Statement: Data are contained within the article.

Conflicts of Interest: Authors Bindhu Lakshmibai Vasanthakumari and Hans Lee are employees of Kemin Industries. The funders had no role in the design of the study or the collection, analysis, or interpretation of data. The other authors declare no conflicts of interest.

References

1. An, S.H.; Sung, J.Y.; Kong, C. Ileal digestibility and total tract retention of phosphorus in inorganic phosphates fed to broiler chickens using the direct method. *Animals* **2020**, *10*, 2167. [CrossRef] [PubMed]
2. Selim, S.; Abdel-Megeid, N.S.; Khalifa, H.K.; Fakiha, K.G.; Majrashi, K.A.; Hussein, E. Efficacy of various feed additives on performance, nutrient digestibility, bone quality, blood constituents, and phosphorus absorption and utilization of broiler chickens fed low phosphorus diet. *Animals* **2022**, *12*, 1742. [CrossRef] [PubMed]
3. Hernandez, J.R.; Gulizia, J.P.; Adkins, J.B.; Rueda, M.S.; Haruna, S.I.; Pacheco, W.J.; Downs, K.M. Effect of Phytase Level and Form on Broiler Performance, Tibia Characteristics, and Residual Fecal Phytate Phosphorus in Broilers from 1 to 21 Days of Age. *Animals* **2022**, *12*, 1952. [CrossRef] [PubMed]
4. Walters, H.; Coelho, M.; Coufal, C.; Lee, J. Effects of increasing phytase inclusion levels on broiler performance, nutrient digestibility, and bone mineralization in low-phosphorus diets. *J. Appl. Poult. Res.* **2019**, *28*, 1210–1225. [CrossRef]
5. Walk, C.; Bedford, M.; Olukosi, O. Effect of phytase on growth performance, phytate degradation and gene expression of myo-inositol transporters in the small intestine, liver and kidney of 21 day old broilers. *Poult. Sci.* **2018**, *97*, 1155–1162. [CrossRef] [PubMed]
6. Bello, A.; Dersjant-Li, Y.; Korver, D. The efficacy of 2 phytases on inositol phosphate degradation in different segments of the gastrointestinal tract, calcium and phosphorus digestibility, and bone quality of broilers. *Poult. Sci.* **2019**, *98*, 5789–5800. [CrossRef] [PubMed]
7. Hofmann, P.; Krieg, J.; Francesch, M.; Feuerstein, D.; Rodehutscord, M. Effects of added phytase on growth performance, carcass traits, and tibia ash of broiler chickens fed diets with reduced amino acid, crude protein, and phosphorus concentration. *J. Appl. Poult. Res.* **2022**, *31*, 100258. [CrossRef]
8. Vieira, S.; Anschau, D.; Serafini, N.; Kindlein, L.; Cowieson, A.; Sorbara, J. Phosphorus equivalency of a Citrobracter braakii phytase in broilers. *J. Appl. Poult. Res.* **2015**, *24*, 335–342. [CrossRef]
9. Ribeiro, V., Jr.; Salguero, S.; Gomes, G.; Barros, V.; Silva, D.; Barreto, S.; Rostagno, H.; Hannas, M.; Albino, L. Efficacy and phosphorus equivalency values of two bacterial phytases (*Escherichia coli* and *Citrobacter braakii*) allow the partial reduction of dicalcium phosphate added to the diets of broiler chickens from 1 to 21 days of age. *Anim. Feed Sci. Technol.* **2016**, *221*, 226–233. [CrossRef]
10. Wu, Y.; Xu, S.; Wang, X.; Xu, H.; Liu, P.; Xing, X.; Qi, Z. Phosphorus equivalency of phytase with various evaluation indicators of meat duck. *Poult. Sci.* **2021**, *100*, 101216. [CrossRef]
11. Kwon, C.; Lee, J.; Kim, D.; Kil, D. Determination and validation of available phosphorus equivalency of Escherichia coli-derived phytase in broiler diets. *J. Appl. Poult. Res.* **2022**, *31*, 100298. [CrossRef]
12. ROSS. *308 Broiler Management Handbook*; Aviagen Inc.: Huntsville, AL, USA, 2014.
13. NRC-National Research Council. *Nutrient Requirements of Poultry*; National Academies Press: Washington, DC, USA, 1994.
14. ROSS. *Ross 308 Broiler: Nutrition Specification*; Aviagen Inc.: Huntsville, AL, USA, 2007.
15. AOAC. *Official Methods of Analysis of the Association of Analytical Chemists International*; Official Methods: Gaithersburg, MD, USA, 2005.
16. Oketch, E.O.; Lee, J.W.; Yu, M.; Hong, J.S.; Kim, Y.B.; Nawarathne, S.R.; Chiu, J.W.-C.; Heo, J.M. Physiological responses of broiler chickens fed reduced-energy diets supplemented with emulsifiers. *Anim. Biosci.* **2022**, *35*, 1929–1939. [CrossRef] [PubMed]
17. Phillippy, B.Q.; Bland, J.M. Gradient ion chromatography of inositol phosphates. *Anal. Biochem.* **1988**, *175*, 162–166. [CrossRef]
18. Skoglund, E.; Carlsson, N.-G.; Sandberg, A.-S. Determination of isomers of inositol mono-to hexaphosphates in selected foods and intestinal contents using high-performance ion chromatography. *J. Agric. Food Chem.* **1997**, *45*, 431–436. [CrossRef]
19. Adedokun, S.; Sands, J.; Adeola, O. Determining the equivalent phosphorus released by an Escherichia coli-derived phytase in broiler chicks. *Can. J. Anim. Sci.* **2004**, *84*, 437–444. [CrossRef]
20. Lee, S.A.; Lopez, D.A.; Stein, H.H. Mineral composition and phosphorus digestibility in feed phosphates fed to pigs and poultry. *Anim. Biosci.* **2023**, *36*, 167. [CrossRef] [PubMed]
21. Han, J.; Yang, X.; Qu, H.; Xu, M.; Zhang, T.; Li, W.; Yao, J.; Liu, Y.; Shi, B.; Zhou, Z. Evaluation of equivalency values of microbial phytase to inorganic phosphorus in 22-to 42-day-old broilers. *J. Appl. Poult. Res.* **2009**, *18*, 707–715. [CrossRef]
22. De Freitas, H.B.; de Souza Nascimento, K.M.R.; Kiefer, C.; Gomes, G.A.; Dos Santos, T.T.; Garcia, E.R.M.; da Silva, T.R.; Paiva, L.L.; Berno, P.R. Graded levels of phytase on performance, bone mineralization and carcass traits of broiler fed reduced dicalcium phosphate. *Asian-Australas. J. Anim. Sci.* **2019**, *32*, 691. [CrossRef] [PubMed]

23. Ravindran, V.; Morel, P.; Partridge, G.; Hruby, M.; Sands, J. Influence of an Escherichia coli-derived phytase on nutrient utilization in broiler starters fed diets containing varying concentrations of phytic acid. *Poult. Sci.* **2006**, *85*, 82–89. [CrossRef]
24. Abd El-Hack, M.E.; Alagawany, M.; Arif, M.; Emam, M.; Saeed, M.; Arain, M.A.; Siyal, F.A.; Patra, A.; Elnesr, S.S.; Khan, R.U. The uses of microbial phytase as a feed additive in poultry nutrition—A review. *Ann. Anim. Sci.* **2018**, *18*, 639. [CrossRef]
25. Jendza, J.; Dilger, R.N.; Sands, J.; Adeola, O. Efficacy and equivalency of an Escherichia coli-derived phytase for replacing inorganic phosphorus in the diets of broiler chickens and young pigs. *J. Anim. Sci.* **2006**, *84*, 3364–3374. [CrossRef] [PubMed]
26. Rousseau, X.; Létourneau-Montminy, M.-P.; Meme, N.; Magnin, M.; Nys, Y.; Narcy, A. Phosphorus utilization in finishing broiler chickens: Effects of dietary calcium and microbial phytase. *Poult. Sci.* **2012**, *91*, 2829–2837. [CrossRef] [PubMed]
27. Selle, P.H.; Ravindran, V. Microbial phytase in poultry nutrition. *Anim. Feed Sci. Technol.* **2007**, *135*, 1–41. [CrossRef]
28. Gautier, A.; Walk, C.; Dilger, R. Effects of a high level of phytase on broiler performance, bone ash, phosphorus utilization, and phytate dephosphorylation to inositol. *Poult. Sci.* **2018**, *97*, 211–218. [CrossRef] [PubMed]
29. Li, W.; Angel, R.; Kim, S.-W.; Brady, K.; Yu, S.; Plumstead, P. Impacts of dietary calcium, phytate, and phytase on inositol hexakisphosphate degradation and inositol phosphate release in different segments of digestive tract of broilers. *Poult. Sci.* **2017**, *96*, 3626–3637. [CrossRef] [PubMed]
30. Pontoppidan, K.; Glitsoe, V.; Guggenbuhl, P.; Quintana, A.P.; Nunes, C.S.; Pettersson, D.; Sandberg, A.-S. In vitro and in vivo degradation of myo-inositol hexakisphosphate by a phytase from Citrobacter braakii. *Arch. Anim. Nutr.* **2012**, *66*, 431–444. [CrossRef] [PubMed]
31. Zeller, E.; Schollenberger, M.; Witzig, M.; Shastak, Y.; Kühn, I.; Hoelzle, L.E.; Rodehutscord, M. Interactions between supplemented mineral phosphorus and phytase on phytate hydrolysis and inositol phosphates in the small intestine of broilers. *Poult. Sci.* **2015**, *94*, 1018–1029. [CrossRef]

animals

MDPI

Article

The Effect of Low Temperature on Laying Performance and Physiological Stress Responses in Laying Hens

Da-Hye Kim, Ju-Yong Song, Jina Park, Byung-Yeon Kwon and Kyung-Woo Lee *

Department of Animal Science and Technology, Konkuk University, 120 Neungdong-ro, Gwangjin-gu, Seoul 05029, Republic of Korea; kdh142536@naver.com (D.-H.K.); 0jysong970@gmail.com (J.-Y.S.); jinaa97@konkuk.ac.kr (J.P.); byung64@konkuk.ac.kr (B.-Y.K.)
* Correspondence: kyungwoolee@konkuk.ac.kr; Tel.: +82-2-450-0495

Simple Summary: Cold stress is considered an environmental stress and is an important managemental factor in regions where winter temperatures drop below 18 °C. Although a low temperature lowered the productive performance of laying hens, there are still unanswered questions as to the physiological changes that occur when they are exposed to low temperature, especially changes in antioxidants and stress indicators, which prompted us to set up the current study. Laying hens reared at low temperature (12 ± 4.5 °C) showed impaired laying performance compared with laying hens reared at normal temperature (24 ± 3 °C). However, eggshell color was intensified in laying hens exposed to low temperature. During the early stage of exposure to low-temperature stress, malondialdehyde levels in the egg yolk increased in laying hens raised at low temperature vs. normal temperature. However, yolk corticosterone, an indicator of stress responses, remained unchanged. Low temperature had an impact on total cholesterol and triglyceride levels in serum increases in laying hens reared at low temperature. In essence, low temperature in laying hens altered antioxidant systems and lipid metabolism without inducing stress responses.

Abstract: The present study investigated the effect of low temperature on laying performance, egg quality, body temperature, yolk malondialdehyde, yolk corticosterone, and serum biochemistry in laying hens. A total of 40 laying hens (Hy-Line Brown) aged 36 weeks were housed in one of two environmental chambers kept at 12 ± 4.5 °C (low temperature) or 24 ± 3 °C (normal temperature) for 4 weeks. Low vs. normal temperature significantly increased ($p < 0.05$) live body weight, feed intake, and feed conversion ratio in laying hens. Skin surface temperature, but not rectal temperature, was decreased in laying hens exposed to low vs. normal temperature. Hens exposed to low temperature laid an intense eggshell color compared with those raised in a normal temperature. Malondialdehyde concentrations in yolk were increased in low-temperature-exposed laying hens compared with those at normal temperature conditions, but this effect was only noted on day 7, post the low-temperature exposure ($p = 0.04$). Finally, low vs. normal temperature increased the concentrations of total cholesterol and triglyceride in serum. Collectively, this study indicates that exposure to low temperature in laying hens initially disrupted antioxidant system and altered lipid metabolism in laying hens without inducing stress responses.

Keywords: cold stress; egg production; egg quality; lipid metabolism; body temperature; laying hens

check for updates

Citation: Kim, D.-H.; Song, J.-Y.; Park, J.; Kwon, B.-Y.; Lee, K.-W. The Effect of Low Temperature on Laying Performance and Physiological Stress Responses in Laying Hens. *Animals* **2023**, *13*, 3824. https://doi.org/10.3390/ani13243824

Academic Editors: Velmurugu Ravindran and Janghan Choi

Received: 23 October 2023
Revised: 9 November 2023
Accepted: 5 December 2023
Published: 11 December 2023

1. Introduction

The future of poultry husbandry will be focused on improved animal welfare, with less reliance on preventative medical interventions. In the poultry husbandry environment, animals are likely to encounter various kinds of simultaneous environmental stressors. In emphasizing the latter, authors have tried to understand the physiological and stress responses of laying hens exposed to various stressors, including stocking density, transport, or ambient temperature [1]. It is generally understood that the thermal neutral zone for the optimal metabolic and productive activity of poultry ranges from around 18 to 23.9 °C [2] with a relative humidity between 50% and 70% during the laying period [3]. Heat stress is well-explored and acknowledged when laying hens are raised beyond the thermal neutral zone [1]. In contrast to heat stress, our understanding of the impact of low temperature (i.e., cold stress) on laying hens is still limited.

It has been reported that the ambient environmental temperature below 16 °C results in negative effects on poultry production performance, such as egg production, egg mass, and egg quality [4–6]. Cold stress is considered an environmental stress and can be considered an important managemental factor in regions where environmental temperatures during the winter periods drop below 18 °C. Earlier studies reported that cold exposure could influence the antioxidant and immune system functions of the host [7–9]. During low-temperature exposure, the hypothalamus, which acts as the control center for temperature regulation, triggers homeostatic mechanisms to maintain body temperature by generating heat. Also, upon experiencing stress, the activation of the hypothalamic–pituitary–adrenal (HPA) axis leads to the elevated secretion of corticosterone (CORT) from the adrenal glands. This process is intended to stimulate gluconeogenesis within various tissues and lipolysis in adipose tissue, thereby augmenting both glucose accessibility and the metabolic rate of energy [10].

In general, low temperature, as a physical environmental stressor, is known to increase feed intake to meet energy requirements [11]. Low ambient temperatures also result in decreased insulin and increased corticosterone (CORT) concentrations in the serum of laying hens [12,13]. It is well-known that insulin and CORT are influenced by temperature and these hormones play an important role in controlling nutrient metabolism [14,15]. In addition, a low temperature can disrupt the balance in an antioxidant system, leading to oxidative damage of several tissues by altering the enzymatic and non-enzymatic antioxidant status, lipid peroxidation, antioxidant vitamins, and reactive oxygen species production [16]. It has been reported that malondialdehyde (MDA) was increased in chickens exposed to low temperature, when compared to normal-temperature-reared counterparts [17,18]. Nonetheless, studies of the effects of low temperature on the egg quality, body temperature (rectal and surface), MDA in yolk, and CORT in yolk of laying hens that engage on a weekly basis are scarce, which prompted us to set up the current experiment. Thus, this present study aimed to address the biological and physiological responses of laying hens exposed to low ambient temperature.

2. Materials and Methods

2.1. Birds, Diets, and Experimental Design

A total of forty 36-week-old Hy-Line Brown laying hens with the average body weight of 1871 ± 123 g were used in this study. The birds were housed (one bird/cage) in one of two experimental chambers, which had one tier of 20 cages, 1 m high from the floor. Each cage, measuring 41 × 37 × 40 cm (length × depth × height), had nipples and a trough-type feeder. Two adjacent cages sharing one feeder were considered a replicate (n = 2/replicate, 10 replicates/treatment). The sample size (n = 10 per treatment) was calculated based on a Type I error of 5%, 80% statistical power, an effect size of 7.6%, and a coefficient of variation of 6%. In addition, to reduce variation within replicates, 2 laying hens per replicate were used. Each hen was provided with a 1517 cm^2 floor space. All birds were initially adapted to the chambers for 7 days at an ambient temperature of 24 °C with a relative humidity of 40 ± 4% and a lighting program of 16L/8D. After adaptation periods, hens were exposed to one of two temperature regimes (see Section 2.2 for the temperature regimes) for 28 days. Corn, soybean meal, and dried distillers grains with a solubles-based commercial layer diet were provided (Table 1), with feed and water supplied ad libitum.

Table 1. Ingredients and chemical composition of the basal diet (%, as-fed basis).

Ingredients	g/100 g
Corn	55.122
Corn dried distillers grains with solubles	13.500
Soybean meal, 44% CP	5.000
Wheat, 12% CP	4.000
Corn gluten meal, 60% CP	1.861
Rapeseed meal	1.933
Sesame meal	1.800
Meat meal	3.000
Feather meal	1.000
Full-fat soybeans	0.900
L-lysine-HCl, 78%	0.242
DL-methionine, 98%	0.152
Monocalcium phosphate	0.458
L-Threonine	0.025
Choline, 50%	0.200
Salt	0.230
Limestone	10.387
Vitamin premix [1]	0.120
Mineral premix [2]	0.070
Total	100.00
Calculated values	
AMEn, kcal/kg	2800
Crude protein	18.02
Crude fat	3.82
Ca	3.95
Total phosphorus	0.64
Methionine + Cysteine	0.80
Lysine	0.84
Threonine	0.63
Tryptophan	0.20

[1] Vitamin premix provided following nutrients per kg of diet: vitamin A, 24,000 IU; vitamin D3, 5520 IU; vitamin E, 48 mg; vitamin K3, 7.80 mg; thiamine, 4.32 mg; riboflavin, 9.60 mg; vitamin B6, 6.96 mg; vitamin B12, 0.05 mg; niacin, 72 mg; biotin, 0.30 mg; pantothenic acid, 22.04 mg; folate, 1.68 mg. [2] Mineral premix provided following nutrients per kg of diet: Fe, 49 mg; Mn, 56 mg; Zn, 42 mg; I, 0.7 mg; Se, 0.14 mg; Cu, 5.25 mg; Co, 0.09 mg.

2.2. Temperature Treatments and Temperature Monitor

Two environmental chambers were continuously set at 12 ± 4.5 °C (i.e., low temperature) and 24 ± 3.5 °C (i.e., normal temperature) and the average relative humidity inside the chambers was maintained at 40 ± 4%. Each chamber was equipped with a heater (MCP-300; MAXCON Co., Bucheon, Republic of Korea), air-conditioner (AR07J5174HA, SAMSUNG, Suwon, Republic of Korea), humidifier (MH-601A; mtechwin Co., Gimhae, Republic of Korea), dehumidifier (NED-062P; NAWOOEL Co., Gimpo, Republic of Korea), and the main controller panel. Temperature and humidity loggers (MHT-381SD; Lutron Electronic Enterprise Co., Taipei, Taiwan) were also installed to monitor the temperature and relative humidity at 10 min intervals throughout the experiment.

2.3. Measurements of Rectal Temperature and Skin Surface Temperature

On days 7, 14, 21, and 28 following temperature treatment, the rectal and skin surface temperature of laying hens was measured. One hen per replicate was randomly selected and measured by inserting a rectal thermometer to a depth of 3 cm into the rectum. Skin surface temperature was measured at three different sites (i.e., head, chest, and leg) using a thermal imaging camera (Cat® S60: equipped with an FLIR™ Lepton, FLIR Systems Inc., Wilsonville, OR, USA), as previously described [19]. The hens were handled by wearing latex gloves, to avoid the influences of heat and moisture of the hands on the temperature of the feathers.

2.4. *Egg Production and Egg Quality*

Body weight and feed intake were measured weekly. Eggs were collected daily and weighed per replicate to calculate hen-day egg production. Feed conversion ratio was expressed as kg of feed consumed per kg of eggs produced. Eggs (3 eggs per replicate) collected on the preceding 3 consecutive days, at 7, 14, 21, and 28 days following the temperature treatment, were used to measure egg quality with a digital egg tester (DET-6000, Navel, Kyoto, Japan). Yolk color intensity was evaluated on a scale between 1 and 16, with 1 being a very pale yellow and 16 being a dark orange. Eggshell color was measured using an egg multi-tester made by TSS (QCR, Technical Services and Supplies Ltd., Yolk, UK). Eggshell color intensity was evaluated on a scale between 0 and 100, with 0 being a darkness and 100 being a lightness.

2.5. *Malondialdehyde in Yolk Samples*

Eggs were collected to measure yolk MDA concentrations at days 7, 14, 21, and 28. The eggs were cracked open, and the yolks were separated from the albumin by gently rolling the yolk on filter paper. Three yolks were pooled and homogenized. The yolk MDA was measured using the OxiSelect TBARS Assay kit (Cell Biolabs Inc., San Diego, CA, USA).

2.6. *Corticosterone in Yolk Samples*

Yolk CORT concentrations were measured at 7, 14, 21, and 28 days, and 4 g of pooled yolk were vortexed with an equal volume of PBS. Then, 1 mL of the yolk suspension was mixed with an equal volume of ethanol, incubated at 37 °C for 1 h, and subsequently centrifuged. The 50 μL of supernatants were mixed with 50 μL of ethanol and 50 μL of PBS solution, and these mixtures were analyzed with a CORT ELISA kit (Enzo life science Inc., ADI-901-097, Farmingdale, NY, USA), as previously described [20,21].

2.7. *Serum Parameters*

At 28 days, one bird per replicate was selected to collect blood from a wing vein. Serum was separated by centrifugation at 200 g for 15 min and stored at −20 °C until the analysis. Serum samples were analyzed by using an automatic blood chemical analyzer (Film DRI CHEM 7000i, Fuji film, Tokyo, Japan) to measure for total cholesterol, triglyceride, high-density lipoprotein cholesterol, glutamic oxaloacetic transaminase, glutamic pyruvic transaminase, and uric acid.

2.8. *Statistical Analysis*

Two adjacent cages were considered an experimental unit. All data were analyzed using Student's *t*-test procedure of SAS (SAS Institute Inc., Cary, NC, USA). Results were presented as least square means and standard deviation. Differences were considered significant at $p < 0.05$.

3. Results

Ambient temperature ranged from 21 to 27 °C (normal temperature) and from 7.5 to 16.5 °C (low temperature) in the normal- and low-temperature chambers (Figure 1).

Low temperature increased ($p < 0.05$) the final body weight, feed intake, and FCR of laying hens compared with those raised at a normal temperature (Table 2).

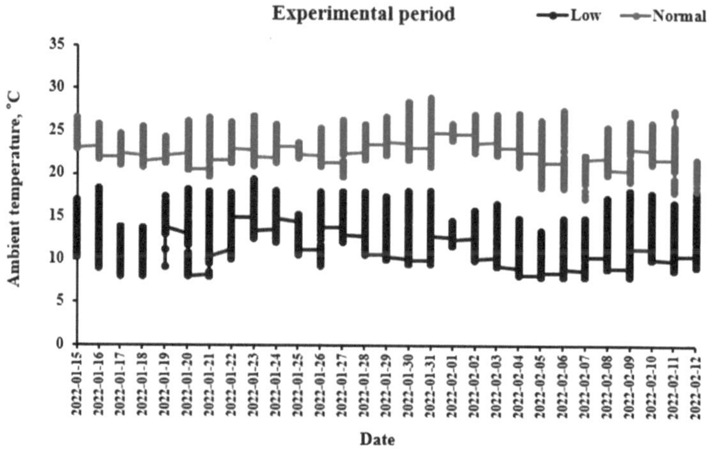

Figure 1. Ambient temperatures in chambers were recorded at an interval of 10 min with a thermometer.

Table 2. The effect of cold stress on production performance [1].

Item	Temperature				*p*-Value
	Low [3]		Normal		
	Mean	SD [4]	Mean	SD	
0 to 28 days					
Initial body weight, kg	1.867	0.128	1.923	0.107	0.302
Final body weight, kg	2.073	0.063	1.993	0.073	0.018
Feed intake, g	133.7	5.273	112.7	8.655	<0.001
FCR [2], kg/kg	2.691	0.451	2.020	0.132	<0.001
Egg weight, g	64.04	1.861	65.28	2.702	0.248
Egg production, %	89.82	8.164	92.68	4.795	0.353
Egg mass	55.52	5.236	57.73	3.023	0.263

[1] n = 10 replicates per treatment. [2] FCR = feed conversion ratio (kg of feed consumed per kg of eggs produced). [3] Low = low temperature (12 ± 4.5 °C); Normal = normal temperature (24 ± 3 °C). [4] SD= standard deviation.

However, egg weight, egg production and egg mass were not affected ($p > 0.05$) by temperature regimes. Rectal temperature during the experimental period was maintained from 41.27 to 41.58 °C and was not different between temperature treatment groups (Table 3).

Table 3. The effect of cold stress on rectal temperature (°C) [1].

Item	Temperature				*p*-Value
	Low [2]		Normal		
	Mean	SD [3]	Mean	SD	
Day 7	41.58	0.175	41.52	0.248	0.507
Day 14	41.44	0.193	41.41	0.242	0.802
Day 21	41.41	0.280	41.27	0.342	0.330
Day 28	41.28	0.190	41.40	0.148	0.133

[1] n = 10 replicates per treatment. [2] Low = low temperature (12 ± 4.5 °C); Normal = normal temperature (24 ± 3 °C). [3] SD = standard deviation.

Table 4 presents the skin surface temperature at 7, 14, 21, and 28 days. The laying hens exposed to low temperature at all ages had lower skin (i.e., head, breast, and leg regions) surface temperature compared with those raised at normal temperature ($p < 0.05$). Skin surface temperature between low- vs. normal-temperature treatments was 5.7 to 12.2 °C lower in the breast surface area, followed by the legs and the head areas.

Table 4. The effect of cold stress on skin surface temperature (°C) [1].

Item	Temperature				p-Value
	Low [2]		Normal		
	Mean	SD [3]	Mean	SD	
Day 7					
Head	36.31	0.693	38.49	0.994	<0.0001
Breast	21.04	0.918	29.10	0.460	<0.0001
Leg	34.89	1.343	39.10	0.330	<0.0001
Day 14					
Head	37.93	0.380	39.75	0.483	<0.0001
Breast	25.39	1.514	31.09	0.437	<0.0001
Leg	38.27	0.853	40.49	0.461	<0.0001
Day 21					
Head	36.26	0.421	39.97	0.447	<0.0001
Breast	21.20	0.511	32.19	0.746	<0.0001
Leg	35.47	1.309	40.72	0.472	<0.0001
Day 28					
Head	36.92	0.684	40.27	0.330	<0.0001
Breast	20.13	0.528	32.29	0.330	<0.0001
Leg	36.16	0.986	41.11	0.372	<0.0001

[1] $n = 10$ replicates per treatment. [2] Low = low temperature (12 ± 4.5 °C); Normal = normal temperature (24 ± 3 °C). [3] SD = standard deviation.

Low temperature significantly increased ($p < 0.05$) eggshell color at days 7 and 28, compared with normal-temperature-raised laying hens (Table 5). However, yolk color, the Haugh unit, shell strength, and shell thickness were not affected by temperature treatments on all days ($p > 0.05$).

Table 5. The effect of cold stress on egg quality [1].

Item	Temperature				p-Value
	Low [4]		Normal		
	Mean	SD [5]	Mean	SD	
Yolk color [2]					
Day 7	8.490	0.239	8.457	0.344	0.521
Day 14	8.486	0.243	8.652	0.528	0.182
Day 21	8.364	0.190	8.479	0.311	0.582
Day 28	8.326	0.325	8.236	0.310	0.252
Haugh unit					
Day 7	90.98	1.924	91.32	2.789	0.807
Day 14	93.43	2.265	94.38	2.999	0.379
Day 21	95.81	1.479	93.83	5.683	0.333
Day 28	92.70	2.616	93.93	2.590	0.535
Shell strength, kgf					
Day 7	5.126	0.323	5.319	0.802	0.748
Day 14	5.247	0.493	5.248	0.855	0.436
Day 21	5.500	0.404	4.984	0.666	0.333
Day 28	5.285	0.459	5.285	0.453	0.306
Shell thickness, mm					
Day 7	0.399	0.024	0.434	0.029	0.490
Day 14	0.410	0.023	0.398	0.026	0.998
Day 21	0.437	0.020	0.424	0.021	0.051
Day 28	0.403	0.024	0.436	0.024	0.997

Table 5. *Cont.*

Item	Temperature				p-Value
	Low [4]		Normal		
	Mean	SD [5]	Mean	SD	
Eggshell color [3]					
Day 7	26.97	2.156	25.83	3.395	0.010
Day 14	27.26	3.041	25.77	3.176	0.295
Day 21	27.83	3.567	25.13	2.161	0.170
Day 28	27.76	2.806	24.71	2.282	0.007

[1] $n = 10$ replicates per treatment. [2] Yolk color was determined using a digital egg tester (DET-6000, Navel, Kyoto, Japan); scale between 1 to 16, with 1 being a very pale yellow and 16 being a dark orange. [3] Eggshell color was measured by an Egg multi tester made by TSS (Technical Services and Sup-plies Ltd., Yolk, UK); scale between 0 to 100, with 0 being a darkness and 100 being a lightness. [4] Low = low temperature (12 ± 4.5 °C); Normal = normal temperature (24 ± 3 °C). [5] SD = standard deviation.

Yolk MDA concentrations were higher in laying hens exposed to low temperature, compared with those in normal temperature at 7 days (Table 6). However, low-temperature-induced decreases in yolk MDA concentrations were not noted ($p > 0.05$) at 14, 21, and 28 days.

Table 6. The effect of cold stress on malondialdehyde in egg yolk (μM) [1].

Item	Temperature				p-Value
	Low [2]		Normal		
	Mean	SD [3]	Mean	SD	
Day 7	47.47	7.612	41.04	4.098	0.040
Day 14	25.41	3.391	24.17	3.440	0.472
Day 21	51.37	6.537	50.78	4.711	0.818
Day 28	61.85	5.820	65.97	6.866	0.272

[1] $n = 10$ replicates per treatment. [2] Low = low temperature (12 ± 4.5 °C); Normal = normal temperature (24 ± 3 °C). [3] SD = standard deviation.

The CORT concentrations were not different in laying hens exposed to low and normal temperature (Table 7). Low vs. normal temperature increased ($p < 0.05$) the concentrations of total cholesterol and triglycerides, but did not affect the concentrations of high-density lipoprotein cholesterol, glutamic oxaloacetic transaminase, glutamic pyruvic transaminase, and uric acid in serum samples (Table 8; $p > 0.05$).

Table 7. The effect of cold stress on corticosterone in egg yolk (ng/g) [1].

Item	Temperature				p-Value
	Low [2]		Normal		
	Mean	SD [3]	Mean	SD	
Day 7	257.7	126.29	236.0	87.19	0.740
Day 14	229.0	87.90	198.1	85.83	0.476
Day 21	108.9	56.32	96.6	49.97	0.622
Day 28	125.5	59.12	103.6	55.90	0.445

[1] $n = 10$ replicates per treatment. [2] Low = low temperature (12 ± 4.5 °C); Normal = normal temperature (24 ± 3 °C). [3] SD = standard deviation.

Table 8. The effect of cold stress on serum biological parameters in serum [1].

Item [2]	Temperature				p-Value
	Low [3]		Normal		
	Mean	SD [4]	Mean	SD	
TCHO, mg/dL	128.3	18.39	103.5	15.44	0.011
Triglyceride, mg/dL	2487	787.1	1832	567.4	0.047
HDL-CHO, mg/dL	13.20	3.225	14.00	3.300	0.590
GOT, IU/L	116.6	15.53	114.0	11.82	0.679
GPT, IU/L	28.30	1.337	28.50	2.321	0.816
Uric acid, mg/dL	5.130	0.811	4.870	1.200	0.577

[1] n = 10 replicates per treatment. [2] TCHO = total cholesterol; HDL-CHO = high density lipoprotein cholesterol; GOT = glutamic oxalacetic transaminase; GPT = glutamic pyruvic transaminase. [3] Low = low temperature (12 ± 4.5 °C); Normal = normal temperature (24 ± 3 °C). [4] SD = standard deviation.

4. Discussion

The present study showed that low vs. normal ambient temperature increased final body weight, feed intake, and FCR in laying hens. This finding is well-established because chickens, as a homeothermic animal, are unable to warm themselves and so overfeed to compensate for heat loss to the environment when they are exposed to low temperature (e.g., 18 °C) [22]. As ambient temperature shifted from 30 to 18 °C, the abdominal fat weight, abdominal fat rate, and subcutaneous fat thickness linearly increased, favoring body fat deposition [23]. For these reasons, the present study indicates that heavier body weight is due to increases in feed intake and concomitant body fat deposition, leading to a higher FCR in laying hens exposed to low vs. normal temperature. It is well-documented that low temperatures stimulate feed intakes in laying hens [6,24], broiler chickens [17,25], quails [4,26] and turkeys [27].

In the present study, rectal temperature at all ages was not altered in birds exposed to a low temperature or those at normal temperature, which agrees with previous studies with turkeys [28] and laying hens [3]. These results suggest that body heat exchange (e.g., production and loss) does not change above or below the normal rectal temperature range when the homeostasis of warm-blooded animals is maintained [29]. On the other hand, skin surface temperature, an important evaluable parameter, quickly shifts in response to environmental changes and serves as an indicator of alterations in peripheral blood flow and heat exchange [30]. Skin surface (e.g., head, chest, and leg) temperatures were significantly decreased in laying hens raised at low temperature, compared with those at normal temperature, during the whole experiment period. In line with our finding, lower skin surface temperatures in broiler chickens [31] and laying hens [29] have been observed when they were exposed to low vs. normal ambient temperature. It is understood as a way of preventing the loss of heat by cold-induced peripheral vasoconstriction, which reduces blood circulation to the body surface [31]. Additionally, the temperatures of the head and legs were kept higher than that of the chest. It is speculated that this is because the head and legs are featherless body surfaces, compared with the breast body surface, which is covered by feathers [30].

We found that yolk color and eggshell strength were not altered in laying hens raised at a low temperature compared with normal-temperature-raised laying hens, although the former ate more. In contrast to our belief [24,32], our finding indicates that hens exposed to low temperatures might not use excess feed-origin carotenoids and essential nutrients (e.g., amino acids, Ca, Mg) to intensify yolk and to increase eggshell strength. It is of interest that low-temperature-raised hens laid more intensified eggshell colors than hens raised in normal ambient temperature. It is known that the endogenously synthesized protoporphyrin IX is the major pigment in brown eggshell, and that stress factors, including stocking density, fear, or molting, often deteriorate eggshell color pigmentation in brown laying hens [33]. As the pigment is known to be synthesized in the shell gland of the oviduct, it seems that low temperature per se may stimulate the synthetic process of pigment or effectively deposit it on the shell layers. Further studies are warranted to reveal the underlying mechanisms that will show how low temperature affects shell color deposition in laying hens.

It is reported that cold stress causes tissue damage [22] via increased metabolic rate, which demands tissue requirement for oxygen in birds [34]. We found that yolk MDA levels were elevated in low- vs. normal-temperature treatments, but this effect was only noted at 7 days post temperature exposure. Rahmani et al. [17] reported that broiler chickens exposed to 15 °C vs. 22 °C had increased plasma MDA concentrations. In addition, MDA levels in plasma and liver samples were significantly higher in broiler chickens in the low temperature group (from 10 to 15 °C) at 21 and 42 days [18].

In contrast to yolk MDA alterations caused by low temperature, no differences in yolk CORT were noted between low-temperature and normal-condition groups. Exposure of chickens to stress increases the secretion of CORT, a major stress hormone in chickens, via the activation of the hypothalamic pituitary adrenal axis [35]. In earlier studies, the levels of CORT in the albumen and yolk of chicken eggs were used as a non-invasive method to measure stress [20,36,37]. The substances or hormone concentrations in the blood during the egg's formation phase are deposited in the yolk composition. Hence, if there is an increase in plasma CORT, it would consequently be transferred into the eggs [36]. Also, catching animals for blood sampling can be stressful and it also causes an increase in CORT [38]–therefore, egg yolks were used for an analysis of CORT concentrations in this study. Conflicted results have been reported regarding plasma CORT in response to low ambient temperature. With various treatments (durations of cold exposure, temperature, and age), increased plasma levels of CORT were reported in young broiler chickens (19 ± 1 °C, 6 h/day, from the first to the seventh day of life) [39], male turkeys [40], chicks (1.2 °C for 3 h) [41], and laying hens (0 °C for 72 h) [42]. On the other hand, Hangalapura et al. [43] reported that low temperature (10.4 ± 0.5 °C) decreased plasma CORT in laying hens and Hu and Cheng [44] reported no effect of low temperature on the CORT level in laying hens. It is not clear at this stage why low temperature failed to affect yolk CORT levels in this study, although the information on the effect of low temperature in yolk CORT is scarce. The unaltered CORT could be due to the negative feedback of CORT on the hypothalamus axis with chronic low-temperature exposure, resulting in an inhibition of adrenocorticotropin secretion [45]. Similarly, previous studies have reported that consistently increased CORT is not commonly seen during chronic stress, possibly due to the negative feedback of CORT on the hypothalamus axis [43,44,46].

The finding that low temperature increased the concentrations of total cholesterol and triglyceride is in agreement with the results of [17,47]. This finding indicates that a low-temperature-induced increase in feed intake, coupled with cold stress response in chickens, could stimulate hepatic lipogenic and hypercholesterolemic metabolic pathways [48,49]. On the other hand, glutamic oxaloacetic transaminase and glutamic pyruvic transaminase were not altered by low temperatures, indicating the maintenance of the hepatic function.

5. Conclusions

In conclusion, low vs. normal ambient temperature stimulated feed intake and increased body weight and FCR. Eggshell color was intensified in laying hens exposed to low temperature. Low temperature elevated skin surface temperature without affecting rectal temperature. Finally, low temperature did not influence stress responses, as manifested by constant yolk CORT concentrations, but altered MDA, total cholesterol, and triglyceride levels. Taken together, our study indicates that the exposure of laying hens to low temperature disrupted the antioxidant system, especially at an early stage of exposure, and altered lipid metabolism (i.e., total cholesterol and triglyceride) without inducing stress responses. The low-temperature-mediated increase in eggshell color seen in this study warrants further study.

Author Contributions: Conceptualization, D.-H.K. and K.-W.L.; methodology, D.-H.K.; formal analysis, J.P., B.-Y.K. and J.-Y.S.; investigation, D.-H.K., J.P., B.-Y.K. and J.-Y.S.; resources, K.-W.L.;

writing—original draft preparation, D.-H.K.; writing—review and editing, D.-H.K. and K.-W.L.; project administration, K.-W.L.; funding acquisition, K.-W.L. All authors have read and agreed to the published version of the manuscript.

Funding: This research received no external funding.

Institutional Review Board Statement: The animal study protocol was approved by the Ethics Committee of Konkuk University (KU22031).

Informed Consent Statement: Not applicable.

Data Availability Statement: The datasets generated or analyzed during the current study are not publicly available but are available from the corresponding author at reasonable request.

Acknowledgments: This paper was supported by Konkuk University in 2023.

Conflicts of Interest: The authors declare no conflict of interest.

References

1. Tilbrook, A.J.; Fisher, A.D. Stress, health and the welfare of laying hens. *Anim. Prod. Sci.* **2020**, *61*, 931–943. [CrossRef]
2. Durmuş, I.; Kamanli, S. Effects of cold and heat stress on egg quality traits of a newly developed native hybrid layer. *Turk. Tarim. Gida. Bilim. Teknol. Derg.* **2015**, *3*, 444–447. [CrossRef]
3. Kamanli, S.; Durmuş, I.; Yalçın, S.; Yıldırım, U.; Meral, Ö. Effect of prenatal temperature conditioning of laying hen embryos: Hatching, live performance and response to heat and cold stress during laying period. *J. Therm. Biol.* **2015**, *51*, 96–104. [CrossRef] [PubMed]
4. Sahin, N.; Sahin, K.; Onderci, M. Vitamin E and selenium supplementation to alleviate cold-stress-associated deterioration in egg quality and egg yolk mineral concentrations of Japanese quails. *Biol. Trace. Elem. Res.* **2003**, *96*, 179–189. [CrossRef] [PubMed]
5. Torki, M.; Akbari, M.; Kaviani, K. Single and combined effects of zinc and cinnamon essential oil in diet on productive performance, egg quality traits, and blood parameters of laying hens reared under cold stress condition. *Int. J. Biometeorol.* **2015**, *59*, 1169–1177. [CrossRef] [PubMed]
6. Li, D.; Tong, Q.; Shi, Z.; Zheng, W.; Wang, Y.; Li, B.; Yan, G. Effects of cold stress and ammonia concentration on productive performance and egg quality traits of laying hens. *Animals* **2020**, *10*, 2252. [CrossRef] [PubMed]
7. Hangalapura, B.N.; Nieuwland, M.G.; de Vries, R.G.; Heetkamp, M.J.; van den Brand, H.; Kemp, B.; Parmentier, H.K. Effects of cold stress on immune responses and body weight of chicken lines divergently selected for antibody responses to sheep red blood cells. *Poult. Sci.* **2003**, *82*, 1692–1700. [CrossRef]
8. Hangalapura, B.N.; Nieuwland, M.G.B.; de Vries Reilingh, G.; Van Den Brand, H.; Kemp, B.; Parmentier, H.K. Durations of cold stress modulates overall immunity of chicken lines divergently selected for antibody responses. *Poult. Sci.* **2004**, *83*, 765–775. [CrossRef]
9. Hangalapura, B.N.; Nieuwland, M.G.B.; Buyse, J.; Kemp, B.; Parmentier, H.K. Effect of duration of cold stress on plasma adrenal and thyroid hormone levels and immune responses in chicken lines divergently selected for antibody responses. *Poult. Sci.* **2004**, *83*, 1644–1649. [CrossRef]
10. Schweiger, M.; Schreiber, R.; Haemmerle, G.; Lass, A.; Fledelius, C.; Jacobsen, P.; Tornqvist, R.; Zechner, R.; Zimmermann, R. Adipose triglyceride lipase and hormone-sensitive lipase are the major enzymes in adipose tissue triacylglycerol catabolism. *J. Biol. Chem.* **2006**, *281*, 40236–40241. [CrossRef]
11. Blahová, J.; Dobšíková, R.; Straková, E.; Suchý, P. Effect of low environmental temperature on performance and blood system in broiler chickens (*Gallus domesticus*). *Acta. Vet. Brno.* **2007**, *76*, 17–23. [CrossRef]
12. Şahin, K.; Önderci, M. Optimal dietary concentrations of vitamin C and chromium for alleviating the effect of low ambient temperature on serum insulin, corticosterone, and some blood metabolites in laying hens. *J. Trace. Elem. Exp. Med.* **2002**, *15*, 153–161. [CrossRef]
13. Siegel, H.S. Stress, strains and resistance. *Br. Poult. Sci.* **1995**, *36*, 3–22. [CrossRef] [PubMed]
14. Sahin, K.; Sahin, N.; Kucuk, O. Effects of chromium, and ascorbic acid supplementation on growth, carcass traits, serum metabolites, and antioxidant status of broiler chickens reared at a high ambient temperature (32 °C). *Nutr. Res.* **2003**, *23*, 225–238. [CrossRef]
15. Şahin, K.; Küçuk, O.; Şahin, N. Effects of dietary chromium picolinate supplementation on performance and plasma concentrations of insulin and corticosterone in laying hens under low ambient temperature. *J. Anim. Physiol. Anim. Nutr.* **2001**, *85*, 142–147. [CrossRef] [PubMed]
16. Sinha, S.; Singh, S.N.; Saha, M.; Kain, T.C.; Tyagi, A.K.; Ray, U.S. Antioxidant and oxidative stress responses of sojourners at high altitude in different climatic temperatures. *Int. J. Biometeorol.* **2010**, *54*, 85–92. [CrossRef]
17. Rahmani, M.; Golian, A.; Kermanshahi, H.; Bassami, M.R. Effects of curcumin or nanocurcumin on blood biochemical parameters, intestinal morphology and microbial population of broiler chickens reared under normal and cold stress conditions. *J. Appl. Anim. Res.* **2018**, *46*, 200–209. [CrossRef]

18. Fathi, M.; Nazer, A.K.; Nezhad, Y.E.; Shahryar, H.A.; Daneshyar, M.; Tanha, T. The role of oxidative stress in development of congestive heart failure (CHF) in broiler with pulmonary hypertension syndrome (PHS). *J. Anim. Vet. Adv.* **2011**, *10*, 2724–2729.
19. Cangar, Ö.; Aerts, J.M.; Buyse, J.; Berckmans, D. Quantification of the spatial distribution of surface temperatures of broilers. *Poult. Sci.* **2008**, *87*, 2493–2499. [CrossRef]
20. Kim, D.H.; Lee, Y.K.; Lee, S.D.; Kim, S.H.; Lee, K.W. Physiological and behavioral responses of laying hens exposed to long-term high temperature. *J. Therm. Biol.* **2021**, *99*, 103017. [CrossRef]
21. Kozlowski, C.P.; Bauman, J.E.; Caldwell Hahn, D. A simplified method for extracting androgens from avian egg yolks. *Zoo Biol.* **2009**, *28*, 137–143. [CrossRef]
22. Dhanalakshmi, S.; Devi, R.S.; Srikumar, R.; Manikandan, S.; Thangaraj, R. Protective effect of triphala on cold stress-induced behavioral and biochemical abnormalities in rats. *Yakugaku Zasshi* **2007**, *127*, 1863–1867. [CrossRef] [PubMed]
23. Liu, Z.L.; Chen, Y.; Xue, J.J.; Huang, X.F.; Chen, Z.P.; Wang, Q.G.; Wang, C. Effects of ambient temperature on the growth performance, fat deposition, and intestinal morphology of geese from 28 to 49 days of age. *Poult. Sci.* **2022**, *101*, 101814. [CrossRef]
24. Zurak, D.; Slovenec, P.; Janječić, Z.; Bedeković, X.D.; Pintar, J.; Kljak, K. Overview on recent findings of nutritional and non-nutritional factors affecting egg yolk pigmentation. *Worlds Poult. Sci. J.* **2022**, *78*, 531–560. [CrossRef]
25. Ipek, A.Y.D.I.N.; Sahan, U. Effects of cold stress on broiler performance and ascites susceptibility. *Asian-Australas. J. Anim. Sci.* **2006**, *19*, 734–738. [CrossRef]
26. Sahin, N.; Sahin, K.; Onderci, M.; Ozcelik, M.; Smith, M.O. In vivo antioxidant properties of vitamin E and chromium in cold-stressed Japanese quails. *Arch. Anim. Nutr.* **2003**, *57*, 207–215. [CrossRef] [PubMed]
27. Yahav, S. Limitations in energy intake affect the ability of young turkeys to cope with low ambient temperatures. *J. Therm. Biol.* **2002**, *27*, 103–108. [CrossRef]
28. Yahav, S. The effect of constant and diurnal cyclic temperatures on performance and blood system of young turkeys. *J. Therm. Biol.* **1999**, *24*, 71–78. [CrossRef]
29. Kim, N.Y.; Kim, S.J.; Oh, M.; Jang, S.Y.; Moon, S.H. Changes in facial surface temperature of laying hens under different thermal conditions. *Anim. Biosci.* **2021**, *34*, 1235–1242. [CrossRef]
30. Nascimento, S.T.; da Silva, I.J.O.; Maia, A.S.C.; de Castro, A.C.; Vieira, F.M.C. Mean surface temperature prediction models for broiler chickens—A study of sensible heat flow. *Int. J. Biometeorol.* **2014**, *58*, 195–201. [CrossRef]
31. Andrade, R.R.; Tinôco, I.F.F.; Baêta, F.C.; Barbari, M.; Conti, L.; Cecon, P.R.; Cândido, M.G.L.; Martins, I.T.A.; Junior Teles, C.G.S. Evaluation of the surface temperature of laying hens in different thermal environments during the initial stage of age based on thermographic images. *Agron. Res.* **2017**, *15*, 629–638.
32. Al-Saffar, A.A.; Rose, S.P. Ambient temperature and the egg laying characteristics of laying fowl. *Worlds Poult. Sci. J.* **2002**, *58*, 317–331. [CrossRef]
33. Samiullah, S.; Roberts, J.R.; Chousalkar, K. Eggshell color in brown-egg laying hens—A review. *Poult. Sci.* **2015**, *94*, 2566–2575. [CrossRef]
34. Guo, J.L.; Zheng, Q.H.; Yin, Q.Q.; Cheng, W.; Jiang, Y.B. Study on mechanism of ascites syndrome of broilers. *Am. J. Anim. Vet. Sci.* **2007**, *2*, 62–65. [CrossRef]
35. Quinteiro-Filho, W.M.; Rodrigues, M.V.; Ribeiro, A.; Ferraz-de-Paula, V.; Pinheiro, M.L.; Sá, L.R.M.; Ferreira, A.J.P.; Palermo-Neto, J. Acute heat stress impairs performance parameters and induces mild intestinal enteritis in broiler chickens: Role of acute hypothalamic-pituitary-adrenal axis activation. *J. Anim. Sci.* **2012**, *90*, 1986–1994. [CrossRef] [PubMed]
36. Downing, J.A.; Bryden, W.L. Determination of corticosterone concentrations in egg albumen: A non-invasive indicator of stress in laying hens. *Physiol. Behav.* **2008**, *95*, 381–387. [CrossRef] [PubMed]
37. Alm, M.; Tauson, R.; Holm, L.; Wichman, A.; Kalliokoski, O.; Wall, H. Welfare indicators in laying hens in relation to nest exclusion. *Poult. Sci.* **2016**, *95*, 1238–1247. [CrossRef] [PubMed]
38. Caulfield, M.P.; Padula, M.P. HPLC MS-MS analysis shows measurement of corticosterone in egg albumen is not a valid indicator of chicken welfare. *Animals* **2020**, *10*, 821. [CrossRef]
39. Borsoi, A.; Quinteiro-Filho, W.M.; Calefi, A.S.; Piantino Ferreira, A.J.; Astolfi-Ferreira, C.S.; Florio, J.C.; Palermo-Neto, J. Effects of cold stress and Salmonella Heidelberg infection on bacterial load and immunity of chickens. *Avian Pathol.* **2015**, *44*, 490–497. [CrossRef] [PubMed]
40. El-halawani, M.E.; Waibel, P.E.; Appel, J.R.; Good, A.L. Effects of temperature stress on catecholamines and corticosterone of male turkeys. *Am. J. Physiol.* **1973**, *224*, 384–388. [CrossRef]
41. Buckland, R.B.; Blagrave, K.; Lague, P.C. Competitive protein-binding assay for corticoids in the peripheral plasma of the immature chicken. *Poult. Sci.* **1974**, *53*, 241–245. [CrossRef]
42. Hester, P.Y.; Muir, W.M.; Craig, J.V.; Albright, J.L. Group selection for adaptation to multiple-hen cages: Hematology and adrenal function. *Poult. Sci.* **1996**, *75*, 1295–1307. [CrossRef] [PubMed]
43. Smith, S.M.; Vale, W.W. The role of the hypothalamic-pituitary-adrenal axis in neuroendocrine responses to stress. *Dialogues Clin. Neurosci.* **2006**, *8*, 383–395. [CrossRef] [PubMed]
44. Hu, J.Y.; Cheng, H.W. Warm perches: A novel approach for reducing cold stress effect on production, plasma hormones, and immunity in laying hens. *Poult. Sci.* **2021**, *100*, 101294. [CrossRef]

45. Vandenborne, K.; de Groef, B.; Geelissen, S.M.E.; Kühn, E.R.; Darras, V.M.; van der Geyten, S. Corticosterone-induced negative feedback mechanisms within the hypothalamo-pituitary-adrenal axis of the chicken. *J. Endocr.* **2005**, *185*, 383–391. [CrossRef] [PubMed]
46. Angelier, F.; Wingfield, J.C. Importance of the glucocorticoid stress response in a changing world: Theory, hypotheses and perspectives. *Gen. Comp. Endocrinol.* **2012**, *190*, 118–128. [CrossRef] [PubMed]
47. Houshmand, M.; Azhar, K.; Zulkifli, I.; Bejo, M.H.; Kamyab, A. Effects of prebiotic, protein level, and stocking density on performance, immunity, and stress indicators of broilers. *Poult. Sci.* **2012**, *91*, 393–401. [CrossRef]
48. Zhou, H.J.; Kong, L.L.; Zhu, L.X.; Hu, X.Y.; Busye, J.; Song, Z.G. Effects of cold stress on growth performance, serum biochemistry, intestinal barrier molecules, and adenosine monophosphate-activated protein kinase in broilers. *Animal* **2021**, *15*, 100138. [CrossRef]
49. Zhang, Z.W.; Bi, M.Y.; Yao, H.D.; Fu, J.; Li, S.; Xu, S.W. Effect of cold stress on expression of AMPKα-PPARα pathway and inflammation genes. *Avian Dis.* **2014**, *58*, 415–426. [CrossRef]

animals

MDPI

Article

Microencapsulated Sodium Butyrate Alleviates Immune Injury and Intestinal Problems Caused by Clostridium Perfringens through Gut Microbiota

Ting Yang [1,†], Yaowei Sun [1,†], Zhenglie Dai [1], Jinsong Liu [2], Shiping Xiao [2], Yulan Liu [2], Xiuxi Wang [1], Shenglan Yang [1], Ruiqiang Zhang [1], Caimei Yang [2] and Bing Dai [1,*]

[1] College of Animal Science and Technology, College of Veterinary Medicine, Key Laboratory of Applied Technology on Green-Eco-Healthy Animal Husbandry of Zhejiang Province, Zhejiang Provincial Engineering Laboratory for Animal Health Inspection and Internet Technology, Zhejiang Agricultural and Forestry University, Hangzhou 311300, China; yangtingly@126.com (T.Y.); 19550177705@163.com (Y.S.); mmddaa0323@163.com (Z.D.); 15261825191@163.com (X.W.); yangshenglan7@163.com (S.Y.); zrq1034@163.com (R.Z.)

[2] Zhejiang Vegamax Biotechnology Co., Ltd., Huzhou 313300, China; huijialjs@163.com (J.L.); xiao1974920@126.com (S.X.); yulanflower@126.com (Y.L.); yangcaimei2012@163.com (C.Y.)

* Correspondence: daibing2008@126.com

† These authors contributed equally to this work.

Simple Summary: Necrotic enteritis is an enterotoxemic disease caused by *Clostridium perfringens*, leading to diarrhea or necrotizing lesions in the intestines of animals, with severe cases leading to death. The butyrate attenuated the inflammatory response and improved intestinal health in piglets challenged with pathogenic bacteria, but it was absorbed by the anterior segment of the gastrointestinal tract. Microencapsulation is a simple and effective method to prevent butyrate from absorption. Thus, we evaluate the effectiveness to two butyrate alleviates clostridium perfringens infections. Our results indicate that dietary supplementation with sodium butyrate or microencapsulated sodium butyrate improves the immune status and morphology of intestinal villi, increases the production of VFAs, and modulates cecal microbiota in chickens challenged with *Clostridium perfringens*. Moreover, microencapsulated sodium butyrate contains less butyrate than sodium butyrate. These findings indicate that microencapsulated sodium butyrate was more effective than sodium butyrate with the same butyrate supplemental amount.

Abstract: Microencapsulated sodium butyrate (MS-SB) is an effective sodium butyrate additive which can reduce the release of sodium butyrate (SB) in the fore gastrointestinal tract. In this study, we assess the protective effects and mechanisms of MS-SB in *Clostridium perfringens* (*C. perfringens*)-challenged broilers. Broiler chickens were pre-treated with SB or MS-SB for 56 days and then challenged with *C. perfringens* three times. Our results indicate that the addition of MS-SB or SB before *C. perfringens* infection significantly decreased the thymus index ($p < 0.05$). Serum IgA, IgY, and IgM concentrations were significantly increased ($p < 0.05$), while pro-inflammatory IL-1β, IL-6, and TNF-α were significantly decreased ($p < 0.05$) under MS-SB or SB supplementation. Compared with SB, MS-SB presented a stronger performance, with higher IgA content, as well as a lower IL-1β level when normal or C. perfringens-challenged. While *C. perfringens* challenge significantly decreased the villus height ($p < 0.05$), MS-SB or SB administration significantly increased the villus height and villus height/crypt depth (V/C ratio) ($p < 0.05$). Varying degrees of SB or MS-SB increased the concentrations of volatile fatty acids (VFAs) during *C. perfringens* challenge, where MS-SB presented a stronger performance, as evidenced by the higher content of isovaleric acid and valeric acid. Microbial analysis demonstrated that both SB or MS-SB addition and *C. perfringens* infection increase variation in the microbiota community. The results also indicate that the proportions of *Bacteroides*, *Faecalibacterium*, *Clostridia*, *Ruminococcaceae*, *Alistipes*, and *Clostridia* were significantly higher in the MS-SB addition group while, at same time, *C. perfringens* infection increased the abundance of *Bacteroides* and *Alistipes*. In summary, dietary supplementation with SB or MS-SB improves the

Citation: Yang, T.; Sun, Y.; Dai, Z.; Liu, J.; Xiao, S.; Liu, Y.; Wang, X.; Yang, S.; Zhang, R.; Yang, C.; et al. Microencapsulated Sodium Butyrate Alleviates Immune Injury and Intestinal Problems Caused by Clostridium Perfringens through Gut Microbiota. *Animals* **2023**, *13*, 3784. https://doi.org/10.3390/ani13243784

Academic Editor: Janghan Choi

Received: 3 November 2023
Revised: 23 November 2023
Accepted: 30 November 2023
Published: 8 December 2023

immune status and morphology of intestinal villi, increases the production of VFAs, and modulates cecal microbiota in chickens challenged with *C. perfringens*. Moreover, MS-SB was more effective than SB with the same supplemental amount.

Keywords: microencapsulated sodium butyrate; broiler chicken; immunity; cecum microflora; *Clostridium perfringens*

1. Introduction

Antibiotic growth promoters have been used to improve animal growth and to control or prevent animal disease during rearing; however, many countries have forbidden the dietary use of antimicrobial agents to avoid the emergence of antibiotic-resistant bacteria [1]. Thus, animal nutritionists have been attempting to discover antibiotic alternatives that are low-cost, widely used, and which have an obvious effect. Butyric acid has a positive effect on the normal intestinal mucosa, which can be attributed to a major energy source for intestinal epithelial cells [2,3]. In the intestine, an appropriate concentration of butyric acid is beneficial in protecting against the invasion of micro-organisms, which may be attributed to the effects of VFAs in terms of maintaining the pH of the intestinal lumen [4]. In a previous study, butyrate significantly ameliorated the mice intestine and intestinal epithelial cell inflammatory response and intestinal epithelium barrier dysfunction caused by 2,4,6-trinitrobenzene sulphonic acid (TNBS) [5]. Butyrate also has an immunomodulatory effect; for example, butyrate attenuated steatohepatitis through restoring the dysbiosis of gut microbiota. As such, butyrate has been considered as a potential gut microbiota modulator and therapeutic substance for non-alcoholic fatty liver disease (NAFLD) [6].

In the context of animal rearing, the harsh environment, stress, harmful bacteria, and other factors may result in intestinal health damage, expressed as intestinal inflammatory responses, intestinal barrier injury, and intestinal flora imbalances. Necrotic enteritis is an enterotoxemic disease caused by *C. perfringens*, leading to diarrhea or necrotizing lesions in the intestines of livestock and humans and, in severe cases, leading to death, with mortality rates of up to 100% in affected piglets [7]. As one of the most economically important diseases affecting poultry worldwide, necrotic enteritis causes $6 billion in annual losses globally [8]. In animals, associated lesions caused by *C. perfringens* have been found in the whole intestine, and were significantly more severe in the jejunum compared to the duodenum and ileum [9]. In healthy chickens, it is hard to find *C. perfringens* spores in the gastrointestinal tract, but the total number of vegetative *C. perfringens* cells increased when necrotic enteritis occurred, presenting a positive correlation between presence in the duodenum, jejunum, and ileum and disease severity [9,10]. In a retrospective study, *C. perfringens* was also found to cause liver morphological lesions, necrotizing hepatitis, congested lungs, and neurological diseases, manifesting as tremors, stargazing, and incoordination [11].

In a previous study, dietary supplementation with butyrate attenuated the inflammatory response and improved intestinal health in piglets challenged with enterotoxigenic Escherichia coli (ETEC) through inhibiting the activation of NF-κB/MAPK and modulating the hindgut microbiota [12]. Another study used an adherent-invasive Escherichia coli (AIEC) challenge model and showed that butyrate augmented AIEC invasiveness, while concurrently bolstering the intestinal epithelial barrier and reducing intestinal inflammation [13]. However, to the best of our knowledge, there have been few studies on the effect of butyrate in mitigating *C. perfringens* infection in broilers. In addition, butyrate given orally is quickly absorbed and used as energy by mucosal cells, absorbed and metabolized by the bird ingluvies and throughout the whole gastrointestinal tract, limiting the amount of butyrate that reaches the hindgut and restricting its practical use in the animal production context [14]. In order to prevent butyrate from absorption and metabolization in the anterior segment of the gastrointestinal tract, encapsulation is a simple and effective method. Previous studies have shown that microencapsulation reduces the release of contents into gastrointestinal fluid [15].

Here, we propose the hypothesis that microencapsulated sodium butyrate has a better protective effect than sodium butyrate in broilers challenged with *C. perfringens*. Consequently, we systematically assess the protective and underlying mechanisms of sodium butyrate (SB) and microencapsulated sodium butyrate (MS-SB) against *C. perfringens* infection in broilers. This study possesses considerable theoretical significance in relation to protection against the occurrence of necrotic enteritis and providing a strategy for effectively utilizing gastrointestinal-sensitive biological agents.

2. Materials and Methods

2.1. Animals and Diets

A total of 360 1-day-old male yellow feather broiler chickens were obtained from a commercial hatchery. The chickens were completely randomly allocated into 6 groups, with 6 replicates, and each replicate containing 10 chickens. The study adopted a completely randomized design with a 3×2 factorial pattern (3 kinds of diet and *C. perfringens*-challenged or not). The chickens were reared in a temperature-controlled room and maintained on a 24 h constant light schedule, and allowed ad libitum access to feed and water. The experiment last 57 days. The basal diet was formulated based on the NRC (1994) [16] and nutrient requirements for yellow chickens [17]; see Table 1. During the experimental period, chickens were fed a basal diet (basic), a basal diet with 1000 mg/kg sodium butyrate (SB), or 1000 mg/kg microencapsulated sodium butyrate (MS-SB), respectively. At 53 days old, the challenged group of chickens were challenged with 1 mL *C. perfringens* suspension (10^9 CFU/mL) via intragastric administration every other day, and the non-challenged group of chickens were given 1 mL media in the same way. The *C. perfringens* used in this study were kept in our laboratory. The SB contained 98% sodium butyrate, and was purchase from Wuhan Jiyesheng Chemical Co., Ltd. (Wuhan, China). The MS-SB was coated with a polymer enteral material and contained 40% sodium butyrate, and was provided by Zhejiang Vegamax Biotechnology Co., Ltd. (Huzhou, China).

Table 1. Ingredients and nutrient composition of base diets, as feed basis.

Ingredients (%)	Starter (Days 1–28)	Grower (Days 29–56)	Nutritional Level	Starter (Days 1–28)	Grower (Days 29–56)
Corn	54.4	53	Me (kcal/kg)	2983	3090
Soybean meal	23.6	16	CP (%)	20.4	17.2
Expanded soybean	5	3	Lysine (%)	1.18	0.96
Rice DDGS	5	8	Methionine (%)	0.55	0.44
Rice bran	/	8	Met + Cys (%)	0.90	0.74
Corn bran	/	2	Tryptophan (%)	0.22	0.20
Soybean oil	2.2	4.5	Threonine (%)	0.88	0.78
Limestone	1.5	1.9	Calcium (%)	0.86	0.73
Fermented soybean meal	2.5	/	Total P (%)	0.70	0.71
Corn gluten meal	2.0	/	Available P (%)	0.43	0.44
$CaHPO_4$ ($2H_2O$)	2.0	1.8			
NaCl	0.3	0.3			
Premix [a]	1.5	1.5			
Total	100.00	100.00			

[a] The following substances were supplied per kilogram of diet: vitamin A, 10,000 IU; vitamin D_3, 2500 IU; vitamin E, 20 mg; vitamin B_1, 1.5 mg; vitamin B_2, 3.5 mg; pantothenic acid, 10 mg; vitamin B_{12}, 0.01 mg; folic acid, 1 mg; niacin 30 mg; Choline chloride, 1000 mg; Cu ($CuSO_4 \cdot 5H_2O$), 8 mg; Fe ($FeSO_4 \cdot 7H_2O$), 80 mg; Zn ($MnSO_4 \cdot 7H_2O$), 60 mg; Se ($NaSeO_3$), 0.15 mg; I (KI), 0.2 mg.

2.2. Sample Collection

At 58 days old, blood was collected from the jugular vein after the chickens were starved overnight and weighted. It was centrifuged at $3500\times g$ for 15 min at 4 °C, then stored at −20 °C for future analyses. The chickens were slaughtered by cervical dislocation.

The middle segments of jejunum were collected and fixed with 4% paraformaldehyde. The cecal contents were aseptically collected and assessed for volatile fatty acids (VFAs) and microflora composition.

2.3. Organ Index

The liver, spleen, thymus, and bursa of Fabricius of each sampling broiler were removed, the blood stains were wiped off the surface, then they were weighed. The relative organ weights were calculated as per the following equation:

$$\text{Organ weight indexes} = \text{Organ weight (g)}/\text{Body weight (g)} \times 100$$

2.4. Serum Immune Indicators

The concentrations of serum immunoglobulin A (IgA) (CAS:ANG-E32004C; 10–600 ng/mL), immunoglobulin M (IgM) (CAS:ANG-E32005C; 0.1405–11.25 μg/mL), immunoglobulin Y (IgY) (CAS:ANG-E32209C; 0.062.5–3.75 ng/mL), interleukin-1β (IL-1β) (CAS:ANG-E32031C; 1.875–112.5 ng/L), interleukin-6 (IL-6) (CAS:ANG-E32013C; 1–60 ng/L), and tumor necrosis factor-α (TNF-α) (CAS:ANG-E32030C; 1.25–75 ng/L) were measured using an enzymatic chromatometric method using the ELISA Kits that were purchased from Angle Gene Biotechnology Co., Ltd. (Nanjing, Jiangsu, China), according to the manufacturer's instructions.

2.5. Jejunum Morphology Analysis

The fixed and pruned jejunum was dehydrated with gradient ethanol, followed by cleaning with xylene, waxing, embedding, slicing, and staining with hematoxylin–eosin (HE). Finally, we took pictures using a microscope (Nikon, Tokyo, Japan). Ten intact villi for each sample were randomly selected to measure the villus height (V) and crypt depth (C) using the Image Pro Plus 6.0 software (Rockville, MD, USA), and the ratio of villus height to crypt depth (V/C) was calculated.

2.6. Volatile Fatty Acid (VFA) Analysis

The VFAs in cecum content were determined via gas chromatography according to the method of Yu et al. (2023) [18]. In brief, a sample containing about 0.5 g of cecum content was mixed with pre-cooled water at a mass volume ratio of 1:3 and centrifuged at 12,000× *g* for 10 min at 4 °C. The supernatant was mixed with 25% metaphosphoric acid in a 5:1 ratio and rested for 30 min, followed by centrifugation at 10,000× *g* for 10 min at 4 °C. Then, the supernatant was transferred into a sample bottle for testing using an Agilent Technologies 7890B GC System (column parameter: 30 m × 0.25 mm × 0.25 μm; Agilent Technologies, Santa Clara, CA, USA). Pure acetic acid, propionic acid, isobutyric acid, butyric acid, isovaleric acid, and valeric acid solutions were used as standards to calculate relevant concentrations in the sample.

2.7. Cecum Microflora

Total bacterial DNA of cecum microbiota was extracted using a DNA extraction kit, and the DNA concentration was determined using agarose gel electrophoresis. The V3–V4 regions of 16S rRNA were amplified with universal primers 515F/806R using the Applied Biosystems GENEAMP 9700 (Thermo Fischer Scientific, Waltham, MA, USA). The amplification products were recovered after purification using an AXYGEN DNA Gel Extraction Kit (Union City, CA, USA). The amplification products were sequenced on an Illumina miseq (PE300) platform provided by Majorbio Co., Ltd. (Shanghai, China), after quantitative analysis conducted using a quantifluortm blue fluorescence quantitative system (Promega, WI, USA). In order to obtain the Amplicon Sequence Variant (ASV) variants and feature lists, the DADA2 variants in QIIME2 were used to optimize the data obtained through sequencing. The alpha diversity component, Shannon, Chao, and Simpson indices were adopted to indicate the diversity of microbiota. Principal component analysis (PCA) and principal co-ordinates analysis (PCoA) were conducted to analyze the species indices, including the β-diversity component between different groups. Finally, linear discriminant analysis coupled with effect size (LEfSe) was used to identify microbial differences among all treatment groups.

2.8. Statistical Analysis

The data were tested for normality and homogeneity of variance through the Levene test. Then, two-factor analysis of variance and Tukey's HSD were carried out using the JMP Pro software 13.0 (SAS, Carrey, MS, USA). The model equation included the main effects (sodium butyrate addition and *C. perfringens* challenge) and their interactions. The differences among treatments were tested using Tukey's test when the main effects or interactions were significant. The statistical significance was set to $p < 0.05$. All values are shown as mean ± SEM. The resulting data were plotted using the GraphPad Prism 8.0 software (GraphPad Software, San Diego, CA, USA).

3. Results

3.1. Microencapsulated Sodium Butyrate Alleviated C. perfringens Infection

At 53–57 days old, the broilers were challenged with *C. perfringens* three times through gavage administration every other day. As shown in Table 2, the *C. perfringens* challenge significantly decreased the thymus index and significantly increased the spleen index ($p < 0.05$). The addition of SB significantly decreased the thymus index ($p < 0.05$), and MS-SB had the same effect as sodium butyrate. It is worth noting that the addition of SB and MS-SB decreased the thymus index under *C. perfringens* infection.

Table 2. Effect of microencapsulated sodium butyrate on the organ index of broiler chickens challenged with clostridium perfringens.

Diet	Challenge	Liver Index	Spleen Index	Bursa of Fabricius Index	Thymus Index
Control		16.99	1.37	1.14	2.34 [a]
SB	CON	18.04	1.49	1.23	1.84 [ab]
MS-SB		17.67	1.77	0.76	2.43 [a]
Control		17.77	1.84	0.53	1.46 [bc]
SB	CP	19.74	1.89	0.75	0.98 [c]
MS-SB		17.25	1.62	0.92	0.97 [c]
SEM		0.65	0.11	0.19	0.14
Main effect					
Diet	Control	17.38	1.60	0.83	1.90 [a]
	SB	18.86	1.70	0.99	1.41 [b]
	MS-SB	17.46	1.69	0.84	1.63 [b]
Challenge	CON	17.56	1.54 [b]	0.73	2.24 [a]
	CP	18.25	1.78 [a]	1.04	1.37 [b]
p-value					
Diet		0.057	0.743	0.694	0.005
Challenge		0.221	0.034	0.063	<0.001
Diet × Challenge		0.304	0.051	0.128	0.023

Note. Number with different letters in the same column are statistically significant.

3.2. Microencapsulated Sodium Butyrate Alleviated Reduced Systemic Inflammation Caused by C. perfringens

The levels of inflammatory factors and immunoglobulin in sera are shown in Figure 1. The *C. perfringens* challenge had no influence on inflammatory factor and immunoglobulin content in sera. However, SB addition significantly increased the immunoglobulin content in sera, including IgA, IgM, and IgY ($p < 0.05$). In addition, SB significantly decreased the content of IL-1β, IL-6, and TNF-α ($p < 0.05$). Compared with SB, MS-SB presented a stronger performance, with higher IgA content, as well as a lower IL-1β level when normal or *C. perfringens*-challenged.

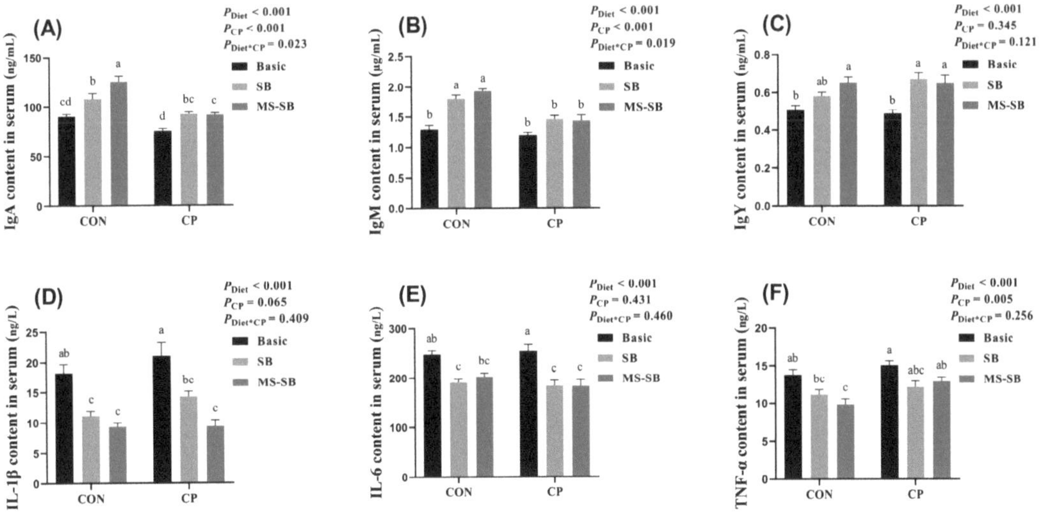

Figure 1. MS-SB modulates immunity through globulins and immune factor content in serum during *C. perfringens* infection. (**A**) IgA content; (**B**) IgM content; (**C**) IgY content; (**D**) IL-1β content; (**E**) IL-6 content; (**F**) TNF-α content. Bars with different letters are statistically significant ($p < 0.05$) in different groups, $n = 6$.

3.3. Microencapsulated Sodium Butyrate Repaired Intestinal Morphology Damaged by C. perfringens

We also measured the parameters of jejunum villi through HE stains, and the results are shown in Figure 2. The results revealed that SB administration significantly increased the villus height and V/C ratio ($p < 0.05$), especially the MS-SB addition. However, *C. perfringens* challenge significantly decreased the villus height ($p < 0.05$) without affecting the V/C ratio.

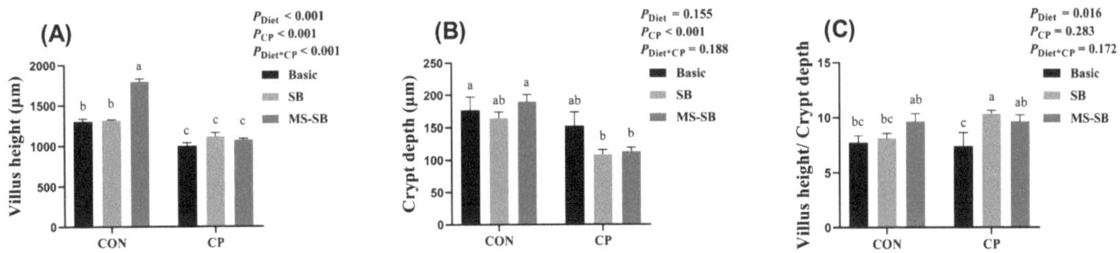

Figure 2. MS-SB repaired intestinal morphology damaged by *C. perfringens*. (**A**) Villus height of jejunum; (**B**) crypt depth of jejunum; (**C**) ratio of the villus height and crypt depth of jejunum. Bars with different letters are statistically significant ($p < 0.05$) in different groups, $n = 6$.

3.4. Microencapsulated Sodium Butyrate Ameliorated VFAs under C. perfringens Challenge

For this study, we also detected the VFA content in the broiler cecum, and the results are shown in Figure 3. The results indicated that the *C. perfringens* challenge had no effect on the VFA content. However, the addition of SB significantly decreased the content of acetic acid ($p < 0.05$) under *C. perfringens* non-challenge; at same time, SB or MS-SB presented varying degrees of increased fatty acid concentrations during *C. perfringens* challenge. It should be noted that MS-SB had a stronger performance, with a higher content of isobutyric acid, isovaleric acid, and valeric acid.

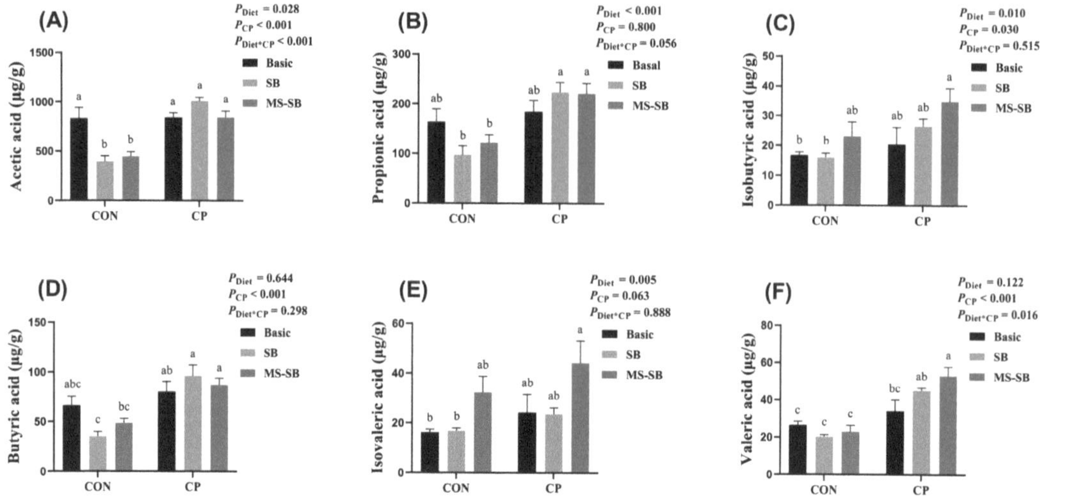

Figure 3. Microencapsulated sodium butyrate modulates the VFA content when challenged with *C. perfringens*. (**A**) Acetic acid; (**B**) propionic acid; (**C**) isobutyric acid; (**D**) butyric acid; (**E**) isovaleric acid; (**F**) valeric acid content in cecum of broiler chickens. Bars with different letters are statistically significant ($p < 0.05$) in different groups, $n = 6$.

3.5. Microencapsulated Sodium Butyrate Modulated Gut Microbiota Community Variation Caused by C. perfringens

The changes in the diversity of cecum microbiota are summarized in Figure 4. A total of 921 ASVs were shared among the five treatment groups based on the Venn diagram, with non-overlapping regions indicating unique OTUs in the CON group ($n = 569$), SB group ($n = 658$), MS-SB group ($n = 743$), CON-CP group ($n = 592$), and MS-SB-CP group ($n = 611$); see Figure 4A. We adopted the Chao index, Shannon index, and Simpson index to assess the microbiota community diversity. The results indicated that *C. perfringens* infection and SB addition individually presented more significant diversity in cecum microbiota when compared to the CON group, but there was no difference between them (Figure 4B–D). Principal component analysis (PCA) and principal coordinate analysis (PCoA) were performed to investigate the differences in species complexity and structural alterations in microbial communities. PCoA and PCA plots revealed the degree of diversity discrepancy in cecum microbiota between different treatments; in particular, the PCoA plot indicated that the CON group was separated from the other groups (Figure 4E,F).

3.6. Microencapsulated Sodium Butyrate Modulated Gut Microbiota Community Composition Caused by C. perfringens

In order to explore the change in cecum microbiota structure, we analyzed the relative abundance of cecum microbiota at phylum to species levels. At the phylum level, about 4 major phyla were detected, including *Firmicutes, Bacteroidota, Verrucomicrobiota,* and *Actinobacteriota* (Figure 5A), and the relative abundance of *Proteobacteria* was higher in the *C. perfringens* infection groups than non-infection groups. The addition of MS-SB significantly increased the relative abundance of *Campilobacteria* (Figure 5D). At the genus level, about 13 major phyla were detected as dominant genera (Figure 5B). The relative abundances of *Bacteroides* in the *C. perfringens* infection groups were higher than in the non-infection groups, and the relative abundances of *Ruminococcus_torques* and *Alistipes* in the MS-SB-CP groups were higher than in other groups (Figure 5E). At the species level, about 11 major phyla were detected as dominant flora (Figure 5C). In addition, the relative abundances of 6 bacteria were higher with MS-SB addition than in the CON group, including *Bacteroides, Faecalibacterium, Clostridia_UCG-014, Ruminococcaceae, Alistipes_inops,* and *Clostridia_vadinBB60*. The *C. perfringens* infection increased the abundance of *Bacteroides* and *Alistipes_inops* (Figure 5F). Finally, we analyzed the association between VFAs and microbiota composition. The results demonstrated that *Ruminococcaceae_torques* and *Alistipes* were positively correlated with isobutyric acid concentration, while *Lachnospiraceae*, *Ruminococcaceae_torques*, and *Alistipes* were positively correlated with valeric acid (Figure 6A). At the species level, *Ruminococcaceae_torques* and *Alistipes* were positively correlated with isobutyric acid, and *Lachnospiraceae*, *Ruminococcaceae_torques*, and *Alistipes* were positively correlated with valeric acid (Figure 6B).

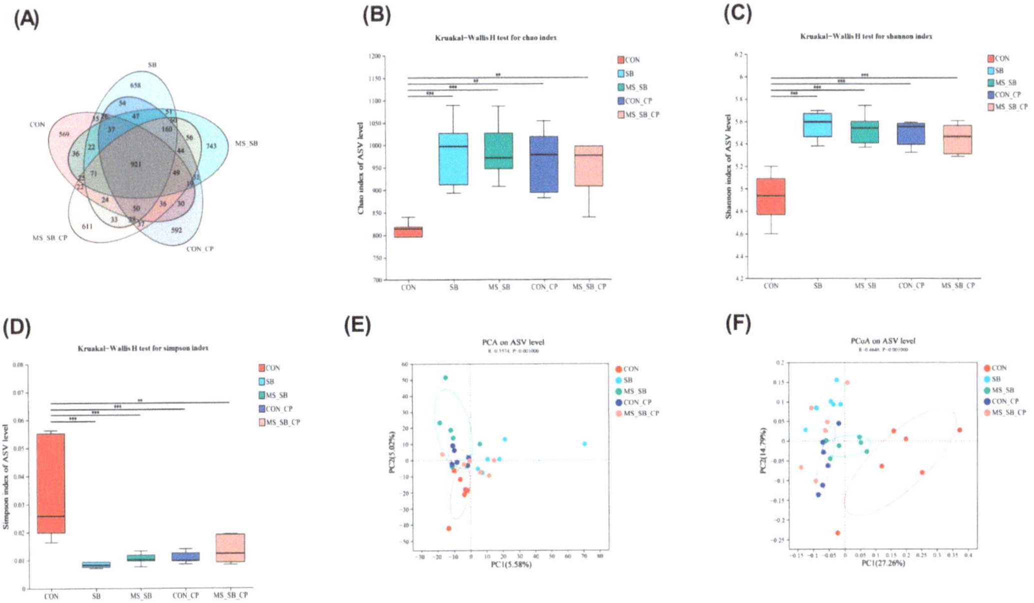

Figure 4. Analysis of the diversity of gut microbiota. (**A**) The Venn diagram summarizing the numbers of common and unique OTUs in cecum microflora community. (**B–D**) The Chao index, Shannon index, and Simpson index reflecting species alpha diversity between groups. (**E**,**F**) The principal component analysis (PCA) and principal co-ordinates analysis (PCoA) reflecting beta diversity within and between groups at the species level. The column with * are statistically significant, ** means $p < 0.01$, *** means $p < 0.001$, $n = 6$.

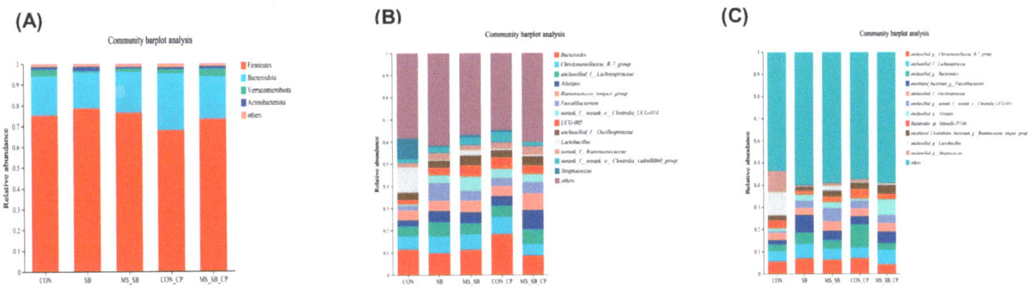

Figure 5. *Cont.*

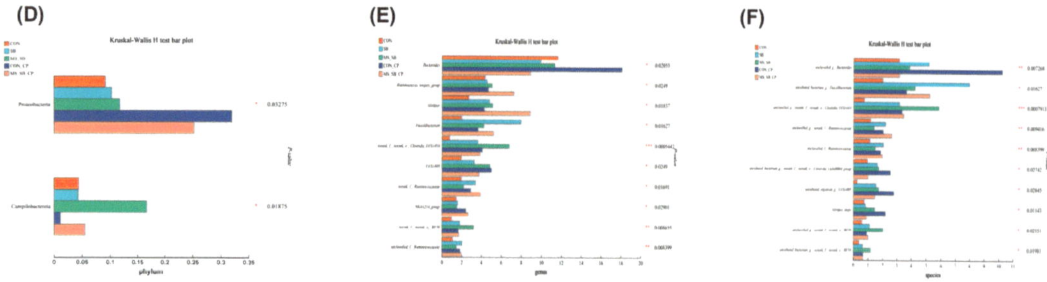

Figure 5. The abundance of the microbial community in cecum content. (**A–C**) The top relative abundance of the microflora community between groups (phylum level, genus level, and species level). (**D–F**) The bacteria with significant differences between groups (phylum level, genus level, and species level). The column with * are statistically significant, * means $p < 0.05$, ** means $p < 0.01$, *** means $p < 0.001$, $n = 6$.

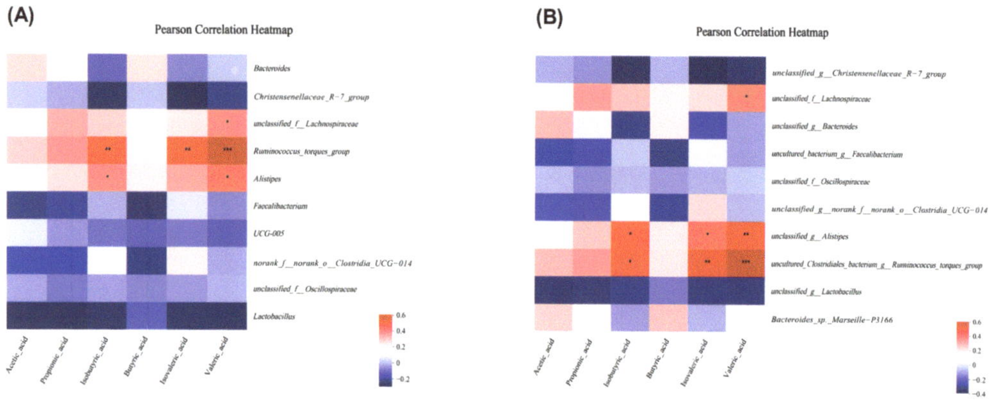

Figure 6. Correlation analysis between gut microbiota and VFAs. (**A**) Genus level and VFAs. (**B**) Species level and VFAs. The column with * are statistically significant, * means $p < 0.05$, ** means $p < 0.01$, *** means $p < 0.001$, $n = 6$.

4. Discussion

In intensive animal production, the animals are faced with a variety of external factors at any time, such as production environment stress and physiological stress. In healthy animals, bacteria and health stand at either end of a pair of scales; once this balance is broken, harmful bacteria in the environment and body can cause damage to animal health, including intestinal inflammation, immune stress, and so on [19]. *C. perfringens* is consumed by chickens from environmental sources during rearing, including contaminated feed, water, and the farm environment [20]. As previously reported, the immune organ changes in response to infection with *C. perfringens*, shown in terms of the morphology and weight of the bursa of Fabricius, spleen, and thymus [21]. In this study, the results indicated that the spleen index increased, which is consistent with previous findings, as well as showing expected changes in the thymus index. This may be attributed to the age and species, as broilers near the end of growth were used in this study. Thus, these results indicate the validity of the *C. perfringens* infection model.

Moreover, we demonstrated that the addition of SB or MS-SB had no promotive or protective effects on the immune organs in this study. The results are not consistent with the results of a previous study, in which the thymus, spleen, and bursa weighed more in the SB addition group compared with control group [22]. Moreover, in a study in quails, dietary supplementation with 1000 mg/kg SB significantly increased the thymus and bursa of Fabricius [23]. As a matter of fact, we have also confirmed that SB and MS-SB could enhance immune organ development throughout the growth period in a previous study (unpublished). Thus, according to the results of this study, SB or MS-SB have no protective effect against the immune organ changes caused by *C. perfringens* infection, evidenced by the fact that SB or MS-SB were unable to reverse the organ changes induced by *C. perfringens*.

The immune organs, immune cells, and immune molecules make up the immune system of animals, and can be categorized into two parts—the innate immune system and the adaptive immune system—according to the manner in which they act against invading pathogens [24]. As our results indicated that SB and MS-SB could enhance immune organ development during the growth period (unpublished), we further determined the levels of immunoglobulins and immunomodulatory cytokines in sera. Immunoglobulins are synthesized and secreted by B-cells after immune cells are activated by antigens, which can bind to specific antigens in order to defend against invading pathogens [25]. In this study, the results showed that *C. perfringens* infection decreased the immunoglobulin levels in sera, while SB and MS-SB improved the secretion of immunoglobulins, consistent with the results of previous studies [26]. However, it should be noted that the enhancement brought by SB and MS-SB was diminished under *C. perfringens* infection, when compared to normal conditions. Immunomodulatory cytokines are produced by immune cells and act on other immune cells, which can be classified as pro-inflammatory or anti-inflammatory according to their function [27]. Pro-inflammatory cytokines (e.g., IL-1α/β, TNF-α/β, and IL-6) up-regulate inflammatory reactions, while anti-inflammatory cytokines (e.g., IL-10) down-regulate inflammatory responses and promote tissue healing [28]. In rat models, necrotic enteritis causes enteric inflammation accompanied by serum proinflammatory cytokine production [29]. Thus, we proceeded to determine the content of inflammatory factors in sera. The results demonstrated that SB or MS-SB exhibited a potent anti-inflammatory effect, evidenced by reductions in IL-1β, IL-6, and TNF-α, especially in the case of the slight increase in inflammatory factors caused by *C. perfringens*. Similar to the results of our study, the study published by Sun et al. (2021) reported that sodium butyrate inhibited intestinal inflammation through the HMGB1-TLR4/NF-κB pathway [3]. Therefore, we speculate that SB or MS-SB may enhance immune function by regulating serum inflammatory cytokines in broiler chickens. Moreover, compared with SB, MS-SB presented a better anti-inflammatory effect.

As an organ for nutrient digestion and absorption, the intestinal tract is also an important barrier to maintain the homeostasis of the internal environment, and is the first barrier to deal with foreign harmful bacteria [30]. Normal intestinal permeability prevents water and electrolyte loss, promotes the absorption of dietary nutrients, and prevents the entry of antigens and micro-organisms into the body, which are dependent on the integrity of the intestinal villi [31]. In a previous study, necrotic enteritis caused by *C. perfringens* was mostly observed in the jejunum, manifesting as intestinal morphological damage and inflammation [9]. In addition, SCFA can improve the proliferation of gut epithelial cells and increase their villi height, which subsequently helps to improve the capacity of the intestine for nutrient absorption [32]. Thus, we detected the morphology of the jejunum using HE staining, and the results indicated that *C. perfringens* negatively affected the intestinal villi morphology, decreasing both the villus height and crypt depth. These results are similar to those previously reported in broilers [33,34] and mice [35]. We cannot ignore that the addition of SB or MS-SB alleviated the morphological damage to the intestinal villi, and the performance of MS-SB was particularly prominent compared with that of SB. Similar results have also been confirmed in a study using a mouse model [5].

In previous research, VFAs have been shown to inhibit pathogenic micro-organisms and increase the absorption of nutrients, which may contribute to a reduction in the luminal pH [36]. Acetic acid is the shortest fatty acid with a carbon chain, which has been shown to be an intermediary involved in *bifidobacteria* inhibiting the proliferation of intestinal pathogens [37]. Propionate and butyrate have been reported to assist in controlling intestinal inflammation by inducing the differentiation of T-regulatory cells, and the inhibition of histone deacetylation may be involved in the relevant regulatory mechanism [2,38]. In addition, recent studies have shown that valerate inhibits the proliferation of *Clostridium difficile* in the intestinal tract, thereby protecting or treating intestinal diseases [39]. Moreover, butyric acid is usually produced in the large intestine by intestinal bacteria and plays important roles, such as fueling intestinal epithelial cells and increasing mucin production, which may result in changes in bacterial adhesion and improved tight-junction integrity [40,41]. Thus, VFAs seem to play an important role in the maintenance of gut barrier function [42]. We also measured the concentrations of VFAs in the cecum, such as acetic acid, butyric acid (isobutyric acid), and valeric acid (isovaleric acid). The results indicated that *C. perfringens* infection has no effect on the VFA content in the cecum. However, SB or MS-SB increased the concentrations of fatty acids to varying degrees under the *C. perfringens* challenge. Our results are partially consistent with previous findings in broilers [43], and the dietary composition and butyric acid originating from foregut may be the cause of this result, as a result of butyric acid and other VFAs being produced through the bacterial fermentation of unabsorbed carbohydrates or food scraps [44]. Meanwhile, MS-SB was used in our study, which can significantly delay the enteric release of butyric acid, thus reducing small intestinal absorption and enhancing colonic delivery [44–46]. This may explain the difference between our results and those reported in previous studies.

Numerous studies have revealed that the gut microbiota that improve growth and metabolism promote host nutrient absorption and modulate the immune system, which are all behaviors providing irreplaceable functionality [47,48]. Furthermore, the diversity of the microbial community helps to maintain the homeostasis of the intestinal microbiome and improve resistance to pathogens in the host [49]. Our results on alpha diversity and beta diversity in the cecum content revealed a degree of diversity discrepancy in the cecum microbiota. Both the *C. perfringens* challenge and addition of SB improved the diversity of bacterial flora. This result is not consistent with previous studies, such as that of Pammi et al. (2017), in which the intestinal flora diversity in necrotizing enterocolitis patients was lower than that in unaffected patients, such as the lower relative abundance of *Firmicutes* and *Bacteroides* and higher relative abundance of *Proteobacteria* [50]. Meanwhile, our results were consistent with those of Zhang et al. (2018), who revealed the α-diversity index of broiler gut microbial community after *C. perfringens* infection [51]. This may be explained by the fact that *C. perfringens* infection can destroy the ecological balance of intestinal flora, resulting in intestinal ecological imbalance [52]. In addition, *C. perfringens* strains and dietary components, as well as the timing and duration of the *C. perfringens* challenge, may have contributed to the observed discrepancy [53]. In this study, the results indicated that the proportions of *Bacteroides, Faecalibacterium, Clostridia, Ruminococcaceae, Alistipes,* and *Clostridia* were significantly higher in the MS-SB addition group; at the same time, *C. perfringens* infection increased the abundance of *Bacteroides* and *Alistipes*. In previous studies, the genera *Alistipes* and *Bacteroides* have been identified as butyrate producers in the gut and have demonstrated good anti-inflammatory effects through butyrate [54]. The results of the correlation analysis considering VFAs and microbiota indicated that the isobutyric acid content increased and was positively correlated with the abundance of *Alistipes* and *Clostridia*. Thus, our findings demonstrate that SB and MS-SB exhibit a protective role to suppress *C. perfringens*-induced intestinal damage and microbiota disturbances.

5. Conclusions

In summary, sodium butyrate ameliorated *C. perfringens* infection by reducing inflammation, repairing intestinal damage, and modulating the cecum microbiota. Compared to sodium butyrate, microencapsulated sodium butyrate presented a better effect. This study highlights the effectiveness of microencapsulated sodium butyrate and provides a novel strategy for protection against *C. perfringens* infection through the effective utilization of gastrointestinal-sensitive biological agents.

Author Contributions: B.D. designed the experiments. S.X., Y.L. and X.W. performed the experiments. C.Y., J.L., S.X. and Z.D. analyzed the data. T.Y. and Y.S. wrote the main manuscript. Z.D., S.Y. and R.Z. revised the main manuscript. All authors have read and agreed to the published version of the manuscript.

Funding: This work was supported by the Leading Innovation and Entrepreneurship Team Project of Zhejiang Province (2020R01015), Zhejiang Provincial Key R&D Program of China (2021C02008), the Natural Science Foundation of China (32260847, 32202687), Jiangxi Modern Agricultural Research Collaborative Innovation Project (JXXTCXBSJJ202208), and Hangzhou Key Projects for Agricultural and Social Development (202203A08).

Institutional Review Board Statement: The animal study protocol was approved by the Animal Health and Care Committee of Zhejiang Agricultural and Forestry University (No. ZAFUAC2022005, 15 October 2022).

Informed Consent Statement: Not applicable.

Data Availability Statement: The data that support the findings of this study were not deposited in an official repository, but they are available from the authors upon request.

Conflicts of Interest: Jinsong Liu, Shiping Xiao, Yulan Liu, and Caimei Yang are employees of Zhejiang Vegamax Biotechnology. This paper reflects the views of the scientists, and not the company. We certify that there is no conflict of interest with any financial organization regarding the material discussed in the manuscript, and the authors declare no conflict of interest.

References

1. Li, Z.; Wang, W.; Liu, D.; Guo, Y. Effects of Lactobacillus acidophilus on the growth performance and intestinal health of broilers challenged with Clostridium perfringens. *J. Anim. Sci. Biotechnol.* **2018**, *9*, 25. [CrossRef]
2. Salvi, P.S.; Cowles, R.A. Butyrate and the Intestinal Epithelium: Modulation of Proliferation and Inflammation in Homeostasis and Disease. *Cells* **2021**, *10*, 1775. [CrossRef]
3. Sun, Q.; Ji, Y.-C.; Wang, Z.-L.; She, X.; He, Y.; Ai, Q.; Li, L.-Q. Sodium Butyrate Alleviates Intestinal Inflammation in Mice with Necrotizing Enterocolitis. *Mediat. Inflamm.* **2021**, *2021*, 6259381. [CrossRef]
4. Manrique Vergara, D.; González Sánchez, M.E. Short chain fatty acids (butyric acid) and intestinal diseases. *Nutr. Hosp.* **2017**, *34*, 58–61. [CrossRef]
5. Chen, G.; Ran, X.; Li, B.; Li, Y.; He, D.; Huang, B.; Fu, S.; Liu, J.; Wang, W. Sodium Butyrate Inhibits Inflammation and Maintains Epithelium Barrier Integrity in a TNBS-induced Inflammatory Bowel Disease Mice Model. *EBioMedicine* **2018**, *30*, 317–325. [CrossRef]
6. Zhou, D.; Pan, Q.; Xin, F.Z.; Zhang, R.N.; He, C.X.; Chen, G.Y.; Liu, C.; Chen, Y.W.; Fan, J.G. Sodium butyrate attenuates high-fat diet-induced steatohepatitis in mice by improving gut microbiota and gastrointestinal barrier. *World J. Gastroenterol.* **2017**, *23*, 60–75. [CrossRef]
7. Posthaus, H.; Kittl, S.; Tarek, B.; Bruggisser, J. Clostridium perfringens type C necrotic enteritis in pigs: Diagnosis, pathogenesis, and prevention. *J. Vet. Diagn. Investig. Off. Publ. Am. Assoc. Vet. Lab. Diagn. Inc.* **2020**, *32*, 203–212. [CrossRef]
8. Wade, B.; Keyburn, A. The true cost of necrotic enteritis. *World Poult.* **2015**, *31*, 16–17.
9. Hustá, M.; Tretiak, S.; Ducatelle, R.; Van Immerseel, F.; Goossens, E. Clostridium perfringens strains proliferate to high counts in the broiler small intestinal tract, in accordance with necrotic lesion severity, and sporulate in the distal intestine. *Vet. Microbiol.* **2023**, *280*, 109705. [CrossRef]
10. Daneshmand, A.; Kermanshahi, H.; Mohammed, J.; Sekhavati, M.H.; Javadmanesh, A.; Ahmadian, M.; Alizadeh, M.; Razmyar, J.; Kulkarni, R.R. Intestinal changes and immune responses during Clostridium perfringens-induced necrotic enteritis in broiler chickens. *Poult. Sci.* **2022**, *101*, 101652. [CrossRef]
11. Marcano, V.; Gamble, T.; Maschek, K.; Stabler, L.; Fletcher, O.; Davis, J.; Troan, B.V.; Villegas, A.M.; Tsai, Y.Y.; Barbieri, N.L.; et al. Necrotizing Hepatitis Associated with Clostridium perfringens in Broiler Chicks. *Avian Dis.* **2022**, *66*, 337–344. [CrossRef]
12. Tian, M.; Li, L.; Tian, Z.; Zhao, H.; Chen, F.; Guan, W.; Zhang, S. Glyceryl butyrate attenuates enterotoxigenic Escherichia coli-induced intestinal inflammation in piglets by inhibiting the NF-κB/MAPK pathways and modulating the gut microbiota. *Food Funct.* **2022**, *13*, 6282–6292. [CrossRef]
13. Pace, F.; Rudolph, S.E.; Chen, Y.; Bao, B.; Kaplan, D.L.; Watnick, P.I. The Short-Chain Fatty Acids Propionate and Butyrate Augment Adherent-Invasive Escherichia coli Virulence but Repress Inflammation in a Human Intestinal Enteroid Model of Infection. *Microbiol. Spectr.* **2021**, *9*, e0136921. [CrossRef]
14. Kaczmarek, S.A.; Barri, A.; Hejdysz, M.; Rutkowski, A. Effect of different doses of coated butyric acid on growth performance and energy utilization in broilers. *Poult. Sci.* **2016**, *95*, 851–859. [CrossRef]
15. Pourjafar, H.; Noori, N.; Gandomi, H.; Basti, A.A.; Ansari, F. Viability of microencapsulated and non-microencapsulated Lactobacilli in a commercial beverage. *Biotechnol. Rep.* **2020**, *25*, e00432. [CrossRef]

16. Pesti, G.M. *Nutrient Requirements of Poultry*; National Academy Press: Washington, DC, USA, 1994.
17. Sciences, G.A.o.A. *Nutrient Requirements of Yellow Chickens*; Ministry of Agriculture and Rural Affairs, PRC: Beijing, China, 2020.
18. Yu, X.; Dai, Z.; Cao, G.; Cui, Z.; Zhang, R.; Xu, Y.; Wu, Y.; Yang, C. Protective effects of Bacillus licheniformis on growth performance, gut barrier functions, immunity and serum metabolome in lipopolysaccharide-challenged weaned piglets. *Front. Immunol.* **2023**, *14*, 1140564. [CrossRef]
19. Cong, J.; Zhou, P.; Zhang, R. Intestinal Microbiota-Derived Short Chain Fatty Acids in Host Health and Disease. *Nutrients* **2022**, *14*, 1977. [CrossRef]
20. He, W.; Goes, E.C.; Wakaruk, J.; Barreda, D.R.; Korver, D.R. A Poultry Subclinical Necrotic Enteritis Disease Model Based on Natural Clostridium perfringens Uptake. *Front. Physiol.* **2022**, *13*, 788592. [CrossRef]
21. Villagrán-de la Mora, Z.; Vázquez-Paulino, O.; Avalos, H.; Ascencio, F.; Nuño, K.; Villarruel-López, A. Effect of a Synbiotic Mix on Lymphoid Organs of Broilers Infected with Salmonella typhimurium and Clostridium perfringens. *Animals* **2020**, *10*, 886. [CrossRef]
22. Sikandar, A.; Zaneb, H.; Younus, M.; Masood, S.; Aslam, A.; Khattak, F.; Ashraf, S.; Yousaf, M.S.; Rehman, H. Effect of sodium butyrate on performance, immune status, microarchitecture of small intestinal mucosa and lymphoid organs in broiler chickens. *Asian-Australas. J. Anim. Sci.* **2017**, *30*, 690–699. [CrossRef]
23. Elnesr, S.; Ropy, A.; Abdel-Razik, A. Effect of dietary sodium butyrate supplementation on growth, blood biochemistry, haematology and histomorphometry of intestine and immune organs of Japanese quail. *Anim. Int. J. Anim. Biosci.* **2019**, *13*, 1234–1244. [CrossRef]
24. Ferreira, N.S.; Tostes, R.C.; Paradis, P.; Schiffrin, E.L. Aldosterone, inflammation, immune system, and hypertension. *Am. J. Hypertens.* **2021**, *34*, 15–27. [CrossRef]
25. Hand, T.W.; Reboldi, A. Production and function of immunoglobulin A. *Annu. Rev. Immunol.* **2021**, *39*, 695–718. [CrossRef]
26. Zhang, R.; Qin, S.; Yang, C.; Niu, Y.; Feng, J. The protective effects of Bacillus licheniformis against inflammatory responses and intestinal barrier damage in broilers with necrotic enteritis induced by Clostridium perfringens. *J. Sci. Food Agric.* **2023**, *13*, 6958–6965. [CrossRef]
27. Azad, M.A.K.; Sarker, M.; Wan, D. Immunomodulatory effects of probiotics on cytokine profiles. *BioMed Res. Int.* **2018**, *2018*, 8063647. [CrossRef]
28. Crawford, C.K.; Lopez Cervantes, V.; Quilici, M.L.; Armién, A.G.; Questa, M.; Matloob, M.S.; Huynh, L.D.; Beltran, A.; Karchemskiy, S.J.; Crakes, K.R. Inflammatory cytokines directly disrupt the bovine intestinal epithelial barrier. *Sci. Rep.* **2022**, *12*, 14578. [CrossRef]
29. Yu, R.; Jiang, S.; Tao, Y.; Li, P.; Yin, J.; Zhou, Q. Inhibition of HMGB1 improves necrotizing enterocolitis by inhibiting NLRP3 via TLR4 and NF-κB signaling pathways. *J. Cell. Physiol.* **2019**, *234*, 13431–13438. [CrossRef]
30. Di Tommaso, N.; Gasbarrini, A.; Ponziani, F.R. Intestinal barrier in human health and disease. *Int. J. Environ. Res. Public Health* **2021**, *18*, 12836. [CrossRef]
31. Bischoff, S.C.; Barbara, G.; Buurman, W.; Ockhuizen, T.; Schulzke, J.-D.; Serino, M.; Tilg, H.; Watson, A.; Wells, J.M. Intestinal permeability–a new target for disease prevention and therapy. *BMC Gastroenterol.* **2014**, *14*, 189. [CrossRef]
32. Pérez-Reytor, D.; Puebla, C.; Karahanian, E.; García, K. Use of short-chain fatty acids for the recovery of the intestinal epithelial barrier affected by bacterial toxins. *Front. Physiol.* **2021**, *12*, 650313. [CrossRef]
33. Fasina, Y.O.; Lillehoj, H.S. Characterization of intestinal immune response to Clostridium perfringens infection in broiler chickens. *Poult. Sci.* **2019**, *98*, 188–198. [CrossRef]
34. Ibrahim, D.; Ismail, T.A.; Khalifa, E.; Abd El Kader, S.A.; Mohamed, D.I.; Mohamed, D.T.; Shahin, S.E.; Abd El-Hamid, M.I. Supplementing Garlic Nanohydrogel Optimized Growth, Gastrointestinal Integrity and Economics and Ameliorated Necrotic Enteritis in Broiler Chickens Using a Clostridium perfringens Challenge Model. *Animal* **2021**, *11*, 2027. [CrossRef]
35. Navarro, M.A.; Li, J.; McClane, B.A.; Morrell, E.; Beingesser, J.; Uzal, F.A. NanI sialidase is an important contributor to Clostridium perfringens type F strain F4969 intestinal colonization in mice. *Infect. Immun.* **2018**, *86*, 10–1128. [CrossRef]
36. Dąbek-Drobny, A.; Kaczmarczyk, O.; Woźniakiewicz, M.; Paśko, P.; Dobrowolska-Iwanek, J.; Woźniakiewicz, A.; Piątek-Guziewicz, A.; Zagrodzki, P.; Zwolińska-Wcisło, M. Association between Fecal Short-Chain Fatty Acid Levels, Diet, and Body Mass Index in Patients with Inflammatory Bowel Disease. *Biology* **2022**, *11*, 108. [CrossRef]
37. Fukuda, S.; Toh, H.; Hase, K.; Oshima, K.; Nakanishi, Y.; Yoshimura, K.; Tobe, T.; Clarke, J.M.; Topping, D.L.; Suzuki, T.; et al. Bifidobacteria can protect from enteropathogenic infection through production of acetate. *Nature* **2011**, *469*, 543–547. [CrossRef]
38. Yang, J.; Wei, H.; Zhou, Y.; Szeto, C.H.; Li, C.; Lin, Y.; Coker, O.O.; Lau, H.C.H.; Chan, A.W.H.; Sung, J.J.Y.; et al. High-Fat Diet Promotes Colorectal Tumorigenesis Through Modulating Gut Microbiota and Metabolites. *Gastroenterology* **2022**, *162*, 135–149.e132. [CrossRef]
39. McDonald, J.A.; Mullish, B.H.; Pechlivanis, A.; Liu, Z.; Brignardello, J.; Kao, D.; Holmes, E.; Li, J.V.; Clarke, T.B.; Thursz, M.R. Inhibiting growth of Clostridioides difficile by restoring valerate, produced by the intestinal microbiota. *Gastroenterology* **2018**, *155*, 1495–1507.e1415. [CrossRef]
40. Pietrzak, A.; Banasiuk, M.; Szczepanik, M.; Borys-Iwanicka, A.; Pytrus, T.; Walkowiak, J.; Banaszkiewicz, A. Sodium Butyrate Effectiveness in Children and Adolescents with Newly Diagnosed Inflammatory Bowel Diseases-Randomized Placebo-Controlled Multicenter Trial. *Nutrients* **2022**, *14*, 3283. [CrossRef]

41. Jung, T.-H.; Park, J.H.; Jeon, W.-M.; Han, K.-S. Butyrate modulates bacterial adherence on LS174T human colorectal cells by stimulating mucin secretion and MAPK signaling pathway. *Nutr. Res. Pract.* **2015**, *9*, 343–349. [CrossRef]
42. Liu, P.; Wang, Y.; Yang, G.; Zhang, Q.; Meng, L.; Xin, Y.; Jiang, X. The role of short-chain fatty acids in intestinal barrier function, inflammation, oxidative stress, and colonic carcinogenesis. *Pharmacol. Res.* **2021**, *165*, 105420. [CrossRef]
43. Mallo, J.J.; Sol, C.; Puyalto, M.; Bortoluzzi, C.; Applegate, T.J.; Villamide, M.J. Evaluation of sodium butyrate and nutrient concentration for broiler chickens. *Poult. Sci.* **2021**, *100*, 101456. [CrossRef]
44. Krix-Jachym, K.; Onyszkiewicz, M.; Swiatkiewicz, M.; Fiedorowicz, M.; Sapierzynski, R.; Grieb, P.; Rekas, M. Evaluation of butyric acid as a potential supportive treatment in anterior uveitis. *Ophthalmol. J.* **2022**, *7*, 117–126. [CrossRef]
45. Pham, V.T.; Fehlbaum, S.; Seifert, N.; Richard, N.; Bruins, M.J.; Sybesma, W.; Rehman, A.; Steinert, R.E. Effects of colon-targeted vitamins on the composition and metabolic activity of the human gut microbiome—A pilot study. *Gut Microbes* **2021**, *13*, 1875774. [CrossRef]
46. Makowski, Z.; Lipiński, K.; Mazur-Kuśnirek, M. The Effects of sodium butyrate, coated sodium butyrate, and butyric acid glycerides on nutrient digestibility, gastrointestinal function, and fecal microbiota in turkeys. *Animals* **2022**, *12*, 1836. [CrossRef]
47. Xu, Y.; Yu, Y.; Shen, Y.; Li, Q.; Lan, J.; Wu, Y.; Zhang, R.; Cao, G.; Yang, C. Effects of Bacillus subtilis and Bacillus licheniformis on growth performance, immunity, short chain fatty acid production, antioxidant capacity, and cecal microflora in broilers. *Poult. Sci.* **2021**, *100*, 101358. [CrossRef]
48. Yao, H.; Zhang, D.; Yu, H.; Yuan, H.; Shen, H.; Lan, X.; Liu, H.; Chen, X.; Meng, F.; Wu, X. Gut microbiota regulates chronic ethanol exposure-induced depressive-like behavior through hippocampal NLRP3-mediated neuroinflammation. *Mol. Psychiatry* **2023**, *28*, 919–930. [CrossRef]
49. Li, C.; Wang, S.; Chen, S.; Wang, X.; Deng, X.; Liu, G.; Chang, W.; Beckers, Y.; Cai, H. Screening and characterization of Pediococcus acidilactici LC-9-1 toward selection as a potential probiotic for poultry with antibacterial and antioxidative properties. *Antioxidants* **2023**, *12*, 215. [CrossRef]
50. Pammi, M.; Cope, J.; Tarr, P.I.; Warner, B.B.; Morrow, A.L.; Mai, V.; Gregory, K.E.; Kroll, J.S.; McMurtry, V.; Ferris, M.J. Intestinal dysbiosis in preterm infants preceding necrotizing enterocolitis: A systematic review and meta-analysis. *Microbiome* **2017**, *5*, 31. [CrossRef]
51. Zhang, B.; Lv, Z.; Li, Z.; Wang, W.; Li, G.; Guo, Y. Dietary l-arginine Supplementation Alleviates the Intestinal Injury and Modulates the Gut Microbiota in Broiler Chickens Challenged by Clostridium perfringens. *Front. Microbiol.* **2018**, *9*, 1716. [CrossRef]
52. Gong, L.; Wang, B.; Zhou, Y.; Tang, L.; Zeng, Z.; Zhang, H.; Li, W. Protective effects of Lactobacillus plantarum 16 and Paenibacillus polymyxa 10 against Clostridium perfringens infection in broilers. *Front. Immunol.* **2021**, *11*, 628374. [CrossRef]
53. Liu, Y.; Zhang, S.; Luo, Z.; Liu, D. Supplemental Bacillus subtilis PB6 improves growth performance and gut health in broilers challenged with Clostridium perfringens. *J. Immunol. Res.* **2021**, *2021*, 2549541. [CrossRef] [PubMed]
54. Wu, M.; Cao, W.; Chen, X.; She, J. Intelligent Optimization and Control for Reheating Furnaces. In *Intelligent Optimization and Control of Complex Metallurgical Processes*; Springer: Singapore, 2020; pp. 223–271. [CrossRef]

animals MDPI

Technical Note

A Technical Report on the Potential Effects of Heat Stress on Antioxidant Enzymes Activities, Performance and Small Intestinal Morphology in Broiler Chickens Administered Probiotic and Ascorbic Acid during the Hot Summer Season

Victory Osirimade Sumanu [1,*], Vinny Naidoo [2], Marinda Oosthuizen [3] and Joseph Panashe Chamunorwa [1]

1 Department of Anatomy and Physiology, Faculty of Veterinary Science, University of Pretoria, Onderstepoort, Pretoria P.O. Box 14679, South Africa; joseph.chamunorwa@up.ac.za
2 Department of Paraclinical Sciences, Faculty of Veterinary Science, University of Pretoria, Onderstepoort, Pretoria P.O. Box 14679, South Africa; vinny.naidoo@up.ac.za
3 Department of Veterinary Tropical Diseases, Faculty of Veterinary Science, University of Pretoria, Onderstepoort, Pretoria P.O. Box 14679, South Africa; marinda.oosthuizen@up.ac.za
* Correspondence: u20741139@tuks.co.za; Tel.: +27-617873794

Simple Summary: Thermal stress is an environmental factor that negatively affects poultry production globally. It elicits behavioural and physiological changes in broiler chickens, hence the need to find ways of ameliorating its detrimental effects which are mainly expressed as oxidative stress. This study was designed as an intervention on the effect of heat stress during the hot summer season in broiler chickens' production using probiotic and ascorbic acid as anti-stress agents. From the results, probiotic and/or ascorbic acid were effective in enhancing the antioxidant enzyme activities and performance of the broiler chickens. This study stands as a basis for application in animal production trials with a larger sample size.

Abstract: Oxidative stress negatively affects the welfare of broiler chickens leading to poor productivity and even death. This study examined the negative effect of heat stress on antioxidant enzyme activities, small intestinal morphology and performance in broiler chickens administered probiotic and ascorbic acid during the hot summer season, under otherwise controlled conditions. The study made use of 56 broiler chickens; which were divided into control; probiotic (1 g/kg); ascorbic acid (200 mg/kg) and probiotic + ascorbic acid (1 g/kg and 200 mg/kg, respectively). All administrations were given via feed from D1 to D35 of this study. Superoxide dismutase, glutathione peroxidase and catalase activities were highly significant ($p < 0.0001$) in the treatment groups compared to the control. Performance indicators (water intake and body weight gain) were significantly higher ($p < 0.05$) in the probiotic and probiotic + ascorbic acid group. The height of duodenal, jejunal and ileal villi, and goblet cell counts of broiler chickens were significantly different in the treatment groups. In conclusion, the study showed that heat stress negatively affects the levels of endogenous antioxidant enzymes, performance and the morphology of small intestinal epithelium, while the antioxidants were efficacious in ameliorating these adverse effects.

Keywords: ascorbic acid; probiotic; antioxidant enzymes; performance; small intestinal morphology

Citation: Sumanu, V.O.; Naidoo, V.; Oosthuizen, M.; Chamunorwa, J.P. A Technical Report on the Potential Effects of Heat Stress on Antioxidant Enzymes Activities, Performance and Small Intestinal Morphology in Broiler Chickens Administered Probiotic and Ascorbic Acid during the Hot Summer Season. *Animals* **2023**, *13*, 3407. https://doi.org/10.3390/ani13213407

Academic Editors: Janghan Choi and Jeffrey Downing

Received: 5 August 2023
Revised: 24 October 2023
Accepted: 31 October 2023
Published: 2 November 2023

1. Introduction

Thermal stress is a considerable problem experienced globally, especially during the summer in the sub-tropics and tropics [1,2]. Heat stress in poultry production severely affects the profitability of both small- and large-scale farms due to lower productivity and growth in the stressed birds [3,4]. High ambient temperature (AT) induces an increase in the production of reactive oxygen species (ROS) above the levels that the body can tolerate, with resultant oxidative stress [5]. Oxidative stress results from an imbalance between free radicals and antioxidant enzymes in the body system [6]. Oxidative stress induces DNA mutations, damages the respiratory chain and alters the permeability of the membrane in the mitochondria. Of the various effects of oxidative stress, one is lipid peroxidation where the free radicals tend to extract electrons from lipids in the cell membrane which leads to damage in all cell types [7].

Endogenous antioxidants like catalase (CAT), glutathione peroxidase (GPx) and superoxide dismutase (SOD), are important enzymes that improve the immune status of individuals [8,9]. While the concentrations of these enzymes are adequate for preserving normal cellular functionality under normal physiology; however, under increased ROS production such as during heat stress, they are often depleted necessitating the need for exogenous antioxidants [10,11] or for precursors to be administered [12,13]. The small intestine is a vital organ involved with the role of digestion and absorption of nutrients from the diet, alterations in its function negatively impact other organ systems' functions in an individual. Oxidative stress adversely affects the intestinal integrity of broiler chickens necessitating the administration of anti-stress agents to protect the morphology of the intestine [14,15]. Although some osmolytes (betaine), phytochemicals (lycopene), electrolytes (bicarbonate ions), minerals and vitamins were reported to be effective in mitigating heat stress, not all were found to be effective in the gut [16].

Probiotics generally provide protection against gut pathogens, improve the intestinal microbial balance, and modulate the immune system [17,18]. A probiotic is a living organism that elicits an antioxidant, anti-stress and a growth-promoting effect in a host. They are rich in vitamin B, protein, fat and enzymes such as phytase and cellulase [19–21]. Ascorbic acid, a water-soluble vitamin, serves as both an antioxidant and anti-stress agent, it protects the body against ROS's deleterious effects by preventing the formation of excessive free radicals above what the body can cope with [22].

Our aim was to intervene in broiler chicken production by alleviating the adverse effects of heat stress during summer under semi-controlled conditions via the administration of probiotic and/or ascorbic acid in their feed. This study is interventionist by design and allowed the animals to be closely monitored, repeatedly sampled and to minimize the potential of additional stressors associated with large group housing.

2. Materials and Methods

2.1. Experimental Animals, Management, and Environmental Conditions

A total of 56 (day-old) male and female broiler chicks (Ross 308) with an average weight of 40 g that were apparently healthy according to chick classification methods, were purchased from Gauteng poultry farmers Alfa chicks (215 Loerie Street, Haakdoornboom, Pretoria, South Africa). The sample size was calculated using G*Power software version 3.1.9.7, with an effect size f value of 0.59, critical F of 2.78 and power of 0.95.

The broiler chicks were kept under an intensive management system with prevailing natural environmental conditions at the Onderstepoort Veterinary Animal Research Unit, Faculty of Veterinary Science, University of Pretoria, South Africa. The guidelines of the Research Ethics Committee and Animal Ethics Committee of the University of Pretoria, South Africa (REC050-20) were adhered to. Electric sensors (Hobo) were used to monitor the ambient temperature of the pens. Infra-red bulbs were used as a source of warmth during brooding for 14 days at 32 °C. The relative humidity and ambient temperature of the pen were 75–80% and 30–36 °C, respectively. The broiler chicks were allowed access to feed (Epol commercial feed, Gauteng, South Africa) and water ad libitum. The broiler chicks were divided into 4 groups of 14 each and they were fed with broiler starter (D1–D14), broiler grower (D15–D25) and broiler finisher (D26–D35). The feed analysis is presented in Table 1.

The pen was made of concrete floor, cement blocks, with tunnel ventilation and an aluminium roof. Foot baths (F10SC, Health and Hygiene (Pty) Ltd., Roodepoort, South Africa), protective clothing and footwear were provided for the assistants to enhance biosecurity.

Table 1. Proximate composition of feed.

Feed Composition	Starter	Grower	Finisher
Proximate analysis (%)			
Protein	18.22	18.23	18.24
Total lysine	1.67	1.68	1.69
Total methionine	0.63	0.64	0.65
Fat	3.45	3.46	3.47
Fibre	5.54	5.55	5.56
Calcium	0.58	0.59	0.6
Phosphorus	0.52	0.53	0.54
Metabolizable energy (kcal/kg)	2910	2980	3000

2.2. *Experimental Design*

The broiler chickens (n = 56) were divided into control; probiotic Saccharomyces cerevisiae (1 g/kg of feed) [20]; ascorbic acid (200 mg/kg of feed) [22]; probiotic (Saccharomyces cerevisiae); and ascorbic acid groups (1 g/kg of feed and 200 mg/kg of feed), respectively. These agents were mixed with the feed of the chickens during the study, taking the above dose rates into consideration. Colour markings and wing tags were used as a means of identification for each broiler chicken.

2.3. *Measurement of Oxidative Stress Biomarkers*

2.3.1. Superoxide Dismutase Activity

Seven broiler chickens were randomly selected and fasted at D35 of this study. They were sacrificed via euthanasia with the use of a gas mixture (35% CO_2, 35% N_2 and 30% O_2). The breast muscle tissues were severed, and 5 g of the tissue was placed in tubes and homogenised. SOD activity was determined using an ELISA kit (BIOCOM Africa, Clubview, Gauteng, South Africa). Briefly, reagent 1 (1 mL) was added to 2 mL of the samples in each well. Reagent 2 (0.1 mL), reagent 3 (0.1 mL) and reagent 4 (0.1 mL) were added to the wells. The mixture was gently mixed by shaking and incubated for 40 min at 37 °C. Chromogenic agent (2 mL) was added to each well and mixed for 10 min. The optical density (OD) was determined at 550 nm using a spectrophotometer (UV-VIS, Perlong Medical Equipment Co., Ltd., Nanjing, China).

2.3.2. Catalase Activity

The activity of catalase (CAT) was determined with the aid of a CAT ELISA kit (BIOCOM Africa, Clubview, Gauteng, South Africa). Briefly, reagent 1 (200 µL) was added to about 20 µL of the samples in each well and incubated for 5 min at 37 °C. Reagent 2 (20 µL) was added to the wells and mixed for 1 min; 200 µL of reagent 3 application solution and reagent 4 (20 µL) were added, respectively, to each well. It was mixed and incubated for 10 min (at room temperature). OD was measured at 405 nm with a spectrophotometer.

2.3.3. Glutathione Peroxidase Activity

The activity of glutathione peroxidase (GPx) was determined with the aid of a GPx ELISA kit (BIOCOM Africa). Briefly, 50 µL of the sample was added to the wells and biotinylated detection antibodies (50 µL) were added to each well immediately. The mixture was incubated for 45 min at 37 °C; after that, it was aspirated and washed thrice. HRP conjugate (100 µL) was added to the wells and incubated (for 30 min at 37 °C). The mixture was washed and 90 µL of substrate reagent was added. Stop solution (50 µL) was also added to each well and the OD was determined at 450 nm immediately after, with the aid of a spectrophotometer.

2.4. Measurement of Performance Parameters

2.4.1. Measurement of Feed Intake

Broiler chickens' feed intake was measured daily at 07:00 h. The feed weight before placement, and the remnant feed after 24 h of feeding were measured using a Digital Precision (Mettler Toledo®) weighing balance (Greifensee, Switzerland). Absolute feed intake for each day was calculated as the difference between the amount of left-over feed and the amount of feed supplied to the broiler chickens.

2.4.2. Measurement of Water Intake

A graduated cylinder (Rutland Industries, 8 Theodore Road, Benrose, Johannesburg, South Africa) was used to measure the quantity of water before placement and after consumption (24 h later).

2.4.3. Measurement of Body Weight Gain

Upon arrival at the poultry pen, each broiler chicken was weighed at D1 and these values served as the initial body weight. The broiler chickens' weights were measured once weekly to determine the average weekly body weight gain. The study was terminated after the broiler chickens were euthanized via the use of a gas mixture (35% CO_2, 35% N_2 and 30% O_2).

2.5. Measurement of Small Intestinal Morphology

At D35 of the experiment, seven broiler chickens were selected at random from various groups and fasted (for 12 h). About 3 cm length of the middle portion of the ileum and jejunum and the descending portion of the duodenum for each broiler chicken were taken and stored in a tube containing 10% solution of buffered neutral formaldehyde (pH 7.2–7.4). After dehydration, the samples were placed in paraffin blocks, they were sectioned and stained with periodic acid–Schiff reagent according to the method of Luna [23]. The crypt width and height, villus width and height, villus and crypt surface areas and goblet cells of the small intestinal epithelium were measured at 100× magnification (Sigma Scan, Jandel Scientific, San Rafael, CA, USA).

Transmission electron microscopy was used to evaluate the morphology of the goblet cells. 0.5 × 0.5 × 0.5 cm tissue blocks of the ileum, jejunum and duodenum were taken and inserted into a tube containing 2.5% glutaraldehyde and processed according to the standard method of Cheville and Stasko [24]. Briefly, tiny pieces of the small intestinal tissue were placed into glutaraldehyde and fixed at room temperature for 1 h, and then were post fixed in osmium tetroxide. The fixed cells were embedded in agar and afterward were dehydrated in ethanol-graded concentrations (with propylene oxide). A glass knife was used to cut the tissue sections from blocks for thinning. The thin sections were placed on copper grids; uranyl acetate and lead citrate were used for impregnation and scoping. Stained sections were visualised using a JEOL 1400 electron microscope operated at 80 kV.

2.6. Data Analyses

Antioxidant enzymes and morphological data were subjected to one-way analysis of variance (ANOVA), while the performance parameters were subjected to two-way ANOVA followed by Tukey's multiple comparison post hoc test to compare differences between the treatment and control groups' means. The data obtained were expressed as mean ± standard error of the mean (Mean ± SEM). Version 27 (Armonk, NY, USA: IBM Corp.) SPSS Statistics for Windows software was used for the analysis. Values of $p < 0.05$ were considered significant.

3. Results

3.1. Superoxide Dismutase Enzyme Activity

There were no differences ($p > 0.05$) in SOD activity obtained between the probiotic, probiotic + ascorbic acid and ascorbic acid groups. The activity of SOD was higher

($p < 0.0001$) in the probiotic, probiotic + ascorbic acid and ascorbic acid groups of broiler chickens when compared to the control group (Figure 1).

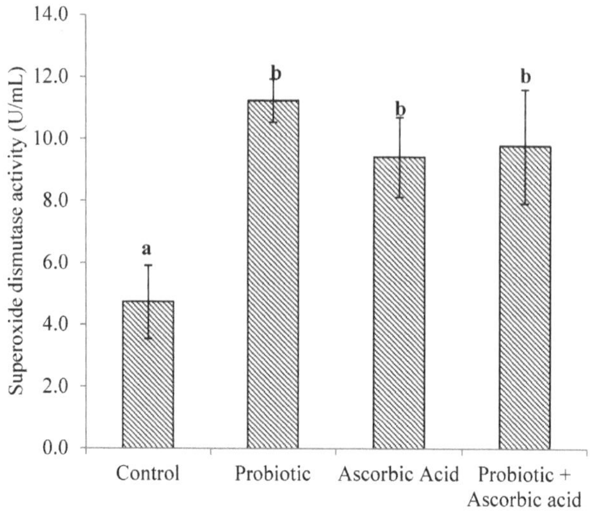

Figure 1. Superoxide dismutase activity in broiler chickens' muscle tissue exposed to heat stress and administered probiotic and ascorbic acid. $p < 0.001$. Data are expressed as Mean ± SEM. Vertical bars with different superscript letters [a,b] are significantly different at $p < 0.05$.

3.2. Catalase Enzyme Activity

There was a significant difference ($p < 0.05$) in CAT activity obtained between the probiotic and ascorbic acid groups. The activity of CAT was higher ($p < 0.0001$) in the probiotic, probiotic + ascorbic acid and ascorbic acid groups of broiler chickens in comparison with the control group (Figure 2).

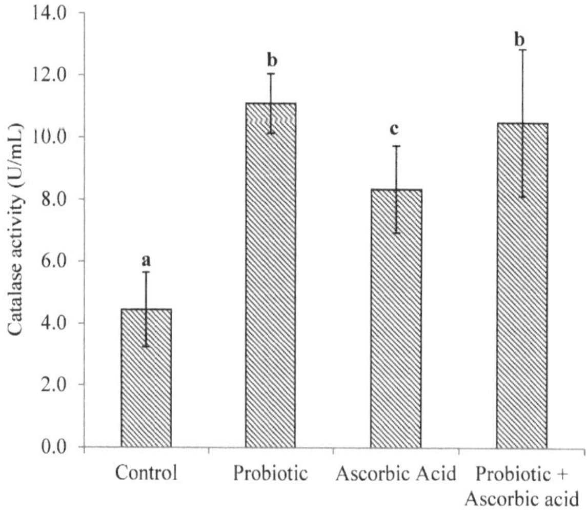

Figure 2. Catalase activity (U/mL) in broiler chickens muscle tissue exposed to heat stress and administered probiotic and ascorbic acid. $p < 0.01$. Data are expressed as Mean ± SEM. Vertical bars with different superscript letters [a,b,c] are significantly different at $p < 0.05$.

3.3. Glutathione Peroxidase Enzyme Activity

The activity of GPx was higher ($p < 0.0001$) in the probiotic + ascorbic acid, probiotic and ascorbic acid groups of broiler chickens when compared to the control. There was no difference ($p > 0.05$) in GPx activity obtained between the treatment groups (Figure 3).

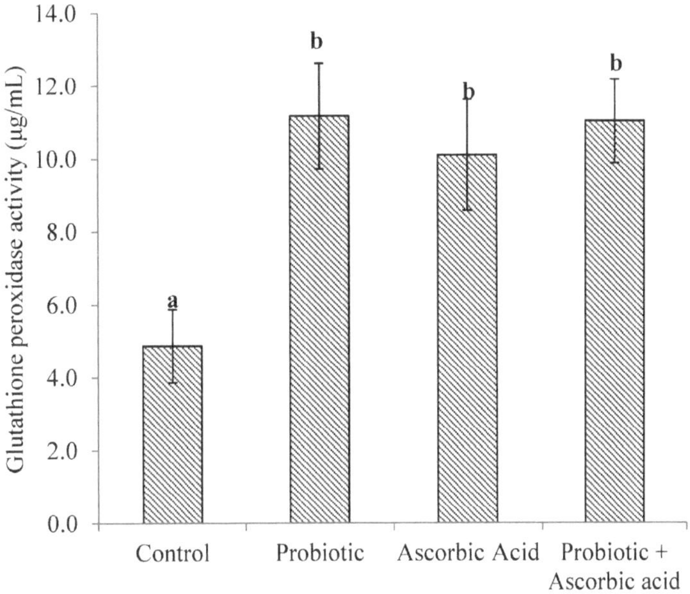

Figure 3. Glutathione peroxidase activity in broiler chickens' muscle tissue exposed to heat stress and administered probiotic and ascorbic acid. $p < 0.002$. Data are expressed as Mean ± SEM. Vertical bars with different superscript letters ^{a,b} are significantly different at $p < 0.05$.

3.4. Feed and Water Intake, and Body Weight Gain

Values of feed intake recorded in the treatment groups were not different ($p > 0.05$) when compared with the control group. Water intake recorded in the ascorbic acid and probiotic + ascorbic acid groups was higher ($p < 0.05$) when compared with the values obtained in the control group at D 21 to D 35. Body weight gain of broiler chickens in the probiotic + ascorbic acid and ascorbic acid groups were not significantly different in comparison with the control, while that of the probiotic group was higher ($p < 0.05$) when compared with the control group at D 28 to D 35 (Table 2).

Table 2. Performance indicators in broiler chickens treated with probiotic and ascorbic acid.

	Group	Day 0–7	Day 8–14	Day 15–21	Day 22–28	Day 29–35
Feed intake (g)	Control	675.14 ± 157.20	1027.57 ± 58.95	1836.71 ± 80.19	2505.43 ± 70.01	2656.86 ± 45.88
	Probiotic	674.43 ± 137.40	981.57 ± 64.21	1923.00 ± 134.98	2821.86 ± 41.10	3126.71 ± 56.54
	Ascorbic acid	708.71 ± 153.98	1045.00 ± 64.21	2327.71 ± 175.52	2626.00 ± 31.61	2792.29 ± 70.02
	Prob + AA	672.29 ± 139.07	1322.29 ± 71.42	2066.57 ± 87.43	2571.43 ± 16.12	2920.71 ± 62.78
Water intake (mL)	Control	1400.00 ± 293.58^{a}	2542.86 ± 184.98^{a}	4585.71 ± 280.67^{a}	5928.57 ± 184.80^{a}	7350.00 ± 227.26^{a}
	Probiotic	2000.00 ± 243.98^{b}	2971.43 ± 176.90^{a}	4900.00 ± 325.87^{a}	6828.57 ± 164.34^{b}	8628.57 ± 316.77^{a}
	Ascorbic acid	1942.86 ± 220.23^{a}	3200.00 ± 211.57^{b}	5142.86 ± 348.37^{b}	7057.14 ± 165.99^{b}	8771.43 ± 222.23^{b}
	Prob + AA	1928.57 ± 229.61^{a}	2971.43 ± 178.24^{a}	5085.71 ± 317.30^{b}	6871.43 ± 171.43^{b}	7471.43 ± 153.86^{b}

Table 2. *Cont.*

	Group	Day 0–7	Day 8–14	Day 15–21	Day 22–28	Day 29–35
Body weight gain (g)	Control	181.50 ± 2.24	505.86 ± 15.14	1031.21 ± 29.89	1756.50 ± 48.23ᵃ	2138.50 ± 68.02ᵃ
	Probiotic	183.93 ± 1.16	515.57 ± 19.33	1289.07 ± 90.23	1822.41 ± 41.00ᵇ	2730.79 ± 55.26ᵇ
	Ascorbic acid	190.07 ± 1.95	521.14 ± 16.33	1063.14 ± 30.99	1768.86 ± 45.56ᵃ	2321.71 ± 58.36ᵃ
	Prob + AA	185.07 ± 2.93	522.21 ± 15.89	1077.64 ± 34.02	1629.02 ± 54.73ᵃ	2432.64 ± 58.82ᵇ

Mean ± SEM with different superscript letters [a,b] within the same column are significantly different at $p < 0.01$. Prob + AA = Probiotic + Ascorbic acid, n = 14.

3.5. Morphological Analysis

Morphological analysis of the villus width and height revealed that villus width and height were different ($p < 0.001$) in the probiotic and probiotic + ascorbic acid groups in comparison with the control group. The crypt width and depth of the ileum, jejunum and duodenum were greater ($p < 0.0001$) in the probiotic and probiotic + ascorbic acid groups in comparison with the control. The probiotic and probiotic + ascorbic acid groups had significantly different surface areas of villi and crypts of the small intestine when compared to the control group. The number of goblet cells of the duodenum, ileum and jejunum was greater ($p < 0.001$) in the treatment groups in comparison with the control group during this study (Table 3).

Table 3. Histomorphometry of the duodenum, ileum and jejunum of broiler chickens treated with probiotic and/or ascorbic acid during heat stress.

	Parameters	Control	Probiotic	Ascorbic Acid	Probiotic + AA
Duodenum	Villus height (µm)	652.18 ± 39.94^{c}	1706.92 ± 129.15^{a}	1427.69 ± 66.69^{b}	1230.55 ± 81.85^{b}
	Villus width (µm)	117.33 ± 6.98^{b}	288.21 ± 53.41^{a}	173.55 ± 8.23^{b}	278.15 ± 44.77^{a}
	Crypt depth (µm)	95.61 ± 12.29^{c}	329.45 ± 37.26^{a}	267.20 ± 53.10^{b}	324.23 ± 43.89^{b}
	Crypt width (µm)	48.16 ± 4.86^{d}	159.57 ± 16.87^{a}	125.07 ± 20.47^{c}	140.95 ± 25.55^{b}
	Villus height/crypt depth	2.40 ± 0.10^{c}	4.03 ± 0.62^{a}	3.24 ± 1.24^{b}	3.03 ± 0.66^{b}
	Villus SA (mm^2)	5.40 ± 1.29^{c}	16.00 ± 2.98^{a}	7.50 ± 3.21^{b}	10.80 ± 1.32^{b}
	Crypt SA (mm^2)	2.80 ± 0.66^{c}	12.80 ± 2.13^{a}	3.80 ± 1.24^{c}	8.40 ± 1.03^{b}
	Goblet cells	58.00 ± 13.99^{c}	101.00 ± 4.92^{b}	108.20 ± 9.32^{a}	102.80 ± 12.14^{b}
Ileum	Villus height (µm)	284.58 ± 56.72^{d}	1025.07 ± 66.80^{a}	678.08 ± 58.14^{c}	865.22 ± 68.37^{b}
	Villus width (µm)	84.70 ± 19.19^{c}	236.10 ± 41.28^{a}	107.93 ± 8.03^{b}	230.98 ± 19.38^{a}
	Crypt depth (µm)	98.95 ± 5.09^{d}	159.73 ± 25.54^{b}	135.13 ± 21.12^{c}	231.73 ± 35.60^{a}
	Crypt width (µm)	40.80 ± 5.94^{c}	88.70 ± 10.96^{a}	65.38 ± 5.05^{b}	92.28 ± 7.71^{a}
	Villus height/crypt depth	1.22 ± 0.01^{c}	2.41 ± 0.10^{a}	1.49 ± 5.01^{b}	2.00 ± 1.01^{b}
	Villus SA (mm^2)	3.60 ± 0.93^{b}	10.40 ± 1.44^{a}	7.60 ± 1.60^{a}	9.00 ± 1.22^{a}
	Crypt SA (mm^2)	3.60 ± 0.68^{b}	7.40 ± 1.12^{a}	3.60 ± 1.21^{b}	7.40 ± 0.93^{a}
	Goblet cells	52.40 ± 9.69^{c}	183.20 ± 20.61^{a}	125.40 ± 22.65^{b}	122.20 ± 13.40^{b}
Jejunum	Villus height (µm)	362.95 ± 68.90^{c}	1205.70 ± 105.67^{a}	1010.33 ± 62.26^{b}	1066.50 ± 51.31^{b}
	Villus width (µm)	102.81 ± 11.09^{c}	204.58 ± 20.54^{b}	202.58 ± 37.60^{b}	220.48 ± 35.19^{a}
	Crypt depth (µm)	123.46 ± 11.16^{c}	303.43 ± 36.83^{a}	240.39 ± 39.34^{b}	279.19 ± 26.99^{b}
	Crypt width (µm)	48.47 ± 4.33^{c}	98.71 ± 14.12^{a}	84.21 ± 5.62^{b}	88.41 ± 9.57^{b}
	Villus height/crypt depth	1.10 ± 0.01^{c}	2.44 ± 3.00^{a}	1.90 ± 1.02^{b}	2.00 ± 0.01^{b}
	Villus SA (mm^2)	2.70 ± 0.49^{c}	10.20 ± 1.39^{b}	11.40 ± 3.06^{a}	9.80 ± 0.86^{b}
	Crypt SA (mm^2)	2.80 ± 0.66^{c}	8.20 ± 1.28^{a}	6.40 ± 1.57^{b}	7.80 ± 1.02^{a}
	Goblet cells	56.00 ± 8.15^{c}	150.60 ± 20.92^{a}	93.00 ± 9.95^{b}	102.60 ± 6.37^{b}

Mean ± SEM with different superscript letters [a,b,c,d] along the same row are significantly different. n = 7. AA = ascorbic acid, SA = surface area.

The histological representations of the effect of heat stress on the jejunal epithelium of broiler chickens in the treatment and control groups are shown in Figures 4 and 5.

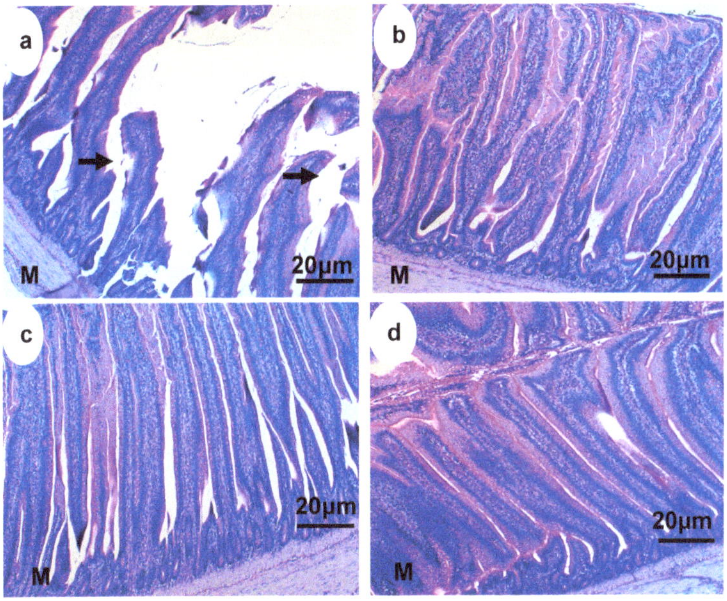

Figure 4. Light microscopy showing the jejunal epithelia of broiler chickens exposed to heat stress: (**a**) control group, (**b**) probiotic, (**c**) ascorbic acid, and (**d**) probiotic and ascorbic acid. Arrows = widening of capillary lacteal due to distortion of epithelia, M = muscularis mucosa.

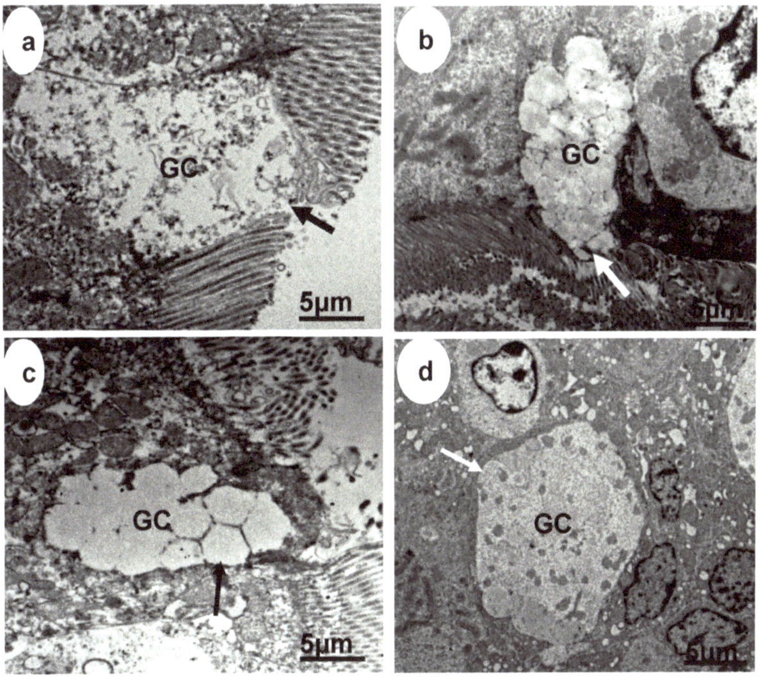

Figure 5. Electron micrographs of jejunum of broiler chickens exposed to heat stress (**a**) without treatment, a distorted goblet cell without a defined contour and absence of mucin (thick black arrow);

(**b**) treated with probiotic, showing the goblet cell and presence of mucin (thick white arrow); (**c**) treated with ascorbic acid, and there was no presence of mucin but presence of dilations within the lobules of the goblet cell (thin black arrow); and (**d**) treated with probiotic and ascorbic acid, in which the goblet cell has a well-defined contour (thin white arrow), GC = goblet cell.

4. Discussion

4.1. Antioxidant Enzyme Activities

The decreased values of SOD, CAT and GPx were evident that the broiler chickens were negatively affected by heat stress. Increased ROS production which occurs during heat stress generally impacts the activities of endogenous enzymes if no exogenous source is used for augmentation in broiler chickens [25]. As shown from previous studies, heat stress stimulates the process of lipid peroxidation which occurs due to the depletion of the antioxidant defence system in broiler chickens [26]. Xue et al. [27] also reported a drastic depletion in the antioxidant activities of broiler chickens exposed to heat stress. They attributed this to the fact that depletion of the antioxidant defence system occurs during heat stress, resulting in cellular damage from rapid oxidation processes. From the treatment groups, the probiotic group had the highest SOD, CAT and GPx activities followed by the combination and the ascorbic acid groups. Antioxidants directly react with free radicals converting them to nonradical products that are stable [28].

According to Winiarska-Mieczan et al. [29], exogenous antioxidants induce the activities of SOD, CAT and GSH in the muscle tissue of broiler chickens leading to the protection of the cellular membranes from peroxidation. Kumar et al. [30] stated in their findings that the use of ascorbic acid and Moringa oleifera in broiler chickens raised in the tropics augmented the activities of CAT, SOD and GSH, and lipid peroxidation was reduced significantly. This was based on the ability of the above antioxidants to attach to the cytoplasmic membrane and inhibit oxidases. Deng et al. [31] reported that the supplementation of broiler chickens' diet with yeast probiotic improved their antioxidant status, which is suggestive of their strong antioxidant potential as compared to selenium which had no effect. This was due to the ability of yeast probiotic to scavenge free radicals during normal metabolism in the body.

4.2. Performance Indicators

Generally, it is assumed that the non-significant value in feed intake observed between the control and treatment groups might be due to the degree of high ambient temperature the broiler chickens were exposed to as water intake is prioritized over feed intake during prolonged exposure to thermal stress. The finding agrees with that of Moataz et al. [32], who reported decreased feed intake in layers exposed to heat stress when compared with the groups that were administered the *Bacillus subtilis* probiotic. He associated the decrease to the degree of heat stress the layers were exposed to.

The decreased water intake observed in the control may be based on their low feed intake, as the quantity of feed intake may be directly proportional to water intake to further enhance digestibility. Ascorbic acid-administered group had an increase in water intake, which could be supported by the fact that feed intake stimulates an increase in water intake. Generally, increased water intake obtained in the treatment groups could also be assumed as a thermoregulatory measure to rid the body of excessive heat during the period of study. The increase in water intake recorded in the treatment groups agrees with that of Egbuniwe et al. [33] who reported an increase in water intake in the group of chickens administered ascorbic acid and/or betaine than the control group. They also attributed the increase to a means of thermoregulation.

Body weight gain obtained in the control group was lowest during the study period. This may be in line with the lower feed (though not significantly different) and water intake obtained when compared with the probiotic and probiotic + ascorbic acid groups. It is important to note that the Ross breed of broiler chickens utilized for this study was reported to attain the highest weight of 2.5 kg at D42 under a thermoneutral condition according to the Ross broiler pocket guide [34], but the probiotic-supplemented group had the highest body weight gain (2.8 kg) at D35 (5 weeks) of this study. This increase may be linked to the fact that probiotics are gut effective and increased water intake may further enhance the rate at which ingested feed is utilized for optimum body weight gain Aluwong et al. [18]. Additionally, probiotics could be vital in the process of nutrient digestion due to the optimization of the gut microflora, although its effect on the transportation of nutrients within the body system remains a grey area for further examination.

To the best of our knowledge, most researchers conducted their studies using probiotics to optimize the rate of performance in broiler chickens within 42 days (6 weeks), but our study achieved better outcomes within 35 days with an average weight of 2.7 kg being recorded in the probiotic-administered group. Olnood et al. [35] and Aluwong et al. [36] speculated that the administration of probiotics via gavage is the fastest route that promotes broiler chickens' performance. Nevertheless, we administered the yeast probiotic via feed (oral route) and from the results, improved performance indices were observed in this group of broiler chickens during our study. Wang et al. [37] reported that probiotics (*Bacillus subtilis*) increased the body weight gain of broiler chickens to 2.3 kg at D43 of their study and this was attributed to the gut enhancement effect of the additive. Increased body weight gain attained in the ascorbic acid group might be due to the improved feed and water intake recorded in this group based on its antioxidant effects in improving wellness which could further stimulate an improvement in performance. The finding is in line with that of Egbuniwe et al. [33] who reported an increase in body weight gain in a group of chickens administered ascorbic acid singly during the heat stress, they presumed the increase to be a result of the anti-stress effect of the agent.

4.3. Small Intestinal Morphology

In addition to the changes in the specific markers, the villus height and width, crypt depth and width were shortened, while the villus and crypt surface areas and goblet cell count of the duodenum, jejunum and ileum in the control group were decreased. Intestinal distortion is basically a primary response to heat stress, and this could further compromise the barrier function and integrity of the intestine in broiler chickens as observed in this group in comparison with the treatment group.

Also, the jejunal epithelia lacteal of the control group of broiler chickens was distorted based on the adverse effects of heat stress. The epithelia lacteal is responsible for the absorption of fatty acids and cholesterol in the small intestine, therefore, any form of alteration in their morphology could affect the process of absorption. The gastrointestinal tract has been reported to be more susceptible to any form of stressors, especially heat stress because it impairs the integrity of the epithelium via decreasing crypt depth and villus height [38–40]. Zampiga et al. [41] stated that heat stress induces damage in the small intestinal mucosa of broiler chickens by increasing the permeability of the intestine to endotoxins which subsequently leads to a reduction in their growth performance.

The treatment groups, especially the probiotic-administered broiler chickens had an improved morphology in comparison with the ascorbic acid group. It could be speculated that ascorbic acid was not as effective as probiotics in the gut due to the poor epithelial turnover obtained during the study. Enhanced epithelial turnover directly influences the height of the small intestinal epithelium via the stem cells of the crypts of Lieberkuhn that is responsible for the rapid renewal of the functional villus epithelium as yeast probiotics generally improve the intestinal epithelial cell integrity in broiler chickens [40–42]. In broiler chickens, the stimulation of mixing movements exposes digesta to the enzymes and improves the process of digestion and absorption of nutrients, this could further enhance the growth performance of broiler chickens. The presence of adequate mucins in the goblet cells of broiler chickens treated with probiotics could have influenced their growth rate as they are responsible for protecting the epithelium and regulating the concentration and passage of some immune mediators (antimicrobial peptides), ions and water [43]. Sen et al. [44] suggested that increased intestinal morphology enhances the digestion and absorption capacity of the small intestine in broiler chickens supplemented with *Bacillus subtilis*.

Even though changes to morphology can account for part of the mechanisms of action of probiotics, the specific mechanisms by which it influences transport across the epithelium remain an area for further research.

5. Conclusions

Heat stress negatively affects antioxidant enzymes, performance and small intestinal morphology of the broiler chickens, although probiotic and/or ascorbic acid administrations were efficacious in alleviating the detrimental effects of heat stress during this study. This was evident by an increase in endogenous antioxidant enzyme activities. The probiotic-administered group had the highest body weight gain, followed by the co-administered and ascorbic acid groups, respectively. The morphology of the small intestinal epithelium was enhanced by these treatments in the phase of heat stress exposure. Also, it would be important to determine if the effect of probiotic and/or ascorbic acid evident under controlled conditions would be translatable to open ventilated broiler houses where birds would be exposed to both production conditions and higher thermal conditions.

Author Contributions: Conceptualization, J.P.C., V.N., M.O. and V.O.S.; Methodology, V.O.S., V.N., M.O. and J.P.C.; Investigation, V.O.S., J.P.C., V.N. and M.O.; Resources, J.P.C., V.N. and M.O.; Data Analysis, V.N.; Writing of Manuscript, V.O.S., V.N., J.P.C. and M.O.; Supervision, J.P.C., V.N. and M.O. All authors have read and agreed to the published version of the manuscript.

Funding: This research was financially supported by the University of Pretoria Doctoral Research Bursary and the Department of Anatomy and Physiology, Faculty of Veterinary Science, University of Pretoria, South Africa.

Institutional Review Board Statement: The study was conducted in accordance with the guidelines of the Research Ethics Committee and Animal Ethics Committee of the University of Pretoria, South Africa (REC050-20, 4 June 2021).

Informed Consent Statement: Not applicable.

Data Availability Statement: Data are contained within the article.

Acknowledgments: The technical support by Antoinette Lensink is highly appreciated.

Conflicts of Interest: The authors declare no conflict of interest.

References

1. Elbaz, A.M.; Ahmed, A.M.; Abdel-Maqsoud, A.; Badran, A.M.; Abdel-Moneim, A.M.E. Potential ameliorative role of Spirulina platensis in powdered or extract forms against cyclic heat stress in broiler chickens. *Environ. Sci. Pollut. Res.* **2022**, *29*, 45578–45588. [CrossRef] [PubMed]
2. Madkour, M.; Aboelazab, O.; Abd El-Azeem, N.; Younis, E.; Shourrap, M. Growth performance and hepatic antioxidants responses to early thermal conditioning in broiler chickens. *J. Anim. Physiol. Anim. Nutr.* **2023**, *107*, 182–191. [CrossRef] [PubMed]
3. Goo, D.; Kim, J.H.; Park, G.H.; Delos Reyes, J.B.; Kil, D.Y. Effect of heat stress and stocking density on growth performance, breast meat quality, and intestinal barrier function in broiler chickens. *Animals* **2019**, *9*, 107. [CrossRef]
4. Kim, H.S.; Kim, J.; Choi, Y.H. Characterization of differential gene expression of broiler chicken to thermal stress in discrete developmental stages. *Anim. Cells Syst.* **2022**, *26*, 62–69. [CrossRef] [PubMed]
5. Gopi, M.; Dutta, N.; Rokade, J.J.; Prabakar, G.; Kumar, R.D.; Beulah, P.; Mohan, J. Dietary supplementation of polyphenols alleviates the negative effects of heat stress in broilers. *Biol. Rhythm. Res.* **2019**, *53*, 535–546. [CrossRef]
6. Israr, M.; Chand, N.; Khan, R.U.; Alhidary, I.A.; Abdelrahman, M.M.; Al-Baddani, H.H.; Tufarelli, V. Dietary grape (*Vitis vinifera*) seed powder and zn–gly chelate complex for mitigating heat stress in broiler chickens: Growth parameters, malondialdehyde, paraoxonase-1, and antibody titer. *Agriculture* **2021**, *11*, 1087. [CrossRef]
7. Alagawany, M.; Ashour, E.A.; El-Fakhrany, H.H.H.; Ismail, T.A.; Nasr, M. Early nutrition programming with Astragalus membranaceus polysaccharide: Its effect on growth, carcasses, immunity, antioxidants, lipid profile and liver and kidney functions in broiler chickens. *Anim. Biotechnol.* **2022**, *33*, 362–368. [CrossRef]
8. Abeyrathne, E.S.; Nam, K.; Ahn, D.U. Analytical methods for lipid oxidation and antioxidant capacity in food systems. *Antioxidants* **2021**, *10*, 1587. [CrossRef]

9. Lee, D.; Lee, H.J.; Jung, D.Y.; Kim, H.J.; Jang, A.; Jo, C. Effect of an animal-friendly raising environment on the quality, storage stability, and metabolomic profiles of chicken thigh meat. *Food Res. Int.* **2022**, *155*, 111046. [CrossRef]
10. Jiang, S.; Yan, F.F.; Hu, J.Y.; Mohammed, A.; Cheng, H.W. *Bacillus subtilis*-based probiotic improves skeletal health and immunity in broiler chickens exposed to heat stress. *Animals* **2021**, *11*, 1494. [CrossRef]
11. Attia, Y.A.; Al Khalaifah, H.; Abd El-Hamid, H.S.; Al-Harthi, M.A.; Alyileili, S.R.; El-Shafey, A.A. Antioxidant status, blood constituents and immune response of broiler chickens fed two types of diets with or without different concentrations of active yeast. *Animals* **2022**, *12*, 453. [CrossRef] [PubMed]
12. Gopi, M.; Prabakar, G.; Kolluri, G.; Tamilmani, T.; Tyagi, J.S.; Mohan, J. Effect of organic chromium (chromium picolinate) supplementation on production parameters, egg quality attributes and serum biochemistry in Pearl guinea fowls during second laying cycle. *Indian J. Anim. Sci.* **2018**, *88*, 823–827. [CrossRef]
13. Ebeid, T.A.; Al-Homidan, I.H. Organic acids and their potential role for modulating the gastrointestinal tract, antioxidative status, immune response, and performance in poultry. *World's Poult. Sci. J.* **2022**, *78*, 83–101. [CrossRef]
14. Jahejo, A.R.; Leghari, I.; Sethar, A.; Rao, M.; Nisa, M.; Sethar, G. Effect of heat stress and ascorbic acid on gut morphology of broiler chicken. *Sindh Univ. Res. J.* **2016**, *48*, 829–883.
15. Chaudhary, A.; Mishra, P.; Amaz, S.A.; Mahato, P.L.; Das, R.; Jha, R.; Mishra, B. Dietary supplementation of microalgae mitigates the negative effects of heat stress in broilers. *Poult. Sci.* **2023**, *102*, 102958. [CrossRef]
16. Wasti, S.; Sah, N.; Mishra, B. Impact of heat stress on poultry health and performances, and potential mitigation strategies. *Animals* **2020**, *10*, 1266. [CrossRef]
17. Jiang, S.; Mohammed, A.A.; Jacobs, J.A.; Cramer, T.A.; Cheng, H.W. Effect of synbiotics on thyroid hormones, intestinal histomorphology, and heat shock protein 70 expression in broiler chickens reared under cyclic heat stress. *Poult. Sci.* **2020**, *99*, 142–150. [CrossRef]
18. Aluwong, T.; Kawu, M.; Raji, M.; Dzenda, T.; Govwang, F.; Sinkalu, V.; Ayo, J.O. Effect of yeast probiotic on growth, antioxidant enzyme activities and malondialdehyde concentration of broiler chickens. *Antioxidants* **2013**, *2*, 326–339. [CrossRef]
19. Sumanu, V.O.; Tagang, A.; Ayo, J.O.; Ogbuagu, N.E. Effects of probiotic and fisetin supplementation on performance, carcass characteristics and small intestinal morphology in broiler chickens. *Open Vet. Sci.* **2021**, *2*, 23–32. [CrossRef]
20. Parlat, S.S.; Ozcan, M.; Oguz, H. Biological suppression of aflatoxicosis in Japanese quail (*Coturnixcoturnix japonica*) by dietary addition of yeast (*Saccharomyces cerevisiae*). *Res. Vet. Sci.* **2001**, *71*, 207–211. [CrossRef]
21. Gouda, A.; Amer, S.A.; Gabr, S.; Tolba, S.A. Effect of dietary supplemental ascorbic acid and folic acid on the growth performance, redox status, and immune status of broiler chickens under heat stress. *Trop. Anim. Health Prod.* **2020**, *52*, 2987–2996. [CrossRef] [PubMed]
22. Egbuniwe, I.C.; Uchendu, C.N.; Obidike, I.R. Ameliorative effects of betaine and ascorbic acid on endocrine and erythrocytic parameters of sexually-maturing female Japanese quails during the dry season. *J. Therm. Biol.* **2021**, *96*, 102812. [CrossRef]
23. Luna, L.G. *Manual of Histologic Staining Methods of the Armed Forces Institute of Pathology*; 152.91 M31 Ed.2; McGraw-Hill: New York, NY, USA, 1968; pp. 37–39.
24. Cheville, N.F.; Stasko, J. Techniques in electron microscopy of animal tissue. *Vet. Pathol.* **2014**, *51*, 28–41. [CrossRef] [PubMed]
25. Wang, J.; Ishfaq, M.; Miao, Y.; Liu, Z.; Hao, M.; Wang, C.; Chen, X. Dietary administration of *Bacillus subtilis* KC1 improves growth performance, immune response, heat stress tolerance, and disease resistance of broiler chickens. *Poult. Sci.* **2022**, *101*, 101693. [CrossRef]
26. Ahmed-Farid, O.A.; Salah, A.S.; Nassan, M.A.; El-Tarabany, M.S. Effects of chronic thermal stress on performance, energy metabolism, antioxidant activity, brain serotonin, and blood biochemical indices of broiler chickens. *Animals* **2021**, *11*, 2554. [CrossRef]
27. Xue, B.; Song, J.; Liu, L.; Luo, J.; Tian, G.; Yang, Y. Effect of epigallocatechin gallate on growth performance and antioxidant capacity in heat-stressed broilers. *Arch. Anim. Nutr.* **2017**, *71*, 362–372. [CrossRef]
28. Decker, E.; Beecher, G.; Slavin, J.; Miller, H.E.; Marquart, L. Whole grains as a source of antioxidants. *Cereal Foods World* **2002**, *47*, 370–373.
29. Winiarska-Mieczan, A.; Kwiecien, M.; Mieczan, T.; Kwiatkowska, K.; Jachimowicz, K. The effect of Cu, Zn and Fe chelates on the antioxidative status of thigh meat of broiler chickens. *Animal* **2021**, *15*, 34–44. [CrossRef]
30. Kumar, R.; Kumar, K.; Kumar, A. Nutritional and physiological responses of broiler chicken to the dietary supplementation of *Moringa oleifera* aqueous leaf extract and ascorbic acid in tropics. *Trop. Anim. Health Prod.* **2021**, *53*, 428–435. [CrossRef]
31. Deng, S.; Hu, S.; Xue, J.; Yang, K.; Zhuo, R.; Xiao, Y.; Fang, R. Productive performance, serum antioxidant status, tissue selenium deposition, and gut health analysis of broiler chickens supplemented with selenium and probiotics—A pilot study. *Animals* **2022**, *12*, 1086. [CrossRef]
32. Moataz, F.; Ibrahim, A.; Abdelaziz, A.; Tarek, E.; Osama, A.; Ahmed, A. Effects of dietary probiotic (*Bacillus subtilis*) supplementation on productive performance, immune response, and egg quality characteristics in laying hens under high ambient temperature. *Ital. J. Anim. Sci.* **2018**, *17*, 804–814.
33. Egbuniwe, I.C.; Ayo, J.O.; Kawu, M.U.; Mohammed, A. Behavioral and hematological responses of broiler chickens administered with betaine and ascorbic acid during hot-dry season. *J. Appl. Anim. Welfare Sci.* **2018**, *4*, 334–346. [CrossRef] [PubMed]
34. Ross Broiler Pocket Guide. 2020. Available online: https://en.aviagen.com/assets/Tech_Center/Ross_Broiler/Ross-Broiler-Pocket-Guide-2020-EN.pdf (accessed on 6 September 2022).

35. Olnood, C.G.; Beski, S.S.M.; Iji, P.A.; Choct, M. Delivery routes for probiotics: Effects on broiler performance, intestinal morphology and gut microflora. *Anim. Nutr.* **2015**, *1*, 192–202. [CrossRef] [PubMed]
36. Aluwong, T.; Sumanu, V.O.; Ayo, J.O.; Ocheja, B.; Zakari, F.; Minka, N. Daily rhythms of cloacal temperature in broiler chickens of different age groups administered with zinc gluconate and probiotic during the hot-dry season. *Physiol. Rep.* **2017**, *5*, e13314. [CrossRef] [PubMed]
37. Wang, W.C.; Yan, F.F.; Hu, J.Y.; Amen, O.A.; Cheng, H.W. Supplementation of *Bacillus subtilis*-based probiotic reduces heat stress-related behaviors and inflammatory response in broiler chickens. *J Anim Sci.* **2018**, *96*, 1654–1666. [CrossRef]
38. Nanto-Hara, F.; Kikusato, M.; Ohwada, S.; Toyomizu, M. Heat stress directly affects intestinal integrity in broiler chickens. *J. Poult. Sci.* **2020**, *57*, 284–290. [CrossRef]
39. Mazzoni, M.; Zampiga, M.; Clavenzani, P.; Lattanzio, G.; Tagliavia, C.; Sirri, F. Effect of chronic heat stress on gastrointestinal histology and expression of feed intake-regulatory hormones in broiler chickens. *Animal* **2022**, *16*, 100600. [CrossRef] [PubMed]
40. Song, J.; Jiao, L.F.; Xiao, K.; Luan, Z.S.; Hua, C.H.; Shi, B. Cello-ligosaccharide ameliorates heat stress-induced impairment of intestinal microflora, morphology and barrier integrity in broilers. *Anim. Feed Sci. Technol.* **2013**, *185*, 175–181. [CrossRef]
41. Zampiga, M.; Laghi, L.; Zhu, C.; Mancinelli, A.C.; Mattioli, S.; Sirri, F. Breast muscle and plasma metabolomics profile of broiler chickens exposed to chronic heat stress conditions. *Animal* **2021**, *15*, 100275. [CrossRef]
42. Li, C.; Wang, J.; Zhang, H.; Wu, S.; Hui, Q.; Yang, C.; Fang, R.; Qi, G. Intestinal morphologic and microbiota responses to dietary *Bacillus* spp. in a broiler chicken model. *Front. Physiol.* **2019**, *9*, 1968. [CrossRef]
43. Mohsin, M.; Zhang, Z.; Yin, G. Effect of probiotics on the performance and intestinal health of broiler chickens infected with *Eimeria tenella*. *Vaccines* **2022**, *10*, 97. [CrossRef] [PubMed]
44. Sen, S.; Ingale, S.L.; Kim, Y.W.; Kim, J.S.; Kim, K.H.; Lohakare, J.D. Effect of supplementation of *Bacillus subtilis* LS 1-2 to broiler diets on growth performance, nutrient retention, caecal microbiology and small intestinal morphology. *Res. Vet. Sci.* **2012**, *93*, 264–268. [CrossRef] [PubMed]

animals

MDPI

Article

Evaluation of the Interaction between Gum Arabic Addition and Stocking Density on Growth Performance, Carcass Characteristics, and General Health Parameters of Broiler Chickens

Hani H. Al-Baadani *, Rashed A. Alhotan * and Mahmoud M. Azzam

Department of Animal Production, College of Food and Agriculture Science, King Saud University, P.O. Box 2460, Riyadh 11451, Saudi Arabia; mazzam@ksu.edu.sa

* Correspondence: hsaeed@ksu.edu.sa (H.H.A.-B.); ralhotan@ksu.edu.sa (R.A.A.); Tel.: +966-558829648 (H.H.A.-B.)

Simple Summary: Stocking density in broilers (intensive production) is a method to increase meat production and profitability. However, the use of gum arabic as a prebiotic (soluble fiber) is a possible strategy to maintain performance and gut health and prevent physiological stress in this intensive production. Therefore, the aim of this study was to investigate the interaction between the addition of gum arabic and various stocking densities on performance, intestinal morphology, carcass characteristics, lymphoid organs, and selected blood indices of broiler chickens. In this study, gum arabic as a prebiotic was found to improve growth performance, production efficiency, and intestinal morphology, while high stocking density had negative effects on broilers. Further studies are needed to determine the mechanism under various conditions.

Abstract: The present study aims to investigate the interaction between the addition of gum arabic as a prebiotic and various stocking densities on performance indicators, intestinal morphology, carcass characteristics, lymphoid organs, and selected blood indices of broiler chickens. A total of 816 1-day-old male broilers (Ross 308) were used and randomly divided into six blocks as replicates with eight treatments per block (forty-eight floor pens) based on 4 × 2 factorial arrangements with four dietary treatments containing 0.00% (CONT), 0.12% gum arabic (T1), 0.25% gum arabic (T2), and 0.10% commercial prebiotic (T3) and two stocking densities (normal = 28 kg/m^2; high = 50 kg/m^2). All performance indicators were evaluated during the feeding phases. Blood biochemical indicators were analyzed at 36 days of age. At 37 days of age, carcass characteristics, lymphoid organs, and intestinal morphology were measured. On days 1–36, growth performance indicators were negatively affected at high stocking density, but all growth performance indicators except feed intake improved in chickens receiving T1–T3 compared to CONT ($p < 0.05$). The relative weight of total small intestine and weight-to-length ratio showed a significant interaction between treatments and stocking density ($p < 0.05$). A high stocking density decreased pre-slaughter weight, carcass weight, and dressing yield, while legs and thymus increased ($p < 0.05$). None of the interactions or treatments affected carcass characteristics or lymphoid organs ($p > 0.05$). Indicators of blood biochemistry were not affected by treatments, stocking density, or their interaction ($p > 0.05$), except for uric acid, creatinine, and aspartate aminotransferase, which were higher at a high stocking density ($p < 0.05$). In conclusion, gum arabic as a prebiotic improved growth performance, production efficiency, and intestinal morphology in broilers. In contrast, high stocking density negatively affected performance, production efficiency, some blood indices, carcass weight, dressing yield, and intestinal morphology. Further research is needed to determine the mechanism.

Keywords: broiler; performance; blood; carcass; morphology; gum arabic; prebiotic

Citation: Al-Baadani, H.H.; Alhotan, R.A.; Azzam, M.M. Evaluation of the Interaction between Gum Arabic Addition and Stocking Density on Growth Performance, Carcass Characteristics, and General Health Parameters of Broiler Chickens. *Animals* **2023**, *13*, 3024. https://doi.org/10.3390/ani13193024

Academic Editors: Vincenzo Tufarelli and Janghan Choi

Received: 21 August 2023
Revised: 9 September 2023
Accepted: 21 September 2023
Published: 26 September 2023

1. Introduction

The modern poultry meat industry relies on intensive production in confined housing systems with large numbers of birds. Such intensive production maximizes meat production while minimizing costs. However, birds raised in this manner may become stressed and susceptible to disease [1,2]. Therefore, many previous studies continuously try to achieve the highest possible stocking density of broilers in the smallest possible floor space (more kg live weight per 1 m^2) to reduce production costs and increase profitability while maintaining broiler chickens' performance [3–5]. Vargas-Rodrıguez et al. [6] found that stocking density (10 or 16 chickens/m^2) within the normal range had no effect on broiler performance or health. However, exceeding the stocking density of 28 to 40 kg/m^2 at the end of the growing season (marketing age) could affect broiler health and performance [7]. High stocking density negatively affects the balance of gut microbiota of broilers and causes dysbiosis [8]. In addition, the morphology of the small intestine of broilers was affected by high stocking density, which reduced their ability to absorb nutrients [9]. Gum arabic is one of the most important natural prebiotics because it mainly contains soluble fiber such as rhamnose, arabinose, and galactose, which have a beneficial effect on the host by stimulating the activity of beneficial bacteria through their ability to ferment in the cecum of birds [10,11]. In addition, the use of soluble fiber, including gum arabic and prebiotics, is one of the possible approaches to maintaining performance and gut health and preventing physiological stress in broiler chickens [12–14]. Previous studies have shown that use of gum arabic can effectively improve growth performance, small intestine development, nutrient absorption, and some blood biochemical indicators in broiler chickens [15–19]. Another study found that chickens fed various levels of gum arabic had no effect on carcass characteristics [20].

The hypothesis of the present study is that greater crowding (high stocking density) means a less hygienic environment, and this may be a stress factor for broiler chickens. The additional administration of gum arabic as prebiotic may be beneficial in such conditions to help broilers cope with stress. Therefore, the aim of this study was to investigate the effect of gum arabic on growth performance, small intestine morphology, carcass characteristics, lymphoid organs, and selected blood biochemical indicators of broiler chickens raised in floor pens under various stocking density conditions.

2. Materials and Methods

2.1. Ethics Approval

The procedures and samples used in the present study were approved by the Ethics Committee of the Deanery of Scientific Research, King Saud University, Kingdom of Saudi Arabia (internal reference number: KSU-SE-20-39).

2.2. Feed Additives Used

Gum arabic (*Acacia senegal*) was obtained from an export company in Khartoum State, Sudan. The gum arabic in powder form was chemically analyzed according to the method described by Hani et al. [14]. The prebiotic (*Saccharomyces cerevisiae*) was purchased commercially with a description of the chemical analysis and then used according to the manufacturer's recommendation. The chemical composition of gum arabic and prebiotic is shown in Table 1.

Table 1. Chemical composition of gum arabic and commercial prebiotic on dry matter basis.

Item	Gum Arabic [1]	Commercial Prebiotic [2]
Crude Analysis		
Dry matter	90.68	92.00
Crude protein	2.30	28.90
Lipid	0.10	2.57
Ash	5.13	7.8
Mannanoligosacharide	-	16.00

Table 1. *Cont.*

Item	Gum Arabic [1]	Commercial Prebiotic [2]
β-glucan	-	18.00
Rhamnose	8.40	-
Arabinose	26.00	-
Galactose	40.18	-
Glucuronic acid	18.23	-
Amino acid Analysis		
Threonine	0.14	1.06
Glutamic acid	0.17	1.11
Valine	0.13	1.21
Isoleucine	0.03	1.10
Leucine	0.17	1.66
Tyrosine	0.06	0.88
Phenylalanine	0.12	0.93
Histidine	0.15	0.64
Lysine	0.06	1.39
Arginine	0.03	1.02
Cysteine	0.10	0.39
Mineral Analysis		
Calcium	1.10	0.23
Phosphorus	0.60	3.6
Sodium	0.02	0.50
Magnesium	0.46	0.90
Zinc	0.0002	1.5
Iron	0.75	0.71

[1] The chemical composition analysis of gum arabic was performed according to Al-Baadani et al. [14]. [2] The chemical composition analysis of the commercial prebiotic (*Saccharomyces cerevisiae*) was obtained by the manufacturer of the product.

2.3. Study Design, Birds, and Housing

A total of 816 one-day-old male broilers (Ross 308) were housed in floor pens (0.90 m × 0.90 m) based on market weight at 36 days of age in a completely randomized block design with 4 × 2 factorial arrangements with four dietary treatments and two different stocking densities. All chicks were weighed and randomly assigned to six blocks as replicates with eight treatments per block (forty-eight pens). The dietary treatments were as follows: Control (CONT) broilers received a basal diet without additives, while T1 and T2 received a basal diet with 0.12 and 0.25% gum arabic, respectively; T3 received a basal diet with 0.10% inactivated stabilized *Saccharomyces cerevisiae* as a commercial prebiotic (Milan, Italy). Within each dietary treatment, two different stocking densities were performed as follows: a normal stocking density of 28 kg/m^2 (12 chickens/pen) according to Ross broiler management guidelines in houses with controlled environment and a high stocking density of 50 kg/m^2 (22 chickens/pen).

The basal diet was formulated on form mash in two feeding stages, starter (1–14 days) and grower (15–36 days), to meet the nutrient requirements of Ross 308 (Aviagen, 2019, New York, USA) as indicated in Table 2. Room temperature (°C) and relative humidity (HR%) were set at 35 °C and 22% HR when the chicks arrived (first day) and then decreased by 2 °C every three days until they were constant at nearly 22 °C and 53% HR after 25 days. The lighting program was offered for 24 h (30–40 lux) until 7 days of age, then 23 h of light (minimum 20 lux) and 1 h of darkness from 8 to 36 days. Feed and water were available to all birds throughout the duration of the study. In addition, there were no differences between stocking density factors in terms of space for feeder and drinker [1]. All chicks were vaccinated against Newcastle disease at the hatchery according to the manufacturer's instructions (Fort Dodge Animal Health-USA, Overland Park, KS, USA).

Table 2. Basal diet components and nutrients content on a dry matter basis.

Basal Diet Components, %	Feeding Stages	
	Starter (1–14 Days)	Grower (15–36 Days)
Corn	53.84	55.50
Soybean meal (48% CP)	38.24	35.90
Soy oil	3.43	4.65
Monocalcium phosphate	1.59	1.41
Limestone	1.54	1.39
DL-Methionine	0.37	0.31
L-Lysine HCL	0.25	0.15
L-Threonine	0.16	0.10
Common salt	0.38	0.38
Vitamin Premix [a]	0.10	0.10
Mineral Premix [b]	0.10	0.10
Choline CL 60%	0.003	0.00
Total	100	100
Calculated nutrient, %		
Metabolizable energy, kcal/kg	3000	3100
Dry matter	88.91	88.98
Crude protein	22.58	21.48
Crude fat	6.18	7.41
Crude fiber	2.55	2.49
Calcium	0.96	0.87
Non-phytate P	0.48	0.43
Digestible lysine	1.28	1.15
Digestible methionine	0.67	0.60
Digestible methionine and cysteine	0.95	0.87
Digestible threonine	0.86	0.77
Digestible arginine	1.32	1.26

[a] Contains (in kg of vitamin premix): Vit. A = 2,400,000 IU; Vit. D = 1,000,000 IU; Vit. E = 16,000 IU; Vit. K = 800 mg; Vit. B1 = 600 mg; Vit. B2 = 1600 mg; Vit. B3 = 8000 mg; Vit. B5 = 3000 mg; Vit. B6 = 1000 mg; Vit. B7 = 40 mg; Vit. B9 = 400 mg; Vit. B12 = 6 mg. [b] Contains in kg of mineral premix: Cu = 2000 mg; Fe = 1200 mg; Mn = 18,000 mg; Se = 60 mg; Zn = 14,000 mg; I = 400 mg; Co = 80 mg.

2.4. General Performance Evaluation

All birds were weighed on days 1, 14, and 36 during the study period to determine their average body weight. Weight gain and feed intake were calculated for the starter (1–14 days), grower (15–36 days), and the entire stages (1–36 days) according to Diler et al. [21]. Feed conversion ratio was determined based on weight gain and feed intake during the feeding stage [22]. Production efficiency index as a parameter of economic, productive, and welfare status of broiler chickens was evaluated during the feeding stage [23]. The cumulative mortality rate for each dietary treatment was expressed as a percentage during the feeding stages [24].

2.5. Small Intestine Morphology

After slaughter (37 days of age), 12 chickens from each treatment were randomly selected to remove the small intestine. The small intestine and its fragments (duodenum, jejunum, and ileum) were measured [25]. The weight and length of duodenum, jejunum, and ileum were expressed as a percentage of the total small intestine. In addition, the weight of the total small intestine (SI) was calculated as a percentage of the live weight. The weight (W) and length (L) of the small intestine were used to calculate the weight-to-length ratio (W/L).

2.6. Carcass Characteristics

At 37 days of age, as appropriate to local commercial conditions, 12 chickens from each treatment (2 chickens from each experimental unit (pen) within the dietary treatment) were randomly selected and slaughtered via decapitation after being deprived of feed for 10 h to ensure that the digestive tract was empty. Then, pre-slaughter weight (PSW) was determined using an electronic scale (Adventurer OHAUS, AR3130, Pine Brook, NJ, USA). After slaughter, the carcasses (CW) were weighed.

The breast (pectoralis major and minor) and legs (thigh and drumstick) were weighed separately and expressed as a percentage of live weight [25]. The percentage of dressing yield (DY) was calculated according to Thema et al. [26]. All lymphoid organs (thymus, bursa, and spleen) were removed and weighed separately to calculate the percentage of live weight [27].

2.7. Serum Biochemical Indices

Blood samples (3 mL) were collected at the end of the grower stage (35 days) from 2 chickens per pen (12 chickens per treatment, density) via the wing vein in tubes without heparin. All samples were immediately separated into serum by centrifugation (Thermo Fisher Scientific, Labofuge 200, Dreieich, Germany) at 2500× g for 20 min and then stored at −20 °C until analysis of blood biochemical indicators. Blood biochemical indicators include total protein (TP), albumin (ALB), glucose (GLU), triglycerides (TG), cholesterol (CHO), uric acid (UA), creatinine (CREA), alanine aminotransferase (ALT), and aspartate aminotransferase (AST), which were analyzed with colorimetric kits (Randox Laboratories Limited, London, UK) according to the manufacturer's instructions using an automated spectrophotometric analyzer (Chem Well, Awareness Technology, Palm City, FL, USA).

2.8. Statistical Analysis

The pens were used as the experimental unit for all parameters studied, based on a completely randomized block design. Data were statistically analyzed based on a 2 × 4 factorial analysis of variance (two different SD × four TRT) using the general linear models of the Statistical Analysis System software [28]. The statistical model used was as follows: Observed (Y_{ijb}) = general mean (μ) + stocking density (SD_i; i = low, high) + dietary treatments (TRT_j; j = CONT, T1, T2, T3) + SD × TRT_{ij} + block (B_b) + the random error (e_{ijb}). A normality test (Shapiro–Wilk test) and homogeneity of variances (Levene's test) were performed before the statistical analysis of the data. Tukey's test ($p < 0.05$) was used to detect significant differences between means. On the other hand, all relative data were transformed (data/100) to check for significant differences at $p < 0.05$. Finally, all values were expressed as mean ± standard error of the means (SEM) for each parameter.

3. Results

3.1. Growth Performance Parameters

The effects of dietary treatments and stocking density on growth parameters of broiler chickens are shown in Table 3. The results showed that live weight and weight gain at 14 days and at 1–14 days, respectively, had a significant interaction effect between treatments and stocking density (p = 0.014 and p = 0.014; respectively). In chickens fed 0.12% gum arabic (T1), body weight and weight gain increased at a normal stocking density and decreased at a high stocking density compared with CONT, T2, and T3. Live weight (36 days) and weight gain (15–36 days and 1–36 days) were lower in chickens at high stocking density compared with normal stocking density ($p < 0.001$). In addition, chickens fed diet supplements (T1–T3) had higher live weight and weight gain than chickens fed CONT ($p < 0.001$), while these values were not affected by the interaction between treatments and stocking density ($p > 0.05$).

For the starter stage (1–14 d), there was a significant interaction effect between stocking density and treatment on feed intake (p = 0.008). Feed intake was higher in chickens receiving T1 at normal stocking density, while it decreased at high stocking density compared with the other treatments. On days 15–36 d and 1–36 d, feed intake was decreased at high stocking density compared with normal stocking density ($p < 0.001$). In contrast, feed intake was not affected by the interaction between treatments and stocking density or by the response during this stage ($p > 0.05$).

Table 3. Effect of gum arabic supplementation and stocking density on growth parameters of broiler chickens.

Item		Live Body Weight			Weight Gain			Feed Intake		
		1 d	14 d	36 d	1–14 d	15–36 d	1–36 d	1–14 d	15–36 d	1–36 d
TRT [1]	SD [2]									
CONT	Normal	39.17	397 [b]	2269	358 [b]	1871	2229	457 [b]	2934	3392
T1	Normal	39.19	413 [a]	2477	374 [a]	2064	2438	471 [a]	3017	3489
T2	Normal	39.17	391 [b]	2413	352 [b]	2022	2374	445 [b]	2928	3373
T3	Normal	39.21	400 [b]	2460	361 [b]	2061	2421	454 [b]	2976	3429
CONT	High	39.19	415 [a]	2150	376 [a]	1735	2111	446 [b]	2822	3268
T1	High	39.20	399 [b]	2263	360 [b]	1864	2224	425 [c]	2847	3272
T2	High	39.19	425 [a]	2318	386 [a]	1893	2279	464 [ab]	2845	3308
T3	High	39.20	416 [a]	2317	377 [a]	1900	2277	445 [bc]	2837	3282
SEM [3]		0.02	7.08	28.72	7.12	26.60	28.70	8.90	27.64	31.74
SD										
Normal		39.19	400	2405 [a]	361	2005 [a]	2366 [a]	456	2964 [a]	3421 [a]
High		39.20	414	2262 [b]	374	1848 [b]	2223 [b]	445	2838 [b]	3283 [b]
SEM [3]		0.01	3.54	14.36	3.56	13.30	14.35	4.45	13.82	15.87
TRT										
CONT		39.18	406	2209 [b]	366	1803 [b]	2170 [b]	452	2878	3330
T1		39.20	406	2370 [a]	367	1964 [a]	2331 [a]	448	2932	3381
T2		39.18	408	2365 [a]	369	1958 [a]	2326 [a]	454	2887	3341
T3		39.20	408	2388 [a]	369	1981 [a]	2349 [a]	449	2906	3356
SEM [3]		0.01	5.01	20.31	5.03	18.81	20.29	6.29	19.54	22.44
Source of variance (p-Value)										
SD × TRT		0.864	0.014	0.2061	0.014	0.552	0.207	0.008	0.448	0.137
SD		0.350	0.012	<0.001	0.012	<0.001	<0.001	0.071	<0.001	<0.001
TRT		0.604	0.980	<0.001	0.977	<0.001	<0.001	0.917	0.232	0.428

[a–c] Superscripts above the means for each parameter within column express the significant difference ($p < 0.05$). [1] Dietary treatments (TRT): CONT = the basal diet without supplement, T1 = the basal diets supplemented with 0.12% gum arabic, T2 = the basal diets supplemented with 0.25% gum arabic, and T3 = the basal diets supplemented with 0.10% inactivated stabilized *Saccharomyces cerevisiae* as commercial prebiotic. [2] Stocking density (SD): normal stocking density = 28 kg/m^2 (12 chickens/pen), and high stocking density = 50 kg/m^2 (22 chickens/pens). [3] SEM = Standard error of mean.

The effects of dietary treatments with stocking density on feed efficiency and production of broiler chickens are shown in Table 4. The results of the current study showed that a high stocking density improved the feed conversion ratio and production efficiency index on days 1–14, while there was a negative effect on days 15–36 and 1–36 compared to normal stocking density ($p < 0.001$). In addition, chickens fed dietary supplements (T1–T3) increased the production efficiency index and improved the feed conversion ratio on days 15–36 and 1–36 ($p < 0.001$), whereas it had no effect on days 1–14 ($p > 0.05$), compared with CONT. The interaction between treatments and stocking density had no effect on the feed conversion ratio and production efficiency index ($p > 0.05$).

The mortality rate was the same for all dietary treatments and the control group at a high stocking density (one chicken per treatment) during the grower stage (15–36 days old), while there was no mortality in chickens raised at a low stocking density, so no data were published in this study.

Table 4. Effect of gum arabic supplementation and stocking density on feed efficiency and production of broiler chickens.

Item		Feed Conversion Ratio			Production Efficiency Index		
		1–14 d	15–36 d	1–36 d	1–14 d	15–36 d	1–36 d
TRT [1]	SD [2]						
CONT	Normal	1.28	1.56	1.52	221.7	402.2	414.5
T1	Normal	1.26	1.46	1.43	234.8	470.7	481.0
T2	Normal	1.26	1.45	1.43	221.5	462.8	471.8
T3	Normal	1.26	1.44	1.42	227.3	473.5	483.0
CONT	High	1.19	1.63	1.55	249.5	367.5	386.0
T1	High	1.18	1.53	1.47	241.7	411.7	427.2
T2	High	1.20	1.50	1.45	252.3	428.3	443.5
T3	High	1.18	1.49	1.44	252.2	431.0	446.5
SEM [3]		0.016	0.014	0.012	6.01	8.27	8.22
SD							
Normal		1.26 [a]	1.48 [b]	1.45 [b]	226.3 [b]	452.2 [a]	462.5 [a]
High		1.19 [b]	1.54 [a]	1.48 [a]	248.9 [a]	409.6 [b]	425.7 [b]
SEM [3]		0.008	0.007	0.006	3.00	4.13	4.11
TRT							
CONT		1.23	1.60 [a]	1.53 [a]	235.6	384.8 [b]	400.2 [b]
T1		1.22	1.50 [b]	1.45 [b]	238.2	441.2 [a]	454.1 [a]
T2		1.23	1.48 [b]	1.44 [b]	236.9	445.6 [a]	457.7 [a]
T3		1.22	1.47 [b]	1.43 [b]	239.7	452.2 [a]	464.7 [a]
SEM [3]		0.011	0.009	0.008	4.25	5.84	5.81
Source of variance (p-Value)							
SD × TRT		0.799	0.967	0.965	0.205	0.418	0.377
SD		<0.001	<0.001	0.001	<0.001	<0.001	<0.001
TRT		0.804	<0.001	<0.001	0.911	<0.001	<0.001

[a,b] Superscripts above the means for each parameter within column express the significant difference ($p < 0.05$). [1] Dietary treatments (TRT): CONT = the basal diet without supplement, T1 = the basal diets supplemented with 0.12% gum arabic, T2 = the basal diets supplemented with 0.25% gum arabic, and T3 = the basal diets supplemented with 0.10% inactivated stabilized *Saccharomyces cerevisiae* as commercial prebiotic. [2] Stocking density (SD): normal stocking density = 28 kg/m^2 (12 chickens/pen), and high stocking density = 50 kg/m^2 (22 chickens/pens). [3] SEM = Standard error of mean.

3.2. Small Intestine Morphology

The effects of dietary treatments with stocking density on small intestine morphology of broiler chickens are shown in Table 5. The results show that the relative weight and length of small intestinal fragments (duodenum, jejunum, and ileum) were not affected by the treatments, stocking density, or their interaction ($p > 0.05$). The SI ratio and W:L showed a significant interaction effect between treatments and stocking density ($p = 0.025$ and $p = 0.007$; respectively). Chickens fed the diet treatments (T1–T3) had higher W:L and SI under a normal stocking density compared with CONT and a high stocking density ($p < 0.05$), but it had no effect compared with T3.

Table 5. Effect of gum arabic supplementation and stocking density on relative small intestine morphology of broiler chickens.

Item		Weight [4]				Length [4]			
		Doud.	Jej.	Ile.	Total SI	Doud.	Jej.	Ile.	W:L
TRT [1]	SD [2]								
CONT	Normal	21.70	42.31	35.98	2.96 [b]	16.48	42.91	40.59	0.37 [b]
T1	Normal	22.24	40.37	37.38	4.06 [a]	16.08	41.58	42.33	0.50 [a]
T2	Normal	21.29	40.17	38.53	3.81 [a]	16.80	41.91	41.28	0.48 [a]
T3	Normal	20.52	37.88	41.58	3.59 [a]	16.78	40.16	43.05	0.46 [a]

Table 5. *Cont.*

Item		Weight [4]				Length [4]			
		Doud.	Jej.	Ile.	Total SI	Doud.	Jej.	Ile.	W:L
CONT	High	20.80	42.44	36.75	2.93 [b]	17.10	41.09	41.80	0.37 [b]
T1		22.87	38.88	38.24	2.99 [b]	17.18	41.07	41.73	0.35 [b]
T2		21.61	38.94	39.44	3.07 [b]	16.00	40.77	43.22	0.36 [b]
T3		21.88	40.14	37.96	2.83 [b]	16.29	42.08	41.62	0.34 [b]
SEM [3]		1.31	1.74	2.09	0.16	0.73	0.70	0.75	0.02
SD									
Normal		21.43	40.18	38.37	3.60	16.53	41.64	41.81	0.46
High		21.79	40.10	38.09	2.95	16.64	41.25	42.09	0.36
SEM [3]		0.65	0.87	1.04	0.08	0.37	0.35	0.37	0.01
TRT									
CONT		21.25	42.37	36.36	2.95	16.79	42.00	41.20	0.37
T1		22.55	39.63	37.81	3.53	16.63	41.32	42.03	0.43
T2		21.45	39.55	38.98	3.44	16.40	41.34	42.25	0.42
T3		21.20	39.01	39.77	3.21	16.53	41.12	42.33	0.40
SEM [3]		0.92	1.23	1.48	0.11	0.52	0.49	0.53	0.02
Source of variance (p-Value)									
SD × TRT		0.855	0.698	0.638	0.025	0.527	0.060	0.110	0.007
SD		0.703	0.948	0.854	<0.001	0.839	0.440	0.600	<0.001
TRT		0.705	0.233	0.402	0.006	0.959	0.624	0.426	0.041

[a,b] Superscripts above the means for each parameter within column express the significant difference ($p < 0.05$). [1] Dietary treatments (TRT): CONT = the basal diet without supplement, T1 = the basal diets supplemented with 0.12% gum arabic, T2 = the basal diets supplemented with 0.25% gum arabic, and T3 = the basal diets supplemented with 0.10% inactivated stabilized *Saccharomyces cerevisiae* as commercial prebiotic. [2] Stocking density (SD): normal stocking density = 28 kg/m^2 (12 chickens/pen), and high stocking density = 50 kg/m^2 (22 chickens/pens). [3] SEM = Standard error of mean. [4] Doud = duodenum; Jej = jejunum; Ile = ileum; SI = small intestine; W:L = weight-to-length ratio.

3.3. Carcass Characteristics and Lymphoid Organs

The effects of dietary treatments with stocking density on carcass characteristics and lymphoid organs of broiler chickens are shown in Table 6. The current results show that a high stocking density resulted in a decrease in PSW, CW, DY, and back ($p < 0.001$), while the relative weight of legs and thymus increased compared with normal stocking density ($p = 0.023$ and $p = 0.001$; respectively). The relative weights of breast, bursa, and spleen were not affected by stocking density ($p > 0.05$). The interaction effect or dietary treatments did not affect all parameters of carcass characteristics and lymphoid organs ($p > 0.05$).

Table 6. Effect of gum arabic supplementation and stocking density on carcass characteristics and lymphoid organs of broiler chickens.

Item		Parameters [4]								
		PSW (g)	CW (g)	DY (%)	Breast (%)	Legs (%)	Back (%)	Thymus (%)	Bursa (%)	Spleen (%)
TRT [1]	SD [2]									
CONT	Normal	2597	1968	75.85	28.70	28.09	8.19	0.33	0.15	0.10
T1		2639	1992	75.47	28.99	27.74	8.73	0.41	0.19	0.09
T2		2631	1983	75.39	29.17	26.50	7.49	0.40	0.19	0.09
T3		2652	1974	74.42	29.39	27.76	8.03	0.38	0.19	0.10
CONT	High	2283	1667	73.02	30.46	28.67	7.88	0.45	0.21	0.10
T1		2371	1766	74.49	29.90	29.10	7.94	0.57	0.20	0.10
T2		2446	1811	74.01	29.95	28.22	8.17	0.46	0.18	0.08
T3		2436	1809	74.25	29.44	28.52	8.26	0.46	0.17	0.09
SEM [3]		44.60	33.69	0.47	0.66	0.65	0.25	0.04	0.02	0.01
SD										
Normal		2630 [a]	1979 [a]	75.28 [a]	29.06	27.52 [b]	8.11 [a]	0.38 [b]	0.18	0.10

Table 6. *Cont.*

Item	PSW (g)	CW (g)	DY (%)	Breast (%)	Legs (%)	Back (%)	Thymus (%)	Bursa (%)	Spleen (%)
	Parameters [4]								
High	2384 [b]	1763 [b]	73.94 [b]	29.94	28.63 [a]	8.06 [b]	0.48 [a]	0.19	0.09
SEM [3]	22.30	16.84	0.23	0.33	0.32	0.12	0.02	0.01	0.01
TRT									
CONT	2440	1817	74.43	29.58	28.38	8.04	0.39	0.18	0.10
T1	2505	1879	74.98	29.45	28.42	8.33	0.49	0.20	0.09
T2	2538	1897	74.69	29.56	27.36	7.83	0.43	0.18	0.08
T3	2544	1892	74.33	29.42	28.14	8.15	0.42	0.18	0.09
SEM [3]	31.54	23.82	0.33	0.47	0.46	0.18	0.03	0.01	0.01
	Source of variance (p-Value)								
SD × TRT	0.494	0.177	0.092	0.660	0.790	0.054	0.592	0.105	0.555
SD	<0.001	<0.001	0.001	0.069	0.023	0.001	0.001	0.367	0.426
TRT	0.093	0.088	0.454	0.976	0.426	0.135	0.135	0.682	0.428

[a,b] Superscripts above the means for each parameter within column express the significant difference ($p < 0.05$). [1] Dietary treatments (TRT): CONT = the basal diet without supplement, T1 = the basal diets supplemented with 0.12% gum arabic, T2 = the basal diets supplemented with 0.25% gum arabic, and T3 = the basal diets supplemented with 0.10% inactivated stabilized *Saccharomyces cerevisiae* as commercial prebiotic. [2] Stocking density (SD): normal stocking density = 28 kg/m^2 (12 chickens/pen), and high stocking density = 50 kg/m^2 (22 chickens/pens). [3] SEM = Standard error of mean. [4] PSW = pre-slaughter weight; CW = carcass weight; DY = dressing yield. Body components and lymphoid organs were computed as a ratio to live weight.

3.4. Blood Biochemical Indices

The effects of dietary treatments with stocking density on blood biochemistry indicators of broiler chickens are shown in Table 7. The results show that the blood biochemistry indicators including the concentrations of TP, GLU, TG, CHO, UA, CREA, ALT, and AST were not affected by the treatments and the interaction between treatments and stocking density ($p > 0.05$). Albumen concentration (ALB) showed a significant interaction effect between treatments and stocking density ($p = 0.001$). A high stocking density resulted in higher UA, CREA, and AST concentrations compared with a normal stocking density ($p = 0.003$, $p = 0.008$, and $p = 0.001$; respectively). Other blood biochemical indicators (TP, ALB, GLU, TG, CHO, and ALT) had no effect on the stocking density ($p > 0.05$).

Table 7. Effect of gum arabic supplementation and stocking density on blood biochemistry indicators of broiler chickens.

Item		TP (g/dL)	ALB (g/dL)	GLU (mg/dL)	TG (mg/dL)	CHO (mg/dL)	UA (mg/dL)	CREA (mg/dL)	ALT (U/L)	AST (U/L)
		Parameters [4]								
TRT [1]	SD [2]									
CONT	Normal	3.39	1.95 [b]	169.6	93.4	103.85	4.53	1.09	12	40
T1	Normal	3.40	1.72 [c]	152.3	99.0	97.10	4.57	0.93	16	39
T2	Normal	3.39	1.70 [c]	158.3	100.6	87.03	4.55	1.05	13	41
T3	Normal	3.23	2.13 [a]	152.8	92.3	70.81	4.26	0.75	14	42
CONT	High	3.31	1.87 [b]	155.2	105.0	78.95	4.94	1.02	15	57
T1	High	3.51	2.02 [a]	163.6	110.4	96.91	4.64	1.17	16	49
T2	High	3.63	2.06 [a]	150.7	106.7	91.95	4.80	1.25	15	45
T3	High	3.51	1.62 [c]	154.8	96.5	92.98	5.03	1.28	14	51
SEM [3]		0.17	0.11	8.70	9.79	6.92	0.17	0.12	2.03	3.23
SD										
Normal		3.35	1.88	158.2	96.3	89.70	4.48 [b]	0.95 [b]	14	41 [b]
High		3.49	1.89	156.1	104.6	90.20	4.85 [a]	1.18 [a]	15	51 [a]
SEM [3]		0.09	0.05	4.35	4.89	3.46	0.09	0.06	1.01	1.61
TRT										

Table 7. *Cont.*

Item	Parameters [4]								
	TP (g/dL)	ALB (g/dL)	GLU (mg/dL)	TG (mg/dL)	CHO (mg/dL)	UA (mg/dL)	CREA (mg/dL)	ALT (U/L)	AST (U/L)
CONT	3.35	1.91	162.4	99.2	91.40	4.73	1.05	14	49
T1	3.45	1.87	158.0	104.7	97.01	4.60	1.05	16	45
T2	3.51	1.89	154.5	103.7	89.49	4.67	1.153	14	44
T3	3.37	1.89	153.8	94.4	81.90	4.65	1.01	15	47
SEM [3]	0.12	0.08	6.15	6.92	4.89	0.12	0.08	1.43	2.28
Source of variance (*p*-Value)									
SD × TRT	0.732	0.001	0.486	0.974	0.015	0.232	0.095	0.789	0.288
SD	0.270	0.847	0.730	0.238	0.919	0.003	0.008	0.401	0.001
TRT	0.786	0.983	0.748	0.709	0.200	0.894	0.668	0.702	0.398

[a–c] Superscripts above the means for each parameter within column express the significant difference ($p < 0.05$). [1] Dietary treatments (TRT): CONT = the basal diet without supplement, T1 = the basal diets supplemented with 0.12% gum arabic, T2 = the basal diets supplemented with 0.25% gum arabic, and T3 = the basal diets supplemented with 0.10% inactivated stabilized *Saccharomyces cerevisiae* as commercial prebiotic. [2] Stocking density (SD): normal stocking density = 28 kg/m^2 (12 chickens/pen), and high stocking density = 50 kg/m^2 (22 chickens/pens). [3] SEM = Standard error of mean. [4] TP = total protein, ALB = albumin, GLU = glucose, TG = triglycerides, CHO = cholesterol, UA = uric acid, CREA = creatinine, ALT = alanine aminotransferase, and AST = aspartate aminotransferase.

4. Discussion

Greater crowding (high stocking density) means a less hygienic environment, and this can be a stressor on broiler performance. The additional administration of gum arabic as a prebiotic could be beneficial in such conditions to help broilers cope with stress. Accordingly, this study was conducted to investigate this objective by using two different stocking densities with the addition of gum arabic or prebiotics to broiler diets and measuring their effects on growth performance, intestine morphology, carcass yield, lymphoid organs, and selected blood indices. The current results showed that weight gain and production efficiency increased and feed conversion improved, but there was no difference in feed intake when chickens were raised at high stocking density in the starter stage (1–14 days of age) compared to normal stocking density. This could indicate that high stocking density during this period has a positive effect on chick performance by maintaining thermoregulation, which is incomplete at 1–14 days of age, which could also be due to their smaller size and space requirements. On the other hand, our results showed that body weight, weight gain, feed conversion ratio, and production efficiency were negatively affected when chickens were raised at a high stocking density at 15–36 and 1–36 days of age. These results are consistent with those of Heidari et al. [29] and Miao et al. [30], who confirmed that a high stocking density (18 and 20 chickens/m^2) at 35 days of age resulted in a decrease in weight gain and feed consumption compared to chickens with normal stocking density. Ghanima et al. [2] reported that a high stocking density has a negative impact on performance indicators of broiler chickens. A high stocking density deteriorates litter and air quality by increasing litter moisture and the volatilization of ammonia, which negatively affects growth performance during the grower stage due to the increasing size of chickens [31]. In addition, the disruption of gut microbial ecology that supports digestion, absorption, and production of beneficial molecules may contribute to poor growth performance of broiler chickens raised at 22–42 and 1–42 days of age at high stocking density [32]. Therefore, high stocking density is one of the stressors that negatively affect growth performance such as body weight, weight gain, and feed intake [9]. Production efficiency and feed conversion ratio are used as indicators of the economic status of broiler production [33]. Thus, a higher production efficiency index and lower feed conversion indicate better performance and feed efficiency when chickens are kept at normal stocking density (28 kg/m^2).

The current results show that gum arabic and prebiotics (T1–T3) improved growth performance parameters by increasing live weight, weight gain, and production efficiency, while feed intake did not differ during grower and overall stage (15–36 and 1–36 days of age). These results are consistent with those of Khan et al. [34,35], who found that the administration of gum arabic as a prebiotic improved overall performance indicator in broilers. A prebiotic-enriched diet improved weight gain, feed intake, and carcass weight in broilers [36]. Zhen et al. [37] reported that a *Saccharomyces cerevisiae*-derived prebiotic increased the body weight gain and feed intake and improved the feed conversion ratio in broilers. The improvement in the feed conversion ratio via dietary treatments could be attributed to the increase in body weight gain by improving the activity of gut microbiota and feed digestibility. Gum arabic has the ability to ferment through the activity of commensal bacteria in the gut, which could reflect its positive effect on the performance indicators of broilers [38]. In contrast, Tabidi and Ekram [39] reported that gum arabic had no effect on weight gain, feed intake, and feed conversion in broilers. Houshmand et al. [40] showed that the use of mannooligosaccharide derived from the cell wall of *Saccharomyces cerevisiae* as a feed additive had no effect on growth performance parameters of broilers. Morphological measurements of the intestine are used as indicators to evaluate the function of the small intestine and its absorption capacity [41]. Chickens fed T1 to T3 had a higher relative W:L ratio and higher SI compared to CONT at a normal stocking density. However, gum arabic as a natural prebiotic may improve nutrient absorption in broilers [16,42]. This suggests that dietary supplementation improves intestinal morphology, which is reflected in performance parameters and carcass yield by increasing the surface area for nutrient absorption in broilers. The relative weight of SI and the W:L ratio decreased when chickens were kept at a high stocking density compared to a normal stocking density. This suggests that the decrease in the relative weight of the W:L ratio and the total weight of SI leads to a decrease in nutrient uptake ability and thus negatively affects growth performance at a high stocking density during the grower stage [9,43].

Selected biochemical indicators in blood are among the tests performed to evaluate the body metabolism of broiler chickens [44]. Therefore, selected blood biochemical indicators were determined to investigate the effects of feed additives and various stocking densities on the health and nutritional status of broiler chickens. Our results showed that the concentrations of TP, GLU, TG, CHO, and ALT were not affected by the treatments, stocking density or their interaction. Singh et al. [45] reported that the biochemical indicators in the blood of broiler chickens were not affected by high stocking density. The gum arabic-fed groups (T1 and T2) had lower ALB concentrations, while the addition of a prebiotic (T3) at a normal stocking density gave a higher concentration of ALB than that in CONT and other treatments at a high stocking density. Normally, a high albumin concentration is one of the physiological stress indicators in broiler chickens [46]. Therefore, the ALB concentration was probably increased in T1 and T2 at a high stocking density as a homeostatic response to stress. A high stocking density resulted in higher UA, CREA, and AST concentrations compared to normal stocking density. These results suggest that some changes in blood biochemical indicators (UA, CREA, and AST) may be due to broiler stocking density. These results are in agreement with those of Ghanima et al. [2] and Nasr et al. [47], who reported that a high stocking density (40 kg/m^2) at 35 days of age resulted in a decrease in the concentrations of UA, CREA, and AST compared to chickens with a normal stocking density. The high AST activity in broiler chickens kept at a high stocking density could be due to increased competition for feed and water, which increases muscle injury, which is the reason for the higher AST activity in blood [48]. The addition of gum arabic had no effect on the ALT and AST activity of rabbits [19].

Dressing yield and carcass components are important criteria for evaluating the carcass traits and slaughter value of broiler chickens [49]. The results of the current study showed that carcass traits and lymphoid organs were not affected by treatments or by the interaction between treatments and stocking density. In contrast to our results, chickens fed prebiotics had a higher carcass and breast meat yield [50]. Pre-slaughter weight (PSW), CW, and DY decreased, while relative leg and thymus weight increased when chickens were kept at a high stocking density compared to a normal stocking density. Similarly, the results of Cengiz et al. [33] showed that a high stocking density decreased the slaughter weight and dressing yield of broiler chickens. In contrast, several studies showed that stocking density had no effect on carcass characteristics, DY, and lymphoid organs [27,51,52]. Other relative weights of lymphoid organs (bursa and spleen) were not affected by treatments, stocking density, or their interaction. These results do not agree with those of Sato et al. [53], who found that the relative weights of lymphoid organs (bursa and spleen) were higher in broiler chickens fed gum arabic. Dietary supplementation with prebiotics (β-glucan and mannooligosaccharide) resulted in higher relative weights of lymphoid organs [54]. In contrast, Houshmand et al. [40] found that mannooligosaccharide as a prebiotic supplement at rate of 0.10% had no effect on the relative weight of lymphoid organs.

5. Conclusions

In conclusion, the present results provide useful evidence that the use of gum arabic as a prebiotic improves growth performance, production efficiency, and small intestinal morphology without negatively affecting blood biochemical indicators, carcass characteristics, or lymphoid organs to the same extent as the prebiotic (*Saccharomyces cerevisiae*), based on the results of the entire experimental period (1–36 d) in broiler chickens. At the same time, they do not provide a solution for chickens kept at a high stocking density. Higher stocking density (50 kg/m^2) of broilers negatively affected growth performance, production efficiency, some blood indices (UA, CREA and AST), carcass weight, dressing yield, and intestinal morphology. Further studies are needed to determine the potential mechanism of this supplementation by testing for stress indicators and the intestinal ecosystem.

Author Contributions: Conceptualization, methodology, formal analysis, and writing—original draft: H.H.A.-B.; visualization, data curation, investigation, writing—review, and editing: R.A.A. and M.M.A. All authors have read and agreed to the published version of the manuscript.

Funding: The authors extend their appreciation to the Deputyship for Research and Innovation, Ministry of Education in Saudi Arabia for funding this research work through Project No. IFKSUOR3-415-1.

Institutional Review Board Statement: The Scientific Ethics Committee of King Saud University in Saudi Arabia approved all procedures used in this study (internal reference number: KSU-SE-20-39, 27 May 2021).

Data Availability Statement: All data presented in this study are available by all authors.

Conflicts of Interest: The authors declare no conflict of interest.

References

1. Goo, D.; Kim, J.H.; Choi, H.S.; Park, G.H.; Han, G.P.; Kil, D.Y. Effect of stocking density and sex on growth performance, meat quality, and intestinal barrier function in broiler chickens. *Poult. Sci.* **2019**, *98*, 1153–1160. [CrossRef] [PubMed]
2. Ghanima, M.M.A.; Swelum, A.A.; Shukry, M.; Ibrahim, S.A.; Abd El-Hack, M.E.; Khafaga, A.F.; Younis, M.E. Impacts of tea tree or lemongrass essential oils supplementation on growth, immunity, carcass traits, and blood biochemical parameters of broilers reared under different stocking densities. *Poult. Sci.* **2021**, *100*, 101443. [CrossRef] [PubMed]
3. Dozier, W.A.; Thaxton, J.P.; Purswell, J.L.; Olanrewaju, H.A.; Branton, S.L.; Roush, W.B. Stocking density effects on male broilers grown to 1.8 kilograms of body weight. *Poult. Sci.* **2006**, *85*, 344–351. [CrossRef] [PubMed]
4. Kryeziu, A.J.; Kamberi, M.; Muji, S.; Mestani, N.; Berisha, S. Carcass traits of broilers as affected by different stocking density and sex. *Bulg. J. Agric. Sci.* **2018**, *24*, 1097–1103.
5. Gholami, M.; Chamani, M.; Seidavi, A.; Sadeghi, A.A.; Aminafschar, M. Effects of stocking density and environmental conditions on performance, immunity, carcass characteristics, blood constitutes, and economical parameters of cobb 500 strain broiler chickens. *Italian J. Anim. Sci.* **2020**, *19*, 524–535. [CrossRef]
6. Vargas-Rodrıguez, L.M.; Duran-Melendez, L.A.; Garcıa-Masıas, J.A.; Arcos-Garcıa, J.L.; Joaquın-Torres, B.M.; Ruelas-Inzunza, M.G. Effect of probiotic and population density on the growth performance and carcass characteristics in broiler chickens. *Int. J. Poult. Sci.* **2013**, *12*, 390–395. [CrossRef]
7. Estevez, I. Density allowances for broilers: Where to set the limits? *Poult. Sci.* **2007**, *86*, 1265–1272. [CrossRef] [PubMed]
8. Sohail, M.U.; Ijaz, A.; Yousaf, M.S.; Ashraf, K.; Zaneb, H.; Aleem, M.; Rehman, H. Alleviation of cyclic heat stress in broilers by dietary supplementation of mannan oligosaccharide and Lactobacillus-based probiotic: Dynamics of cortisol, thyroid hormones, cholesterol, C-reactive protein, and humoral immunity. *Poult. Sci.* **2010**, *89*, 1934–1938. [CrossRef]

9. Kridtayopas, C.; Rakangtong, C.; Bunchasak, C.; Loongyai, W. Effect of prebiotic and synbiotic supplementation in diet on growth performance, small intestinal morphology, stress, and bacterial population under high stocking density condition of broiler chickens. *Poult. Sci.* **2019**, *98*, 4595–4605. [CrossRef]
10. Khalid, S.A.; Musa, A.M.; Saeed, A.M.; Abugroun, E.A.; Ahmed, E.O.S.; Ghalib, M.B.; Phillips, G.O. Manipulating dietary fibre: Gum Arabic making friends of the colon and the kidney. *Bioact. Carbohydr. Diet Fibre* **2014**, *3*, 71–76. [CrossRef]
11. Ahmed, A.A.; Musa, H.H.; Fedail, J.S.; Sifaldin, A.Z.; Musa, T.H. Gum Arabic suppressed diet induced obesity by alteration the expression of mRNA levels of genes involved in lipid metabolism in mouse liver. *Bioact. Carbohydr. Diet Fibre* **2016**, *7*, 15–20. [CrossRef]
12. Sohail, M.U.; Hume, M.E.; Byrd, J.A.; Nisbet, D.J.; Ijaz, A.; Sohail, A.; Shabbir, M.Z.; Rehman, H. Effect of supplementation of prebiotic mannan-oligosaccharides and prebiotic mixture on growth performance of broilers subjected to chronic heat stress. *Poult. Sci.* **2012**, *91*, 2235–2240. [CrossRef] [PubMed]
13. Korcz, E.; Kerényi, Z.; Varga, L. Dietary fibers, prebiotics, and exopolysaccharides produced by lactic acid bacteria: Potential health benefits with special regard to cholesterol lowering effects. *Food Funct.* **2018**, *9*, 3057–3068. [CrossRef] [PubMed]
14. Al-Baadani, H.H.; Al-Mufarrej, S.I.; Azzam, M.M.; Alharthi, A.S.; Al-Garadi, M.A.; Al-Gabri, N.A.; Alhidary, I.A. Evaluation of gum Arabic (Acacia Senegal) as a natural prebiotic to improve growth performance and health status of broiler chickens. *Trop. Anim. Health Prod.* **2022**, *54*, 244. [CrossRef] [PubMed]
15. Kamal, E.; Kaddam, L.A.; Dahawi, M.; Osman, M.; Salih, M.A.; Alagib, A.; Saeed, A. Gum Arabic fibers decreased inflammatory markers and disease severity score among rheumatoid arthritis patients, Phase II Trial. *Int. J. Rheumatol.* **2018**, *1*, 4197537. [CrossRef] [PubMed]
16. Liu, S.D.; Song, M.H.; Yun, W.; Lee, J.H.; Kim, H.B.; Cho, J.H. Effect of carvacrol essential oils on immune response and inflammation-related genes expression in broilers challenged by lipopolysaccharide. *Poult. Sci.* **2019**, *98*, 2026–2033. [CrossRef] [PubMed]
17. Al-Asmakh, M.; Sohail, M.U.; Al-Jamal, O.; Shoair, B.M.; Al-Baniali, A.Y.; Bouabidi, S.; Bawadi, H. The effects of gum acacia on the composition of the gut microbiome and plasma levels of short-chain fatty acids in a rat model of chronic kidney disease. *Front. Pharmacol.* **2020**, *11*, 1896. [CrossRef]
18. Amber, K.; Abd El-Nabi, F.M.; Morsy, W.A.; Shama, S.H.A. Gum Arabic as prebiotic in growing rabbit's diet. *Glob. Vet.* **2017**, *19*, 465–471.
19. El-Ratel, I.T.; Ismail, R.F.; Fouda, S.F. Productive performance, carcass traits, lipid profile, antioxidants and immunity of growing rabbits treated with gum Arabic under Egyptian summer condition. *Egypt. J. Nutr. Feeds.* **2019**, *22*, 143–154. [CrossRef]
20. Al-Baadani, H.H.; Alhotan, R.A.; Azzam, M.M.; Suliman, G.M.; Alharthi, A.S.; Fazea, E.H.; Alhidary, I.A. Effects of gum Arabic (*Acacia senegal*) powder on characteristics of carcass and breast meat quality parameters in male broiler chickens. *Foods* **2023**, *12*, 2526. [CrossRef]
21. Diler, Ö.; Özil, Ö.; Bayrak, H.; Yiğit, N.Ö.; Özmen, Ö.; Saygın, M.; Aslankoç, R. Effect of dietary supplementation of sumac fruit powder (*Rhus coriaria* L.) on growth performance, serum biochemistry, intestinal morphology and antioxidant capacity of rainbow trout (*Oncorhynchus mykiss*, Walbaum). *Anim. Feed Sci. Technol.* **2021**, *278*, 114993–115002. [CrossRef]
22. Dai, D.; Qiu, K.; Zhang, H.J.; Wu, S.G.; Han, Y.M.; Wu, Y.Y.; Wang, J. Organic acids as alternatives for antibiotic growth promoters alter the intestinal structure and microbiota and improve the growth performance in broilers. *Front. Microbiol.* **2021**, *11*, 618144. [CrossRef] [PubMed]
23. Zhang, S.; Zhong, G.; Shao, D.; Wang, Q.; Hu, Y.; Wu, T.; Shi, S. Dietary supplementation with Bacillus subtilis promotes growth performance of broilers by altering the dominant microbial community. *Poult. Sci.* **2021**, *100*, 100935. [CrossRef] [PubMed]
24. Itafa, B.T.; Mohamed, A.S.; Abate, W.H.; Derseh, M.A.; Woldegiorgiss, W.E. Effect of reciprocal crossing Koekoek and Sasso chickens on growth performance, feed efficiency, carcass yield, mortality rate, and genetic components. *J. Appl. Poult. Res.* **2021**, *30*, 100214. [CrossRef]
25. Alkhulaifi, M.M.; Alqhtani, A.H.; Alharthi, A.S.; Al Sulaiman, A.R.; Abudabos, A.M. Influence of prebiotic yeast cell wall extracts on growth performance, carcass attributes, biochemical metabolites, and intestinal morphology and bacteriology of broiler chickens challenged with Salmonella typhimurium and Clostridium perfringens. *Ital. J. Anim. Sci.* **2022**, *21*, 1190–1199. [CrossRef]
26. Thema, K.K.; Mnisi, C.M.; Mlambo, V. Stocking density induced changes in growth performance, blood parameters, meat quality traits, and welfare of broiler chickens reared under semi-arid subtropical conditions. *PLoS ONE* **2022**, *17*, e0275811. [CrossRef] [PubMed]
27. Karar, E.M.H.; Atta, A.R.M.M.; Gharib, H.B.A.; El-Menawey, A.R.A.H. Impact of prebiotic supplementation on productive performance, carcass traits, and physiological parameters of broiler chickens under high stocking density condition. *J. World Poult. Res.* **2023**, *13*, 48–60. [CrossRef]
28. SAS Institute. *SAS Users Guide: Statistics*; SAS Institute Inc.: Cary, NC, USA, 2008.
29. Heidari, S.; Toghyani, M. Effect of stocking density and methionine levels on growth performance and immunity of broiler chicks. *Iranian J. Appl. Anim. Sci.* **2018**, *8*, 483–489.

30. Miao, Z.Q.; Dong, Y.Y.; Qin, X.; Yuan, J.M.; Han, M.M.; Zhang, K.K.; Shi, S.R.; Song, X.Y.; Zhang, J.Z.; Li, J.H. Dietary supplementation of methionine mitigates oxidative stress in broilers under high stocking density. *Poult. Sci.* **2021**, *100*, 101231. [CrossRef]
31. Cengiz, O.; Koksal, B.H.; Tatlı, O.; Kuter, E.; Ahsan, U.; Güven, G.; Bilgili, S.F.; Önol, A.G. Supplemental boric acid does not prevent the development of footpad dermatitis in broilers subjected to high stocking density. *Poult. Sci.* **2018**, *97*, 4342–4350. [CrossRef]
32. Cengiz, Ö.; Köksal, B.H.; Tatlı, O.; Sevim, Ö.; Ahsan, U.; Üner, A.G.; Önol, A.G. Effect of dietary probiotic and high stocking density on the performance, carcass yield, gut microflora, and stress indicators of broilers. *Poult. Sci.* **2015**, *94*, 2395–2403. [CrossRef] [PubMed]
33. Huff, G.R.; Huff, W.E.; Jalukar, S.; Oppy, J.; Rath, N.C.; Packialakshmi, B. The effects of yeast feed supplementation on turkey performance and pathogen colonization in a transport stress/Escherichia coli challenge. *Poult. Sci.* **2013**, *92*, 655–662. [CrossRef] [PubMed]
34. Siham, A.; Khadiga, A.; Huwaida, E.; Bakheit, M.D.; Khalid, M.E. Effect of dietary inclusion of gum Arabic (*Acacia senegal*) on performance and blood chemistry of broiler chicks. *Wayamba J. Anim. Sci.* **2015**, *3*, 305–310.
35. Khan, S.; Chand, N.; Hafeez, A.; Ahmad, N. Effect of gum arabic on overall growth performance, visceral and lymphoid organs along with intestinal histomorphology and selected pathogenic bacteria of broiler chickens. *J. Anim. Health Prod.* **2022**, *10*, 73–80. [CrossRef]
36. Mohammed, M.Y.; Kareem, K.Y. A comparison study of probiotic, postbiotic and prebiotic on performance and meat quality of broilers. *Tikrit J. Agric. Sci.* **2022**, *22*, 24–32. [CrossRef]
37. Zhen, W.; Zhu, T.; Wang, P.; Guo, F.; Zhang, K.; Zhang, T.; Ma, Y. Effect of dietary Saccharomyces-derived prebiotic refined functional carbohydrates as antibiotic alternative on growth performance and intestinal health of broiler chickens reared in a commercial farm. *Poult. Sci.* **2023**, *102*, 102671. [CrossRef] [PubMed]
38. Lockyer, S.; Stanner, S. Prebiotics–an added benefit of some fiber types. *Nutr. Bull.* **2019**, *44*, 74–91. [CrossRef]
39. Tabidi, M.H.; Ekram, K.A. Effect of feeding gum Arabic with or without commercial xylem enzyme 500 on the performance of broiler chicks. *World J. Pharm. Pharm. Sci.* **2015**, *4*, 1863–1872.
40. Houshmand, M.; Azhar, K.; Zulkifli, I.; Bejo, M.H.; Kamyab, A. Effects of prebiotic, protein level, and stocking density on performance, immunity, and stress indicators of broilers. *Poult. Sci.* **2012**, *91*, 393–401. [CrossRef]
41. Zhang, T.; Li, Q.; Cheng, L.; Buch, H.; Zhang, F. *Akkermansia muciniphila* Is a promising probiotic. *Microb. Biotechnol.* **2019**, *12*, 1109–1125. [CrossRef]
42. Jha, R.; Kim, S.W. Nutritional intervention for the intestinal health of young monogastric animals. *Front. Vet. Sci.* **2021**, *8*, 668536. [CrossRef] [PubMed]
43. Guardia, S.; Konsak, B.; Combes, S.; Levenez, F.; Cauquil, L.; Guillot, J.F.; Moreau-Vauzelle, C.; Lessire, M.; Juin, H.; Gabriel, I. Effects of stocking density on the growth performance and digestive microbiota of broiler chickens. *Poult. Sci.* **2011**, *90*, 1878–1889. [CrossRef] [PubMed]
44. Raju, S.; Gurram, S.; Bora, S.; Kumar, B.S. Effect of herbal immunomodulators on immune status, haematological and serum biochemical parameters of Japanese quails (*Coutrinix japanica*) during summer stress. *Indian J. Anim. Sci.* **2019**, *89*, 575–577. [CrossRef]
45. Singh, K.D.; Pramanik, P.S.; Kumar, R. Effect of stocking density on stress reaction and mortality in broiler chickens. *Int. J. Curr. Microbiol. Appl. Sci.* **2018**, *7*, 182–189.
46. Nwaigwe, C.U.; Ihedioha, J.I.; Shoyinka, S.V.; Nwaigwe, C.O. Evaluation of the hematological and clinical biochemical markers of stress in broiler chickens. *Vet. World.* **2020**, *13*, 2294. [CrossRef]
47. Nasr, M.A.; Alkhedaide, A.Q.; Ramadan, A.A.; Abd-El Salam, E.H.; Hussein, M.A. Potential impact of stocking density on growth, carcass traits, indicators of biochemical and oxidative stress and meat quality of different broiler breeds. *Poult. Sci.* **2021**, *100*, 101442. [CrossRef] [PubMed]
48. Ozbey, O.; Yildiz, N.; Aysondu, M.H.; Ozmen, O. The effects of high temperature on blood serum parameters and the egg productivity characteristics of Japanese quails (*Coturnix coturnix japonica*). *Int. J. Poult. Sci.* **2004**, *7*, 485–489.
49. Shewita, R.S.; Taha, A.E. Influence of dietary supplementation of ginger powder at different levels on growth performance, haematological profiles, slaughter traits and gut morphometry of broiler chickens. *S. Afr. J. Anim. Sci.* **2018**, *48*, 998–1008. [CrossRef]
50. Falaki, M.; Shargh, M.S.; Dastar, B.; Zrehdaran, S. Effects of different levels of probiotic prebiotic on performance and carcass characteristics of broiler chickens. *J. Anim. Vet. Adv.* **2011**, *10*, 378–384. [CrossRef]
51. Sekeroglu, A.; Sarica, M.U.S.A.; Gulay, M.S.; Duman, M. Effect of stocking density on chick performance, internal organ weights and blood parameters in broilers. *J. Anim. Vet. Adv.* **2011**, *10*, 246–250. [CrossRef]
52. Adeyemo, G.O.; Fashola, O.O.; Ademulegun, T.I. Effect of stocking density on the performance, carcass yield and meat composition of broiler chickens. *Br. Biotechnol. J.* **2016**, *14*, 1–7. [CrossRef]

53. Sato, K.; Takahashi, K.; Tohno, M.; Miura, Y.; Kamada, T.; Ikegami, S.; Kitazawa, H. Immunomodulation in gut-associated lymphoid tissue of neonatal chicks by immunobiotic diets. *Poult. Sci.* **2009**, *88*, 2532–2538. [CrossRef]
54. Usama, T.M.; Manal, A.M.; Hosnia, S.A.; Omar, A.A. AGRIMOS® Prebiotics: Effect on Behavior, Performance, Cecal Microbial Population and Humeral Immunity in Broiler Chickens. *J. Adv. Vet. Res.* **2018**, *8*, 49–59.

MDPI AG
Grosspeteranlage 5
4052 Basel
Switzerland
Tel.: +41 61 683 77 34

Animals Editorial Office
E-mail: animals@mdpi.com
www.mdpi.com/journal/animals

www.ingramcontent.com/pod-product-compliance
Lightning Source LLC
Chambersburg PA
CBHW041359210926
51850CB00013B/3126
* 9 7 8 3 7 2 5 8 3 6 6 8 0 *